# Three Mile Island

**Bibliographies and Indexes in Science and Technology**

Publishing Opportunities for Energy Research: A Descriptive Guide to Selective Serials in the Social and Technical Sciences
*Roberta A. Scull, compiler*

Toxic and Hazardous Materials: A Sourcebook and Guide to Information Sources
*James K. Webster, editor*

# Three Mile Island

*A Selectively Annotated Bibliography*

Compiled by
**M. Sandra Wood**
and
**Suzanne M. Shultz**

**Bibliographies and Indexes in Science and Technology, Number 3**

**Greenwood Press**
New York • Westport, Connecticut • London

**Library of Congress Cataloging-in-Publication Data**

Wood, M. Sandra.
Three Mile Island : a selectively annotated bibliography / compiled by M. Sandra Wood and Suzanne M. Shultz.
p. cm. — (Bibliographies and indexes in science and technology, ISSN 0888-7551 ; no. 3)
Includes indexes.
ISBN 0-313-25573-3 (lib. bdg. : alk. paper)
1. Nuclear power plants—Pennsylvania—Harrisburg Region—Accidents—Bibliography. 2. Three Mile Island Nuclear Power Plant (Pa.)—Bibliography. I. Shultz, Suzanne M. II. Title. III. Series.
Z5162.A26W66 1988
[TK1345.H37]
363.1'79—dc19 87-37547

British Library Cataloguing in Publication Data is available.



Library of Congress Catalog Card Number: 87-37547
ISBN: 0-313-25573-3
ISSN: 0888-7551

First published in 1988

Greenwood Press, Inc.
88 Post Road West, Westport, Connecticut 06881

Printed in the United States of America

The paper used in this book complies with the Permanent Paper Standard issued by the National Information Standards Organization (Z39.48-1984).

10 9 8 7 6 5 4 3 2 1

# Contents

# Foreword

Almost eight years ago, on March 28, 1979, an accident occurred that would dramatically change the future of nuclear power in this nation and alert the world to the awesome realities of nuclear technology.

Within hours of the accident, world attention was focused on the soon to become famous cooling towers of the Three Mile Island nuclear power plant in Middletown, Pennsylvania, just ten miles from our capital city of Harrisburg.

The crippled plant was soon swarming with reporters, photographers, government officials, and scientists trying to determine exactly what had happened and what should be done about it. And that was only the beginning.

The extent of the damage and the reasons for the accident were not yet known. The new administration, which had just taken office two months earlier, was in the initial stages of assessing our energy situation. Our thoughts were on revitalizing the state's faltering economy and reinstilling pride and integrity in Pennsylvania state government.

We were to learn quickly. In fact, the accident would come to consume our full time and attention during the next few long days and nights.

Although nuclear fission was discovered in the 1930s, it was not until after World War II that scientists began to see the possibility of using the process as a source of cheap and seemingly limitless power. The American nuclear power industry was actually born in Pennsylvania when a prototype atomic reactor was connected to an electricity distribution network at Shippingport in 1957.

The use of nuclear energy for the generation of electricity continued to expand until the mid-1970s. Then, orders for many new reactors that were made earlier in the decade were canceled. By 1979 the nuclear power industry in the United States had reached a virtual standstill. No new plants have been ordered since 1978, and by the early 1990s all plants currently under construction will have been completed.

It became apparent that there were unresolved cost and safety issues that needed to be addressed if nuclear power was to continue to be a viable energy source. The disaster at Three Mile Island dealt the industry a blow that it is still feeling.

Today, in Pennsylvania, nuclear energy contributes about 19 percent of the state's electrical power. New nuclear capacity has been added at the Limerick Nuclear Generating Station in Chester County and the Beaver Valley Nuclear Plant in Beaver County. The TMI Unit 1 reactor has gone back on line. However, the future is still in question.

Pennsylvania's citizens still are concerned about the safety of these plants and the skyrocketing costs of construction and operation. The disposal of low-level and high-level radioactive waste as yet has not been adequately resolved and emergency planning efforts need to be retested and updated. Regulatory agencies like the Pennsylvania Public Utility Commission and the Federal Energy Regulatory Commission need to provide assurances that nuclear plants are in the best interests of the ratepayers. In short, nuclear energy must prove its worth in order to win back public support.

But on that day in March, I wasn't fully aware of the impact this issue would have on me, on my family, on the goals and programs of this administration, and on the future of this Commonwealth. Nuclear power then was still seen as the wave of the future, the new technological marvel of our times, a source of electricity that would become too cheap to meter.

As chairman of the Pennsylvania Emergency Management Agency and the Governor's Energy Council, it became my immediate responsibility to head the fact-finding efforts during the early stages of the accident. It quickly became obvious, with no precedent for such an event and no state agency responsible for nuclear crisis management, that our task was to take control and prepare for the worst.

Throughout that first day, we received conflicting reports on the severity of the accident. We determined that in order to maintain control of the situation, our first priority was to try to relay accurate and timely information to the public. This proved to be more difficult than we expected.

At a 4:30 p.m. press conference, I reported that the situation at that time was more complex than the company at first led us to believe. This became a continuous problem throughout the ordeal.

I again went to the public at 10 p.m. on that same day to give them our most accurate assessment of the situation and to let them know we were firmly in charge.

Throughout the next few exhausting days, we attempted to maintain this open channel of communication, but overreaction, misinformation, and jumbled communications almost caused the channel to break down. On Friday, the third day after the accident, public confidence reached a low point with the great evacuation scare.

We were able to narrowly avert a massive evacuation effort, which could have caused panic and injuries to many people, by keeping our heads and calmly sorting through the available facts and reports. At that point we insisted on independent on-site experts to help us manage the situation.

The public reacted favorably and by Monday, the sixth day after the accident, we were able to report that the crisis had finally passed. However, the work of searching for the facts and developing plans to avert a future crisis had just begun.

I was called upon to head the Governor's Commission on Three Mile Island, which conducted a seven-month investigation into the consequences of the accident. That report attempted to assess fairly the environmental, economic, health, legal, and social effects of the accident and to evaluate the emergency response procedure.

Through this experience and investigation we learned much about how to manage a nuclear emergency.

We learned to expect the unexpected and make sure the emergency apparatus is adequate at all times.

We learned that when an emergency does occur, it is imperative to have good people in place to handle the crisis.

We learned to resist the temptation to act merely for the sake of action. Calm, thoughtful, well-informed action is very important.

We learned not to try to manage a crisis from anywhere but at the crisis site. The people of this Commonwealth depend on their leaders to be in charge, and to be there when needed.

We learned to search for and evaluate the facts and their sources, and to communicate those facts truthfully and carefully. If we are to maintain public trust we must be sure we know what is happening and let the people know that as well. The TMI crisis, if it was anything, was a lesson in what misinformation and inadequate information can do.

We learned to respect, but not to depend upon the news media. Although the quality of reporting from the more than 400 reporters covering the story was often excellent, we did find some reports to be unreliable, adding to the confusion and erosion of public trust.

We learned to forget partisanship and depend on partnership from both political parties.

We relearned the lesson of valuing and learning from history. We hope this lesson will prove valuable to others.

And finally we learned that even after the immediate crisis is over there are issues that will linger.

Today, because of these efforts, we are better prepared. We know what has to be done; we know what to watch out for; we know how to handle a crisis of this magnitude. We also know that nuclear accidents have far-reaching and long-lasting consequences for the public and for the electric utility industry.

Nuclear energy may still be considered the "wave of the future." But until the questions of safety are adequately addressed and until the problems associated with the disposal of high-level nuclear waste are solved, we must continue to look at all alternatives to meet our electric power needs.

We think there are alternatives that may be less costly and in the long-term best interests of both the utilities and the ratepayers. In 1983, in my role as chairman of the Governor's Energy Council, I convened a blue-ribbon panel of experts to look at such alternatives.

We found that there are opportunities for the electric utility industry to regain financial viability through diversification, decentralization, and reduced reliance on expensive new power plants. If we begin to reward utilities for efficiency and conservation, if we reward utilities for flexibility, and if we continue to maintain an economic climate conducive to growth, we can meet the future power needs of this state and this nation without relying on any one form of energy. It is likely in the future that our power will come from a number of different sources, including coal, oil and gas, solar, and even through energy conservation.

But today the issue of the viability and safety of nuclear power is still very much with us. In October 1985, Three Mile Island's undamaged Unit 1 reactor was restarted after six-and-a-half years of hearings and legal battles. Clean-up efforts continue at the Unit 2 reactor. The issues of safety, competence, and integrity are still of concern to our citizens.

On April 26, 1986 the worst accident in the history of nuclear energy occurred at the Chernobyl nuclear power facility in the Soviet Union. We had hoped the lessons we learned at TMI would never be needed again. In fact, in the interest of sharing our knowledge and experience with the world, the governor had traveled to the Soviet Union in 1979 to meet with that country's top government and scientific leaders. He was told that they regarded nuclear safety as a "solved problem." Obviously they were wrong, and the Soviet people, and indeed, the people of the world are now paying the price.

It is my hope that we can begin to look at all the options, including nuclear energy, realistically. I also hope that we continue to support research into air quality problems, nuclear waste problems, and alternatives to conventional fossil fuel facilities. The future depends on us taking advantage of all our energy resources, and using them wisely for the generations that will follow us.

William W. Scranton III
Lieutenant Governor
January 15, 1987

# Preface

The idea for a comprehensive, annotated bibliography came to us shortly after the March 28, 1979 accident at the Three Mile Island Unit 2 (TMI-2) Nuclear Generating Station. The libraries in which we work, The George T. Harrell Library, The Milton S. Hershey Medical Center, The Pennsylvania State University, and the Medical Library, Polyclinic Medical Center of Harrisburg, were deluged with requests for information on a wide range of subjects related to nuclear power. Although the primary focus of both these institutions is health-related, it became clear quite early that our information services would be taxed beyond that narrow definition.

It was not until January 1986 that we began to seriously assess the volume of material already in our respective libraries and to outline the scope for a bibliography. Because papers are still being produced relative to the accident and its consequences, we arbitrarily chose July 1986 as the end point, although materials published past this date were included if they were readily available. This bibliography brings together a broad base of literature directly related to the TMI-2 accident and its aftermath; it covers the humanities, social sciences, basic and biological sciences, and government publications.

Although the intention was to be as comprehensive as possible, space limitations for the book made it necessary to evaluate each item and set parameters for the inclusion or exclusion of material. The bibliography is comprised of articles, editorials, letters, and news items from professional journals; monographs; federal and Pennsylvania government documents; and congressional hearings. All materials are written in English or are an English translation of a foreign language item. With only a few exceptions (proceedings being an example), all articles/documents were examined by one of the authors. Annotations are provided for journal articles and government documents. Books, congressional hearings, and news items and letters in journals are cited but not annotated. Proceedings (both full-length papers and abstracts) are included only if they appear in regularly published journals. Popular literature has been selectively chosen and, again, not annotated. Excluded are: book reviews; unpublished reports; federal and state laws and statutes; hearings (other than congressional); lawsuits; court decisions; dissertations or theses; brief news items in journals (for example, *Nuclear News* briefs); meeting abstracts/proceedings (if in a non-regularly published journal or unpublished); newspaper articles and reports; pamphlets; drafts or preliminary reports where a final document or report exists; and reports privately produced by utility companies or institutes and not sponsored by a government agency; and audiovisuals.

The bibliography is arranged by subject for the journal articles, with a separate section in the first chapter for books. Because of the overlap between subjects, some entries could legitimately be placed in more than one chapter, resulting in the need to decide on an appropriate placement for an entry. It is hoped that the indexing will provide appropriate access to articles with dual or multiple aspects. Popular literature and proceedings have been placed in separate chapters at the end. Within each chapter, the entries are in alphabetical order by author, or if there is no author, by title.

Government documents are arranged by issuing agency, with chapters for the President's Commission, the Nuclear Regulatory Commission, and the Department of

Energy. Another chapter picks up documents by other federal agencies. In all cases, the order is by department, agency, and division, where appropriate, and then by title of the document. Chapters on Pennsylvania documents and congressional hearings complete the documents chapters.

The bibliography is indexed by author, subject, and government document number, where appropriate. All individual authors are indexed. No attempt has been made to consolidate initials vs. full names, so that some authors may have more than one entry in the author index. Corporate authors and government agencies are not indexed. Entries are indexed by subject except for the proceedings section, which has access by author only. In the subject index, cross references are given to the appropriate index terms; additionally, references refer to other related index terms. For the government document number index, a "primary" number was selected for indexing. In cases where a primary number could not be determined (for example, a GEND and DE number for the same document), both appear in the index. Contract numbers are not indexed. Because of the number of documents, all Department of Energy (DE) numbers are listed last.

Sources consulted in the compilation of this bibliography on Three Mile Island include: *Readers Guide to Periodical Literature*, *Social Science Index*, *Engineering Index*, *Applied Sciences and Technology Index*, *Biological and Agricultural Index*, *Business Periodicals Index*, *General Science Index*, *Index to Legal Periodicals*, *Government Reports and Announcements*, and *Monthly Catalog*. Extensive online sources were searched using three vendors:

1. NLM: MEDLINE and HEALTH file

2. BRS: ABI/INFORM (INFO); AGRICOLA (CAIN); BIOSIS Previews (BIOL); ERIC; HAZARDLINE (HZDB); PAIS; Pollution Abstracts (POLL); PsychINFO (PSYC); Sociological Abstracts (SOCA); and Social SciSearch (SSCI)

3. DIALOG: ENERGYLINE (File 69); ENVIROLINE (File 40); ENVIRONMENTAL BIBLIOGRAPHY (File 68); LEGAL RESOURCE INDEX (File 150); MAGAZINE INDEX (File 47); MANAGEMENT CONTENTS (File 75); SCISEARCH (Files 34, 87, 94); and WATER RESOURCES ABSTRACTS (File 117)

The Nuclear Regulatory Commission supplied a printout of the documents the agency produced in connection with the accident, and the Three Mile Island Communications Library at the Simulator Building, in Middletown, Pennsylvania, provided a printout of documents pertaining to the accident. The Pennsylvania Department of Health supplied copies of documents and studies related to TMI as requested.

Libraries visited during the information gathering stages of the project include the Three Mile Island Communications Library, Simulator Building, Middletown, Pennsylvania; the Three Mile Island Technical Library, on site at Three Mile Island, Middletown, PA; Capital College, The Pennsylvania State University, Middletown, PA; State Library of Pennsylvania, Harrisburg, PA; State Law Library, Harrisburg, PA; and The George T. Harrell Library, The Milton S. Hershey Medical Center, The Pennsylvania State University, Hershey, PA.

Many libraries were called upon to contribute interlibrary loans, either in the form of photocopies of material or loan of original reports or documents. Notable among these were The Pennsylvania State University Libraries, the State Library of Pennsylvania, and the Free Library of Philadelphia.

The authors wish to thank the reference staff at the Milton S. Hershey Medical Center (Virginia Lingle, Dorothy Malcom, Loretta Ulincy, Robin Long and Barbara Chestnut) for their time and assistance in obtaining the many articles on interlibrary loan, and for their perseverance during the compilation of this bibliography; Joan Bernardo, also of Hershey, for her clerical assistance and proofreading; Debbie Phillips, librarian at the Three Mile Island Communications Library, for her many telephone consultations and assistance in locating documents; Joan Slavin, and

Bonnie Lerner, librarians at the Three Mile Island Technical Library, for permitting access to the technical reports on-site at Three Mile Island and for telephone consultation; Charles T. Young, Administrator, General Services Division, Polyclinic Medical Center, for his advice and support during this project; and Ann Hatter, for her clerical assistance and proofreading. Special thanks is given to Peter S. Houts, Ph.D., Department of Behavioral Sciences, the Milton S. Hershey Medical Center, for his technical expertise and advice regarding computer processing and for his assistance in the production of this bibliography.

It is inevitable in a bibliography of this size that some relevant references will have been missed. We have attempted to be as complete as possible but some errors are unavoidable. Additionally, some decisions that we made may have eliminated references that others would have included. "All method is imperfect. Error is all around it, and at the least opportunity invades it... But what can we do? There is no other way." - Charles Nicolle (1866-1936) *Biologie de l'Invention*, Introduction.

# Three Mile Island

# 1

# Accident Overviews, Sequence, and Causes

## GENERAL

0001
Abbott, Howard. "Tracing the Common Threat of Simplicity that Runs Through Four Major Product Disasters." **Risk Management** 29(November 1982):42-44, 46-47, 50, 52. (4 refs.)

TMI is one of four disasters for which causes are traced. In the case of TMI, the disaster included "old-fashioned things like pumps," along with problems in decision-making. Reprinted from *Product Liability International*, June 1982.

*Accident - causes; Human error; Decision making

0002
"The Accident at the Three Mile Island Nuclear Power Plant." **Soviet Atomic Energy** 47(1, 1979):575-578.

This translation from an article in *Atomnaya Energiya* 47(July 1979):61-63 gives a chronology of the accident and discusses the radiation situation.

*Accident sequence; Radiation dosage

0003
Blake, E. Michael. "B&W Presents its Case on TMI, Defends its Hardware." **Nuclear News** 22(July 1979):30-33. (News)

*Accident - causes; Babcock & Wilcox

0004
Blake, E. Michael. "Three Mile Island One Year Later: GPU Perseveres, But to What Future?" **Nuclear News** 23(March 1980):51-54, 56-57.

After one year, major problems still remain for GPU. Discussed here are the cleanup and its costs, interactions with the Pennsylvania Public Utilities Commission regarding rate-hike requests, management problems at GPU, and effects on the industry.

*TMI - anniversary (1); General Public Utilities; Financing; Cleanup; Costs; Nuclear industry

0005
Blake, E. Michael. "Three Mile Island; Public Wariness Remains After Crisis Ends." **Nuclear News** 22(May 1979):32-36. (refs.)

Blake describes the accident, including a chronological sequence of the event, and discusses implications for the nuclear industry in terms of the loss of public confidence.

*Overview; Accident sequence; Public opinion; Nuclear industry

0006
Blake, E. Michael. "TMI Plus Five; On the Island, the Work Goes On." **Nuclear News** 27(April 1984):53-58.

Blake reviews the past five years at TMI. Work continues on the cleanup of unit 2, while unit 1 is ready to be put in operation pending approval by the NRC. The switch from Met Ed to GPU Nuclear management is discussed.

*Overview; TMI - anniversary (5); Cleanup; General Public Utilities; Management

0007
"Bubbles from Three Mile Island." **Economist** 271(May 12, 1979):91. (News)

*Hydrogen bubble

0008
Bunyard, Peter. "Living on a Knife-Edge - The Aftermath of Harrisburg." **Ecologist** 9(June 1979):97-102.

The accident at TMI is overviewed, including the sequence of the accident, the "near melt-down," radiation venting, evacuation, cleanup, risks of nuclear power, and the effect on the nuclear industry.

*Overview; Accident sequence; Nuclear industry

0009
Butler, Paul. "Harrisburg: Nuclear Engineers Still in the Dark." **Engineer** 248(April 5, 1979):7. (News)

*Accident - causes

0010
Castro, W.R., and Cottrell, William B. "Preliminary Report on the Three Mile Island Incident." **Nuclear Safety** 20(July-August 1979):483-490. (29 refs.)

"This preliminary report summarizes the status of the plant and related activities through April 30, primarily on the basis of information from Nuclear Regulatory Commission (NRC) press releases." Included are a detailed incident chronology, a chronology of the system degasification, a preliminary NRC assessment of the causes of the accident, information on NRC instructions to other reactor licensees, and a list of fourteen investigation committees.

*Overview; Accident sequence; Nuclear Regulatory Commission; Accident - causes; Preliminary investigations; Accident investigations

0011
Cole, Richard W. "More on Three Mile Island." **Bulletin of the Atomic Scientists** 35(September 1979):63. (Letter)

*Decay heat; Hydrogen bubble

0012
"Consequences of Three Mile Island." **Medical Research Engineering** 13(July-August 1979):6-7, 26-31. (24 refs.)

The TMI incident is overviewed. Included are information on: the chronology of the incident, the President's Commission, activism related to the incident, safety versus economic concerns, psychological effects, legislative and governmental reactions, and effects on nuclear power.

*Overview; Accident sequence; Legislation; Accident investigations

0013
Crancher, D.W., and Nelmes, R.N. "Accident at the Three Mile Island Nuclear Power Station." **Atomic Energy in Australia** 23(July 1980):2-17.

Unavailable.

*Overview

0014
Culliton, Barbara J., and Waterfall, Wallace K. "Low-dose Radiation." **British Medical Journal** 1(June 9, 1979):1545-1546.

The near-disaster at Three Mile Island "simply capped a winter's growing discontent" over the risks involved in nuclear power. Mounting discontent has resulted from the government's handling of nuclear concerns and disagreement among government committees, scientists, and special task forces on the effects of radiation. The population around TMI will "provide the first big prospective low-level radiation study." Long-term health studies (20 to 30 years) will be carried out by the department of HEW.

*Overview; Nuclear industry; Radiation, low level; Health effects

0015
Denning, R.S. "The Three Mile Island Unit 2 Core: A Post-Mortem Examination." **Annual Review of Energy** 10(1985):35-52. (16 refs.)

The article reviews the sequence of events of the TMI accident, discusses the response of the NRC and the nuclear industry and, in a post accident examination of the plant, describes the cleanup operation and the current state of the TMI-2 core.

*Accident sequence; Industry response; Nuclear Regulatory Commission; Cleanup; Reactor core analysis

0016
Fells, Ian. "Three Mile Island and Its Implications." **Engineering** 220(March 1980):273-278.

The accident sequence at TMI is discussed in detail; illustrations show the accident at successive times, with values for temperatures and pressures. Implications of the Presidential Commission are discussed from the British perspective.

*Accident sequence; President's Commission; Recommendations; Great Britain

0017
FitzGerald, Chris. "Three Mile Island; Everyone's Concern." **Nuclear News** 22(June 1979):40-44, 47.

The accident is overviewed, including bringing the system under control, various commissions appointed, equipment problems, financial problems, public opinion, and legal actions.

*Overview

0018
Gordon, S.; Schmidt, K.H.; and Honekamp, J.R. "An Analysis of the Hydrogen Bubble Concerns in the Three-Mile Island Unit-2 Reactor Vessel." **Radiation Physics and Chemistry** 21(1983):247-258. (23 refs.)

The authors "analyzed all possible sources of non-condensable gases" and concluded that continued growth of the bubble was not possible. Maximum gas volumes are calculated. Computer simulations were used to calculate hydrolytic reactions and maximum amount of oxygen (which was always far below the explosion limit).

*Hydrogen bubble; Chemistry; Computer simulation

0019
Grover, John. "The Nuclear Accident at Harrisburg, Pennsylvania." **Medical Journal of Australia** 1(June 2, 1979):518-519. (Letter)

*Accident sequence; Radiation dosage

0020
Gwynne, Peter. "Happy Anniversary, Three Mile Island." **New Scientist** 85(March 27, 1980):990. (News)

*TMI - anniversary (1)

0021
"The History of Unit 2; A Fateful Chain of Abnormal Incidents." **Engineering News-Record** 202(April 5, 1979):13. (News)

*Accident sequence

0022
"Human Faults Underlie Three Mile Island." **New Scientist** 83(August 9, 1979):428. (News)

*Human error; Accident - causes

0023
Iwler, Louis. "Three Mile Island: Men, Design, Equipment Fail." **Electrical World** 191(May 1, 1979):29-31. (News)

*Overview; Accident sequence; Accident - causes; General Public Utilities; Cleanup; Financing

0024
Jaffee, Leonard. "Report of the Technical Assessment Task Force from Volume I: Technical Staff Analysis Reports Summary." **Progress in Nuclear Energy** 6(1-3, 1980):1-89.

Reprinted from the U.S. government document.

*Accident sequence; Technical assessment

0025
Jaffee, Leonard. "Technical Aspects and Chronology of the Three Mile Island Accident." **Annals of the New York Academy of Sciences** 365(1981):37-47. (3 refs.)

Jaffee discusses the technical aspects and sequence of events of the TMI accident, including possible causes. He also discusses some of the implications for the nuclear industry.

*Accident sequence; Nuclear industry

0026
Krause, Carolyn. "Oak Ridge Report on Three Mile Island: An Improbable Set of Circumstances." **Metal Progress** 116(December 1979):60-61.

Reprinted from the *Oak Ridge National Laboratory Review*, Fall 1979 issue.

*Accident sequence

0027
Kreiling, Janet. "What Happened at TMI-2?" **InTech (Instrumentation Technology)** 26(June 1979):7-8, 10, 13-14, 16 (refs.)

This is a detailed account of the events of the accident at TMI-2, including a chronology as understood by Nuclear Regulatory Commission officials "as reported in Enclosure 1 to the NRC's IE Bulletin 79-05A" and according to testimony given on April 23 by the president of GPU to a Senate committee.

*Overview; Accident sequence

0028
Kuhns, William G. "Three Mile Island Legacy Includes Benefits." **Catalyst for Environment/Energy** 7(4, 1982):8-12.

The positive lessons of TMI include changes in hardware and operations for improved safety, "people lessons," and the need for improved communication. The impact of the accident on the public's health and safety was "negligible."

*Lessons learned; Safety; Communications

0029
Lanouette, William J. "`No Longer Can the NRC Say...'" **Bulletin of the Atomic Scientists** 35(June 1979):6-8. (Commentary)

The focus is on the human and mechanical errors that occurred at TMI, "and what these errors portend for nuclear power's future." From incomplete monitoring of the initial release of radioactivity to the operator's errors to the pump failure and valve malfunction, the accident exposed a lack of emergency preparedness and a failure on the part of the NRC to adequately monitor and regulate the nuclear

industry.
*Nuclear industry; Nuclear Regulatory Commission; Emergency preparedness

0030
"The Lessons Learned at Three Mile Island." **New Scientist** 84(October 4, 1979):7. (News)
*Human error; Operators; Lessons learned

0031
Lewis, Harold W. "The Safety of Fission Reactors." **Scientific American** 242(March 1980):53-65.
The safety of "light water" reactors is discussed along with an evaluation of the Rasmussen Report. The series of events at Three Mile Island is detailed along with diagrams explaining the operation and malfunction of the system. "Three items -- transients, small loss-of-coolant accidents and human errors -- were the central features of the Three Mile Island accident." It is important to assess risk in quantitative terms and to make "sound interpretations of those risks."
*Accident sequence; Safety; Light water reactors; Risk

0032
Lombardo, Thomas G. "TMI Plus 2." **IEEE Spectrum** 18(April 1981):28-44.
The situation at TMI is examined after two years. The changes being made at TMI have affected the nuclear industry and are discussed in four broad categories: engineering changes, including new or updated hardware, instrumentation, and systems designed to discover and prevent future accidents; reducing the chance of human error, including changes in operating procedures, changes in control room design, and improved operator training; improving management attitudes; and revitalizing the Nuclear Regulatory Commission.
*Overview; Nuclear industry; Lessons learned; TMI - anniversary (2)

0033
McQueen, M.M. "Instrumentation Key Fault at TMI." **Nuclear News** 24(May 1981):29-30. (Letter)
*Instrument failure; Accident - causes

0034
Marshall, Eliot. "The Crisis at Three Mile Island: Nuclear Risks are Reconsidered. H2 Bubble is Unexpected Source of Trouble." **Science** 204(April 13, 1979):152-153.
The hydrogen bubble which developed in the nuclear reactor is the focus of this description of the TMI accident. Much of the information quoted here is from Metropolitan Edison or Nuclear Regulatory Commission spokespersons; by March 31, the utility deferred to the NRC "to speak for it on matters of fact." Confusion and lack of preparedness played a major role in reactions to the accident.
*Accident sequence; Communications; Press releases; Hydrogen bubble

0035
Marshall, Eliot. "A Preliminary Report on Three Mile Island." **Science** 204(April 20, 1979):280-281. (News)
*Accident sequence; Accident - causes; Preliminary investigations

0036
Mason, John F. "The Technical Blow-by-Blow." **IEEE Spectrum** 16(November 1979):32-48.
This article gives "details of the Three Mile Island accident as excerpted and edited from interviews with Nuclear Regulatory Commission investigators." A chronolog, in hours, minutes and seconds, details the accident in fairly technical terms; the foldout opposite page 42 is used as a reference for the sequence. A less technical narrative by Rubinstein appears on the same pages as this article.
*Accident sequence

0037
Mathes, J.C. "Three Mile Island: The Management Communication Role." **Engineering Management International** 3(January 1986):261-268. (21 refs.)

The author contends that "the accident at Three Mile Island was precipitated by a management communication failure" that occurred at Babcock & Wilcox (B&W). Communication problems at B&W fall into three areas: inappropriate lines of communication, lack of effective feedback, and informal communication procedures. The inefficient communication meant that a change in operator procedure that was recommended in 1977, and that could have prevented the accident at TMI, was never initiated.

*Accident - causes; Babcock & Wilcox; Management; Communications

0038
Norman, Colin. "Anatomy of an Accident." **Science** 228(April 5, 1985):32. (News)

*Accident sequence

0039
"Not One, but Two LOCAs at Three Mile Island." **Nuclear Engineering International** 24(May 1979):10-11. (News)

*Accident sequence; Radioactive releases

0040
"Odyssey of an `Event'; Ordeal at Three Mile Island." **Mechanical Engineering** 101(June 1979):60-61. (News)

*Accident sequence

0041
"OIE: Operators Could Have Prevented Accident." **Nuclear News** 22(September 1979):32-34. (News)

*Human error; Operators; Accident - causes

0042
"The Ordeal at Three Mile Island." **Nuclear News** 22(April 6, 1979):1-4. (Special Report)

This special report summarizes the accident sequence at TMI, and includes information on radiation release.

*Overview; Accident sequence; Radioactive releases

0043
"The Ordeal at Three Mile Island." **Professional Safety** 24(August 1979):29-32.

Reprinted from an April 6, 1979 Special Report of *Nuclear News*.

*Overview; Accident sequence; Radioactive releases

0044
Pigford, Thomas H. "Reply." **Nuclear News** 24(May 1981):30. (Letter)

*Instrument failure; Accident - causes

0045
"Prelude: The Accident at Three Mile Island." **EPRI Journal** 5(June 1980):6-13.

In the first of four articles, the accident itself is reviewed; the basics of pressurized water reactors are described, along with a detailed accident sequence, the uncovering of the core, and the hydrogen bubble. Diagrams of the reactor show the progression of the accident.

*Overview; Accident sequence; Pressurized water reactors; Reactor core uncovery; Hydrogen bubble

0046
Purcell, Arthur H. "Lessons of the Three Mile Island Nuclear Accident." **Environmental Conservation** 10(Summer 1983):155-157. (2 refs.)

This brief overview of TMI discusses technical causes of the accident, the "systems failure," lessons learned, and the international reaction.

*Overview; Accident - causes; Lessons learned; International reaction

0047
Reed, Adam V. "Who Caused Three Mile Island?" **Reason** 12(August 1980):16-23.

Reed has found the real reason for the TMI accident to be linked with eighty-year-old insurance legislation. Since casualty insurance must be based on prior experience, the nuclear industry was considered a special case, with a consortium of insurers set up for the industry, and the Price-Anderson Act, which limited liability. Given the situation, nuclear plants were not designed with human factors engineering in mind, and the control rooms were not as safe as they should have been.

*Accident - causes; Safety; Casualty insurance; Legislation; Human factors engineering; Control room design

0048
Robinson, Gail. "Bursting the Nuclear Bubble." **Environmental Action** 10(May 1979):23-25.

An overview of the TMI crisis is presented from the perspective of an environmentalist who says that "we have to get out" of nuclear power. Uncertainty on the part of both Met Ed and the NRC in handling the accident, possible inaccuracies in the measurement of the radiation, and the confusion over press releases and public relations are among the concerns pointed out by the author.

*Overview; Public relations; Nuclear industry; Press releases

0049
Rosenthal, Stephen. "What Really Went Wrong at Harrisburg?" **Engineers Australia** 51(May 18, 1979):25-29.

This report on the TMI accident has supplemented material from a congressional testimony to put together an overall account of the sequence of events at TMI. The chronology begins at 4 a.m. on March 28, 1979 and continues through April 3, with a report on cold shutdown on April 30.

*Accident sequence

0050
Rubinstein, Ellis. "The Accident that Shouldn't Have Happened." **IEEE Spectrum** 16(November 1979):32-48.

This is a narrative account of the accident based on reports from many experts. The narration follows a sequence of numbered events on a foldout schematic at the end of the article. Mason's article, on the same pages as this article, gives a more technical account.

*Accident sequence

0051
Schneider, Alfred. "Three Mile Island: The Chemical Engineering Aspects." **Chemical Engineering Progress** 76(September 1980):13-20. (8 refs.)

The author indicates that "much of the problem could have been avoided or mitigated had there been a better recognition of elementary chemical and chemical engineering practices." Some of the problems related to chemical reactions, engineering design or operations are reviewed, including the condensate polishers, the zirconium-steam reaction, and the hydrogen bubble. Brief information about the cleanup is given.

*Accident - causes; Chemistry; Engineering; Chemical process industry; Chemical engineering

0052
Smith, David A. "Three Mile Island: What Has It Taught Us?" **Pennsylvania Medicine** 82(June 1979):7. (Editorial)

Three Mile Island has taught us that there are yet many unanswered questions and that we were unprepared for a disaster of such magnitude. The actions of Harold Denton of the NRC and Governor Thornburgh restored confidence in a "near panic situation." What remains now is the problem of radioactive waste disposal.

*Overview; Lessons learned; Emergency preparedness

0053
"TMI Mishap Laid to Operator Error." **Engineering News-Record** 203(August 9, 1979):10. (News)

*Accident - causes; Operators; Human error

0054
Torrey, Lee. "The Week They Almost Lost Pennsylvania." **New Scientist** 82(April 19, 1979):174-178.

*Overview; Accident sequence

0055
"Valve Failure, Design and Human Error Crippled the Plant." **Engineering News-Record** 202(April 5, 1979):12-15. (News)

*Accident - causes; Valves; Equipment failure; Human error

0056
"We All Live in Harrisburg." **Economist** 271(April 7, 1979):19-20.

"We all live in Harrisburg" because what happened at TMI will affect the rest of the world in terms of energy policy and "less, more expensive nuclear energy" during the 1980s. A proposed solution may be to use smaller, lower-powered nuclear reactors or to fully automate the system, thus eliminating human error.

*Overview

0057
Weiss, Ellyn R. "Three Mile Island: The Loss of Innocence." In: **Accidents Will Happen: The Case Against Nuclear Power**, edited by Lee Stephenson and George R. Zachar, 17-42. New York: Harper and Row, 1980.

TMI is viewed as a watershed for nuclear power; the towers at TMI reflect a loss of innocence with regard to nuclear power. The sequence of events at TMI is described, followed by "warning signs" of the event, and selected quotes from government transcripts that show the "chaotic climate in which decisions were being made...."

*Accident sequence; Information exchange; Activism

0058
"What Happened at Three Mile Island." **New Scientist** 97(January 13, 1983):82-83. (News)

*Accident sequence

0059
"What Happened at Three Mile Island Plant?" **Congressional Quarterly Weekly Report** 37(April 7, 1979):623.

This is a brief overview and description of the accident, including mechanical and human problems.

*Overview; Accident sequence

0060
"The Worst Nuclear Power Plant Accident Yet." **Lancet** 1(April 28, 1979):909-910. (3 refs.) (Editorial)

This editorial succinctly describes the TMI accident, including estimated figures regarding radiation dosage to the general population. Brief reference is made to the Rasmussen and BEIR reports, along with commentary on the nuclear power industry in the United States in general.

*Overview; Radiation dosage; Nuclear industry; Population exposure

**MISCELLANEOUS**

0061
Ackerman, Adolph J. "Three Mile Island." **IEEE Spectrum** 17(February 1980):14. (Letter)
*General

0062
Austin, Nancy. "Diary of Three Mile Island Incident 3/28/79 - 4/4/79." **Social Education** 43(October 1979):458-459.

The diary of Nancy Austin of Camp Hill, Pennsylvania, documents "her day-by-day reactions to the nuclear crisis at Three Mile Island."

*Local residents; Diaries

0063
Bewley, D.K. "Three Mile Island not `a Major Disaster'." **Nature** 282(November 29, 1979):442. (Letter)
*Disasters; General

0064
Blair-Clough, Ida, and Wheeler, Brenda. "In the Shadow of Three Mile Island." **Instructor** 89(September 1979):115-116, 118, 120.

Children in a school near Three Mile Island used a simulation game to gain a better understanding of the accident.

*Simulation; Children; Learning; Education

0065
Bronstein, Daniel A. "Three Mile Island." **Science** 204(May 25, 1979):794-795. (2 refs.) (Letter)
*Hydrogen bubble; General

0066
Bruwer, Andre. "Thoughts After Three Mile Island." **Western Journal of Medicine** 131(July 1979):66-67. (Letter)

In a letter to the editor, Bruwer reflects on nuclear power, humanity, and Three Mile Island.

*General

0067
"Canadian Regulators Clear OH to Export." **Nuclear News** 25(June 1982):49-50. (News)

*General Public Utilities; Electric power purchase; Ontario Hydro; Cleanup; Canada

0068
Cheney, E. "In Nuclear Crisis, RPh's Dispensed Rx's and Advice." **American Druggist** 179(May 1979):37.

Unavailable.

*Pharmacists

0069
Christiansen, Ralph C. "An Analysis of Health Physics Employee Recruitment Activity: 1977-1982." **Health Physics** 44(June 1983):601-608. (2 refs.)

An analysis of the demand for employees in the health physics profession indicated that listings for positions in health physics increased after the Three Mile Island accident. It seems probable that "TMI triggered a major change in demand in the profession" and that "this demand continues even in the face of prolonged recession."

*Nuclear workers; Health physics; Employment

0070
Corey, Gordon R. "TMI in Perspective: Reactor Containment Stands Up, Difficult Decisions Remain." **Electric Perspectives** (4, 1979):3-5.

Unavailable.

*General

0071
Critchley, Joy. "A Jocose Jaunt Around Three Mile Island." **Business and Society Review** no. 58 (Summer 1986):62-64.

Critchley, who proposes turning TMI into "An Historic Park and Planned Community," takes a tongue-in-cheek tour of TMI.

*Satire; Tourism

0072
Davis, R.J. "Three Mile Island Revisited." **Association of Food and Drug Officials. Quarterly Bulletin** 46(3, 1982):205-207.

Unavailable.

*General

0073
"Engineers Committee Helps Evaluate Three Mile Island Nuclear Accident." **Industrial Engineering** 12(February 1980):56. (News)

*Engineers

0074
Foster, L. "Accident at Three Mile Island: How ORNL Responded." **Oak Ridge National Laboratory Review** 12(Fall 1979):1-5, 8-11.

Unavailable.

*Oak Ridge National Laboratory; General

0075
Garfield, Eugene. "Three Mile Island and the Information Explosion on Nuclear Energy." **Current Contents** 1980(15, 1980):5-13.

Unavailable.

*Information; Nuclear power

0076
Gilinsky, Victor. "Three Mile Island and the Hindenberg." **Chemical & Engineering News** 57(October 15, 1979):5. (Editorial)

*Lessons learned

0077
Goldstein, Robert P. "Three Mile Island." **IEEE Sepctrum** 17(February 1980):14, 16. (Letter)

*Engineering

0078
"GPU - OH Cable Link Plan to be Abandoned." **Nuclear News** 25(July 1982):35-36. (News)

*General Public Utilities; Electric power purchase; Ontario Hydro; Cleanup; Canada

0079
"The Great TMI Show." **New Scientist** 97(February 24, 1983):505. (News)
*Tourism; Cleanup

0080
Grover, J.C. "Nuclear Accident at Harrisburg." **Medical Journal of Australia** 2(August 11, 1979):145-146. (2 refs.) (Letter)
*General

0081
Gurdziel, Thomas. "Convert TMI to Coal." **Engineering News-Record** 212(March 29, 1984):17. (Letter)
*Coal

0082
Heller, Herbert H. "TMI and USAB." **IEEE Spectrum** 17(April 1980):8. (Letter)
*U.S. Activities Board

0083
Hugo, Mark. "Pride in TMI." **Mechanical Engineering** 102(December 1980):49. (Letter)
*Engineers

0084
"Investigating the Harrisburg Hoax." **Fusion: Magazine of the Fusion Energy Foundation** 2(1979):8-16.
Unavailable.
*General

0085
Iwler, Louis. "Shortage of Engineers is a TMI Aftermath." **Electrical World** 193(March 1, 1980):17-18. (News)
*Engineers

0086
Jameson, Richard T. "China is Near." **Film Comment** 15(May 1979):28-31.
In a review of the *China Syndrome*, the serendipitous occurrence of TMI is viewed as a "publicity coup" for the movie.
*China Syndrome

0087
Kellar, H.W. "Three-Mile Island Reaction." **Chemical Engineering Progress** 75(September 1979):6. (1 ref.) (Letter)
*General

0088
Levidow, L. "Three Mile Island -- Ideology of Safe Level as a Material Force." **Radical Science Journal** 1979(9, 1979):82-92.
Unavailable.
*General

0089
Lewenz, Lisa. "Close to Home: The View from Three Mile Island." **Environmental Action** 15(March 1984):10-11.

Several photographs of TMI by Lewenz taken from inside neighboring homes are reproduced.

*Photographs; Local residents

0090

Mammano, Nicholas J. "A Chemistry Lesson at Three Mile Island." **Journal of Chemical Education** 57(April 1980):286-287. (2 refs.)

Mammano, who teaches a general chemistry course for pre-technical, non-science majors, used the accident at TMI to develop a presentation on nuclear chemistry, gas laws, Le Chatelier's principle, stoichiometry, and Boyle's Law.

*Teaching; Chemistry

0091

Maness, C. Reid. "Three Mile Island." **Science** 204(May 25, 1979):795. (Letter)

*General

0092

Marshall, Eliot. "Reactor Mishap Raises Broad Questions." **Science** 215(February 12, 1982):877-878. (News)

An emergency shutdown of the Robert E. Ginna reactor near Rochester occurred on January 25, 1982. The emergency and actions of the operators is compared with the accident at Three Mile Island.

*Ginna reactor; Operators; Nuclear workers

0093

Masters, Frank, and Josuweit, Robert. "Meltdown!" **QST** 63(August 1979):57. (News)

*Amateur radio; Local residents

0094

Melov, John. "Nuclear Accident at Harrisburg." **Medical Journal of Australia** 2(August 11, 1979):145.

*General

0095

Milspaw, Yvonne J. "Folklore and the Nuclear Age: `The Harrisburg Disaster' at Three Mile Island." **International Folklore Review** 1(1981):57-65. (42 refs.)

Folklore surrounding the Three Mile Island accident has taken the form of popular beliefs, legends, and humor. Milspaw collected material from students at the Pennsylvania State University-Capitol Campus, from residents in the local area, and from colleagues. The material includes everything from a rumor regarding the prediction of TMI by Jeanne Dixon to jokes, humorous slogans, and sham advertisements/posters about the accident.

*Folklore; Humor

0096

Moss, Thomas H., moderator. "Background of the Three Mile Island Nuclear Accident, I: General Discussion." **Annals of the New York Academy of Sciences** 365(1981):48-53.

General discussion following the initial session on TMI.

*General

0097

Moss, Thomas H., moderator. "Background of the Three Mile Island Nuclear Accident, II: General Discussion." **Annals of the New York Academy of Sciences** 365(1981):86-94.

General discussion following the second session on TMI.

*General

0098
Moss, Thomas H., moderator. "Major Lessons and Issues, I: General Discussion." **Annals of the New York Academy of Sciences** 365(1981):323-333.
A general discussion following the "Major Lessons and Issues" section.
*General; Lessons learned

0099
Moss, Thomas H.; Sills, David L.; and Eisenbud, Merril. "Highlights of the Conference Thus Far, from the Three Session Chairmen." **Annals of the New York Academy of Sciences** 365(1981):275-279.
*General

0100
Myhra, S. "Three Mile Island and Implications for Australian Uranium Mining." **Search** 10(October 1979):350-355. (18 refs.)
The chain of events at TMI are discussed along with the effects on the nuclear power industry. The effect of TMI on nuclear power plants commissioned or on order is evaluated in relation to the projected supply of yellowcake through the 1990s. There will not be a substantial market for Australian uranium before 1990.
*Uranium mining; Nuclear industry; Accident sequence; Australia

0101
"ORNL `Recreates' TMI Incident in Nonnuclear Facility." **Journal of Metals** 32(June 1980):11. (News)
*Simulation; Research and development; Oak Ridge National Laboratory

0102
Pope, Terri, and Wenger, Dennis. "TMI in the Literature: A Partially Annotated Bibliography." **International Journal of Mass Emergencies and Disasters** 2(March 1984):197-211.
This highly selected, partially annotated bibliography contains 60 items related to the TMI accident. The focus is on the social and behavioral aspects of the accident.
*Bibliographies; Social sciences

0103
Pritikin, Roland I. "Feedback: Readers Reply to the July/August Professional Platform on Three Mile Island." **Applied Radiology** 7(November-December 1979):206. (Letter)
*General

0104
Rasmussen, Norman C. "Methods of Hazard Analysis and Nuclear Safety Engineering." **Annals of the New York Academy of Sciences** 365(1981):20-36. (3 refs.)
The author discusses the Reactor Safety Study (Wash 1400) and its ability to predict an event such as TMI. He concludes that the transient tree event did predict TMI with the exception of the hydrogen bubble and the failure to predict the psychological impact. Includes discussion.
*Reactor Safety Study (WASH 1400); Risk

0105
Robertson, W. Davenport. "Three Mile Island." **Bulletin of the Medical Library Association** 68(October 1980):376. (Letter)
*Library services

0106
Snyder, L. "Cooperative Response to TMI." **Association of Food and Drug Officials. Quarterly Bulletin** 45(3, 1981):130-137.

Unavailable.
*General

0107
Straus, Donald B. "Lessons for the Resolution of Technical Disputes." **Annals of the New York Academy of Sciences** 365(1981):334-341.
Straus conducted a discussion of the TMI conference as a whole, and problems and unresolved issues that occurred during the conference.
*General

0108
Swarthe, Eric. "Three Mile Island." **IEEE Spectrum** 17(February 1980):14. (Letter)
*Engineering

0109
"Study Completed on Hydrogen Burn During TMI Accident." **Machine Design** 56(June 21, 1984):10. (News)
*Hydrogen burn; Research and development

0110
"Telco Teamwork Provides Communications in Nuclear Scare." **Telephony** 196(May 21, 1979):24-28, 31.
The teamwork between Bell Telephone Co. of Pennsylvania and United Telephone Co. of Pennsylvania, along with several other telephone companies, during the TMI crisis is described.
*Communications; Telephones

0111
Thom, Gary C. "Three Mile Island." **Science** 204(May 25, 1979):794. (Letter)
*General

0112
"Three Mile Island Alert Gets Library Response." **Library Journal** 104(May 15, 1971):1098. (News)
*Library services

0113
"Three Mile Island -- An Engineering Appraisal." **Mechanical Engineering** 102(July 1980):58-59. (News)
*Engineering; President's Commission

0114
"TMI-2 Hydrogen Burn Studied." **Nuclear Engineering International** 29(July 1984):37-38. (News)
*Hydrogen burn; Research and development; Accident investigations

0115
Tomb, Gordon W. "TMI to Coal?" **Engineering News-Record** 213(November 8, 1984):9. (Letter)
*Coal

0116
Torrey, Lee. "Engineering Solutions to Nuclear Accidents." **New Scientist** 84(November 15, 1979):503.
*Engineering

0117
Wood, M. Sandra; Thevos-Leskawa, Angelica; and Capitani, Cheryl A. "A Medical Library's Response to Three Mile Island." **Bulletin of the Medical Library Association** 68(April 1980):242-244. (5 refs.)

Faced with the need to provide information for patrons on the Three Mile Island accident, radiation injuries, and disaster planning, the reference department of the Hershey Medical Center Library developed a vertical file of materials including: bibliographies, primarily generated from MEDLINE; photocopies of journal articles; news articles from newspapers; and government documents.

*Library services

## BOOKS

0118
Barnes, Kent et al. **Responses of Impacted Populations to the Three Mile Island Nuclear Reactor Accident: An Initial Assessment.** New Brunswick, NJ: Rutgers University, Department of Geography, 1979. 22, 4 p.
*General; Local residents

0119
Bromet, Evelyn. **Three Mile Island: Mental Health Findings.** In collaboration with Parkinson, David; Schulberg, Herbert C.; Dunn, Leslie; and Gondek, Paul C. Pittsburgh, PA: Western Psychiatric Institute and Clinic, 1980. 78 p.
*Health effects - psychological; Mental health

0120
Brunn, Stanley D.; Johnson, James H., Jr.; and Zeigler, Donald J. **Final Report on a Social Survey of Three Mile Island Area Residents.** East Lansing, MI: Michigan State University, Deptartment of Geography, 1979. 218 p.
*Local residents; Social sciences

0121
Cantelon, Philip Louis, and Williams, Robert C. **Crisis Contained: The Department of Energy at Three Mile Island.** Carbondale, IL: Southern Illinois University Press, 1982. 213 p. (0-8093-1079-1)
*General; Department of Energy

0122
Del Tredici, Robert. **The People of Three Mile Island.** San Francisco, CA: Sierra Club Books, 1980. 127 p. (0-87156-237-5)
*Local residents

0123
Ford, Daniel F. **Three Mile Island: Thirty Minutes to Meltdown.** New York, NY: Viking Press, 1982. 271 p. (0-670-70859-3) pbk. (0-14-006048-0) clth.
*General; Accident sequence

0124
Gray, Mike, and Rosen, Ira. **The Warning: Accident at Three Mile Island.** New York, NY: W.W. Norton, 1982. 287 p. (0-393-01522-x)
*General; Nuclear power

0125
Keisling, Bill. **Three Mile Island, Turning Point.** Seattle, WA: Veritas Books, 1980. 184 p. (0-935-80400-5) pbk.
*General; Nuclear power

0126
Leppzer, Robert. **Voices from Three Mile Island: The People Speak Out.** Trumansburg, NY: Crossing Press, 1980. 86 p. (0-89594-041-8) clth. (0-89594-042-6) pbk.
*Local residents

0127
Martin, Daniel. **Three Mile Island: Prologue or Epilogue?** Cambridge, Mass: Ballinger Publishing Company, 1980. 253 p. (0-88410-629-2)
*General; Nuclear power

0128
Sills, David L.; Wolf, C.P.; and Shelanski, Vivien B., editors. **Accident at Three Mile Island: The Human Dimensions.** Boulder, CO: Westview Press, 1982. 258 p. (0-86531-165-x)
*General; Psychosocial effects; Health effects - psychological; Stress; Social sciences

0129
Sills, David L. **Social Science Aspects of the Accident at Three Mile Island.** Boulder, CO: Westview Press, 1980.
*Social sciences

0130
Staley, John C., and Seip, Roger R. **Three Mile Island: A Time of Fear.** Harrisburg, PA: RFJ Inc., 1979. 64 p.
*General; Stress; Health effects - psychological

0131
Starr, Philip, and Pearman, William, compilers. **Three Mile Island Sourcebook: Annotations of a Disaster.** New York, NY: Garland Pub., 1983. 411 p. (0-8240-9184-1)
*Bibliographies

0132
Stephens, Mark. **Three Mile Island.** New York, NY: Random House, 1980. 245 p. (0-394-51092-5)
*Overview

0133
Toth, L.M.; Malinauskas, A.P.; Eidam, G.R.; and Burton, H.M., editors. **Three Mile Island Accident: Diagnosis and Prognosis.** Washington, DC: American Chemical Society, 1986. (ACS Symposium Series 293, 1986). 302 p. (0-8412-0948-0)
*Overview

# 2

# International Commentary and Reaction

0134
Barnaby, Wendy. "First the Election and Then the Referendum." **Bulletin of the Atomic Scientists** 35(November 1979):43-47.

The reaction to the Three Mile Island accident was strongly felt in Sweden, where a referendum will be put to the people on the future of nuclear power in that country.

*International reaction; Sweden; Referendum; Politics

0135
"Britons Hear Lawroski Shed Light on TMI." **Nuclear News** 23(July 1980):54-55. (News)

*International reaction

0136
"Commission Reports on TMI Implications." **Nuclear News** 23(January 1980):54-56. (News)

*International reaction; Sweden

0137
Dunster, H.J. "Some Reactions to the Accident at Three Mile Island." **Nuclear Energy** 19(June 1980):139-146.

Dunster reviews the TMI accident and "identifies those features which ought to encourage changes in UK practices." Discussed are: emergency procedures, siting considerations, and trends in regulatory policy. The paper was given at a BNES meeting in London on March 13, 1980; an extensive discussion is appended.

*United Kingdom; Lessons learned; Safety; Regulation

0138
"Europe Simulates Harrisburg." **Economist** 271(April 14, 1979):100, 102. (News)

*International reaction; Simulation

0139
Greenhalgh, Geoffrey, and Waugh, Alyson. "Harrisburg Overshadows Hamburg Nuclear Conference." **Energy International** 16(July 1979):37-40.

The European Nuclear Conference in Hamburg, held about one month after the TMI accident, was dominated by discussion of the events at Harrisburg. Topics included a description of the accident itself, public relations aspects, and public opinion.

*International reaction; European Nuclear Conference

0140
Halliday, P. "The Three Mile Island Incident -- Reactions from Overseas." **Nuclear Engineer** 23(March-April 1982):44-49. (18 refs.)

Following an overview of the accident and a discussion of the conclusions of the accident investigations, international reaction to the accident is summarized.

Countries overviewed are Germany, France, Japan, Spain, Sweden, Italy, Russia, and Switzerland. All countries have benefited from the accident by improvements in operator training, changes in control room design, and improved safety.

*International reaction; Operators; Control room design; Safety; TMI - anniversary (3)

0141
"Harrisburg Fallout." **Economist** 271(April 14, 1979):45-47.

Nuclear energy policy and production in European countries and the Soviet Union are examined in the wake of Three Mile Island.

*International reaction; Europe; Public policy

0142
"Imperial View." **New Scientist** 83(August 9, 1979):428. (News)

*Great Britain; Accident - causes

0143
Iwler, Louis. "How the World Reacted to the TMI Accident." **Electrical World** 193(January 15, 1980):17-18. (News)

*International reaction

0144
Knox, Richard, and Adkins, Bruce. "Profit and Loss After Three Mile Island." **Nuclear Engineering International** 24(June 1979):15-16. (1 ref.) (Editorial)

The authors discuss some of the lessons of the Three Mile Island accident that were discussed at the recent European Nuclear Conference in Hamburg. The political reaction of calling for international safety talks is predictable, but would only serve to "paper over the cracks" of energy policies. Also discussed was the press coverage of the TMI accident, and the need for improved public confidence in nuclear energy.

*European Nuclear Conference; International reaction; Lessons learned; Politics; Press coverage; Public opinion

0145
Nelkin, Dorothy, and Pollak, Michael. "A Pregnant Pause: The European Response to the Three Mile Island Accident." **Annals of the New York Academy of Sciences** 365(1981):186-195. (20 refs.)

The European response in Europe ranged from a "*de facto* moratorium in Germany to an expansion of the nuclear program in France." TMI added to the political and economic controversy of nuclear power in Europe.

*International reaction; Public policy; Public opinion

0146
"Oil Giants Exaggerated Harrisburg -- Izvestiya." **Nature** 279(May 17, 1979):179. (News)

*Media; International reaction; Soviet Union; Czechoslovakia

0147
Patterson, Walter C. "Harrisburg ist uberall." **Bulletin of the Atomic Scientists** 35(June 1979):9-11.

The TMI accident occurred virtually at the same time as the Gorleben hearings in West Germany, resulting in a "profound" psychological impact in Europe. In addition to the West German impact, reactions in France, Sweden, Australia, Switzerland, Belgium, Denmark, and Great Britain are discussed. The overall perspective is "anti-nuclear"; the author is energy consultant to the Friends of the Earth, London.

*International reaction; Politics; Europe; West Germany

0148
Peterson, P. Owe. "Sweden's Disaster Plans -- After Harrisburg." **Hospitals** 53(November 16, 1979):14.
In the wake of TMI, Sweden, like the United States, has become concerned about the risk of nuclear accidents. Sweden's emergency plan for nuclear disasters is described, with special reference to the Uppsala health service region.
*Emergency planning; Disaster planning; Sweden

0149
Reisch, Frigyes. "Applying the Lessons From Harrisburg in Sweden." **Nuclear Engineering International** 25(March 1980):21-23. (refs.)
As a direct result of the TMI accident, new equipment designed to prevent a similar accident has been installed at the Ringhals 2 plant in Sweden.
*Lessons learned; Sweden; International reaction; Accident prevention

0150
Rippon, Simon. "International Regulators Learn about TMI; but will They Talk?" **Nuclear News** 22(August 1979):55-56. (News)
*International reaction; Regulation

0151
Rippon, Simon. "TMI the Absorbing Topic at Hamburg Conference." **Nuclear News** 22(June 1979):66-67. (News)
*International reaction; European Nuclear Conference

0152
Schwartz, Joe. "`We all Live in Pennsylvania...'." **Nature** 278(April 12, 1979):589-590. (News)
*International reaction; Politics

0153
Spilker, Reinhard. "What Lessons from Harrisburg?" **AMBIO/A Journal of the Human Environment** 10(1981):44-45. (News)
*International reaction; Lessons learned

0154
"Three Mile Island Reactions." **Nuclear News** 22(May 1979):66-67. (News)
*International reaction

0155
"Two Years On." **Nuclear Engineering International** 26(April 1981):11. (News)
*International reaction; Nuclear industry; TMI - anniversary (2)

# 3

# Emergency Preparedness and Disaster Planning

0156
"Area Nurses Faced Moral Dilemma in World's Worst Nuclear Accident." **Pennsylvania Nurse** 34(June 1979):1, 3. (News)
*Emergency preparedness; Emergency medical services; Nurses

0157
Breo, Dennis L. "Nuclear Scare Tests Hospital's Disaster Plan." **Hospitals** 53(May 1, 1979):33-36.
An inside view of emergency procedures and evacuation plans in force at the Hershey Medical Center is provided through interviews with several key individuals at the Hershey hospital. The Hershey facility is the closest hospital to TMI and had been "designated as the receiver of casualties from the Three Mile Island nuclear plant, should a major disaster occur there." Ironically, it had always been assumed that the medical center "would be the provider of emergency medical services" rather than the consumer. The patient population was reduced by two-thirds in anticipation of possible evacuation.
*Hospitals; Disaster planning; Emergency medical services

0158
Carter, Luther J. "Nationwide Protection from Iodine-131 Urged." **Science** 206(October 12, 1979):201-202, 204. (News)
In a report prepared for the Council on Environmental Quality, Jan Beyea and Frank Von Hippel urge nationwide distribution of iodine-131 to populations living within 100 miles of nuclear reactors. The report is being reviewed by the President's Commission on Three Mile Island.
*Potassium iodide; Health effects - cancer; Emergency preparedness

0159
Davis, Ruth W. "Three Mile Island, A Nursing Dilemma." **Supervisor Nurse** 10(December 1979):12-15.
Disaster planning began for York County on Friday following the TMI accident. Three aspects of York County's quickly-conceived TMI disaster plan were: public health, including evacuation paths, mobilizing the necessary personnel and transportation; nursing home evacuation; and the role of the emergency rooms of local hospitals. The development of the plan is described on a personal level from the viewpoint of nurses who were directly involved in the activities.
*Disaster planning; Emergency medical services; Nurses

0160
DeMuth, William E., Jr., and Trautlein, Joseph J. "The Luck of Three Mile Island." **Journal of Trauma** 19(October 1979):792-794. (4 refs.) (Editorial)
In an editorial, DeMuth and Trautlein discuss the lucky aspects of Three Mile Island -- that the accident was no worse than it was and that inadequate emergency procedures did not have to be tested. The Emergency Medical Service system in

effect at the Hershey Medical Center at the time of the accident is described and suggestions are made based on the TMI experience.

*Disaster planning; Emergency medical services; Hospitals

0161
"An Eyewitness Account of Three Mile Island Incident." **American Nurse** 11(May 20, 1979):5, 13. (News)

*Emergency preparedness; Nurses; Emergency medical services

0162
Fischer, David W. "Planning for Large-Scale Accidents: Learning from the Three Mile Island Accident." **Energy** 6(January 1981):93-108. (3 refs.)

The organizations involved, their interactions and connections, and decision-making issues related to the TMI accident are explored. The planning for accident management that has resulted from this in-depth analysis of problems raised by TMI will result in a framework for planning of operations for future accidents.

*Emergency planning; Organizational planning

0163
Gorinson, Stanley M.; Kane, Kevin P.; Harvey, Charles; Dicker, Ruth; Pearson, Eric; Reicher, Dan W.; Blood, Thomas; and Sherman, Dan. "Report of the Office of Chief Counsel: Emergency Preparedness." **Progress in Nuclear Energy** 6(1-3, 1980):361-431. (260 refs.)

Reprinted from the U.S. government document.

*Emergency preparedness; Office of Chief Counsel

0164
Grimes, Brian K.; Ramos, Steve L.; and Weiss, Bernard H. "Emergency Planning and Preparedness Since Three Mile Island." **Progress in Nuclear Energy** 10(3, 1982):363-386. (18 refs.)

Improvements in emergency planning and preparedness that have occurred in the nuclear industry as a direct result of the Three Mile Island accident are detailed. The basic guidance document is NUREG-0654, which was developed immediately following the accident. The role of the NRC versus the Federal Emergency Management Agency (FEMA) is discussed.

*Emergency planning; Emergency preparedness; Lessons learned; Nuclear Regulatory Commission; Federal Emergency Management Agency

0165
Haglund, K. "At Hershey Medical System Near `Failure' During Three Mile Island." **New Physician** 28(June 1979):24-25. (News)

*Disaster planning; Emergency medical services; Hospitals

0166
Henderson, Oran K. "Commonwealth of Pennsylvania Emergency Preparedness and Response: The Three Mile Island Accident." **Annals of the New York Academy of Sciences** 365(1981):316-322.

Henderson, from the Pennsylvania Emergency Management Agency, discusses "Pennsylvania's emergency response mechanism and legal basis in being at the time of the Three Mile Island incident." He discusses the sequence of events affecting the agency, how the agency responded to the incident, and lessons learned from TMI.

*Pennsylvania Emergency Management Agency; Emergency preparedness; Pennsylvania

0167
Hull, Andrew P. "Emergency Preparedness for What? Implications of the TMI-2 Accident." **Nuclear News** 24(April 1981):61-67.

Hull discusses emergency preparedness, including potential offsite doses and the concept of emergency planning zones. He presents fairly detailed dose estimates for

the TMI-2 accident, giving evidence that most of the radioactivity "was successfully retained within the containment building," with no more than 10% of the fission gases escaping.
*Emergency planning; Radiation dosage

0168
"In the Wake of Three Mile Island: FEMA Spearheads Response Plans." **Security Management** 28(September 1984):134-137.
Following the Kemeny Commission report on TMI, executive and legislative actions transferred responsibility for off-site emergency planning around nuclear power plants from the Nuclear Regulatory Commission to the Federal Emergency Management Agency (FEMA). Actions taken by FEMA since TMI regarding emergency plans and preparedness are discussed here.
*Federal Emergency Management Agency; Emergency planning

0169
"Involvement in Team Effort; Nurses on the Scene at Three Mile Island." **American Nurse** 11(April 20, 1979):1, 16. (News)
*Emergency preparedness; Nurses; Disaster planning; Health care

0170
Isler, Charlotte. "Could You Cope With a Nuclear Accident?" **RN** 42(June 1979):66, 71, 73-74, 77.
The accident at TMI triggered this article on disaster planning for radiation injuries. Hospitals should have written procedures for handling radiation emergencies; such a plan is described here. Details are given for preparing the emergency room for the arrival of the patient and what to do after the patient arrives. Also described are the various physical and health effects of radiation following different exposure levels.
*Disaster planning; Emergency medical services; Hospitals

0171
Johnson, James H., Jr. "Role Conflict in a Radiological Emergency: The Case of Public School Teachers." **Journal of Environmental Systems** 15(1986):77-91. (40 refs.)
In a survey of public school teachers near the Diablo Canyon Nuclear Power Plant, nearly one-third of the teachers surveyed indicated that they "would not assist in an emergency evacuation of schools" during a radiological emergency. Findings are consistent with evidence from the accident at TMI.
*Evacuation; Disaster planning; Teachers

0172
Johnston, Donald R. "Crisis Administration of the Hospitalized." **Society** 20(September/October 1983):21-23.
Following an overall disucssion of emergency management/evacuation proceedings for hospitalized patients, the specific case of Three Mile Island is evaluated (p. 23).
*Emergency preparedness; Hospitals; Evacuation; Emergency planning

0173
Kerr, Charles. "Three Mile Island -- the Next Time." **Medical Journal of Australia** 143(July 8, 1985):5. (1 ref.) (Editorial)
In a concise editorial that notes the public anxiety which resulted from official indecision in the Three Mile Island incident, the importance of clear and decisive communication in disaster management is emphasized.
*Disaster planning; Communications

0174
Kmet, Mary Alice. "Planning Put to the Test." **Security Management** 23(June 1979):21-23, 26-28, 74.

The TMI accident is viewed from the perspective of four persons who are involved with management and/or security in three local businesses (AMP, Inc.; United Telephone; and McCoy Electronics) and a hospital (Hershey Medical Center). Reactions and emergency procedures are discussed, along with informal comments concerning reactions of local residents.

*Emergency preparedness; Press releases; Security; Local residents

0175
Kuntz, Esther. "Hospitals Prepare Radiation Plans in Wake of Nuclear Plant Accident." **Modern Healthcare** 9(July 1979):16. (News)

*Evacuation; Emergency planning; Disaster planning; Emergency medical services; Hospitals

0176
Kuntz, Esther. "Radiation Treatment Plans Inadequate for Disasters." **Modern Healthcare** 9(May 1979):15-16. (News)

*Evacuation; Emergency planning; Disaster planning; Emergency medical services; Hospitals

0177
Kuntz, Esther. "Ready to Evacuate Area? Nuclear and Chemical Accidents Test Hospital Disaster Plans." **Modern Healthcare** 9(May 1979):14-15, 19. (News)

*Disaster planning; Hospitals

0178
Lesher, Delores C., and Bomberger, Audrey S. "Experience at Three Mile Island." **American Journal of Nursing** 79(August 1979):1402-1408. (9 refs.)

Using a "Diary of Events" for the eight days following the TMI accident, two nurses from Lebanon County, PA detail the events and their personal reactions to the accident. The psychological effects of a disaster and the biological effects of radiation are briefly described; a table gives the disease classification of the acute radiation syndrome based on doses uniformly received. York County's radiation emergency plan is summarized, and the radiation disaster plan in effect at Lebanon Valley General Hospital is described.

*Emergency planning; Health care; Hospitals; Health effects - psychological

0179
Linnemann, Roger E. "The Three Mile Island Incident in 1979: The Utility Response." In: **The Medical Basis for Radiation Accident Preparedness**, edited by Karl F. Hubner and Shirley A. Fry, 501-509. New York: Elsevier/North Holland, 1980.

The emergency medical services in place at TMI at the time of the accident are reviewed. The resources included use of the Hershey Medical Center as the support hospital and support for regional medical care from the Department of Radiology at the Hospital of the University of Pennsylvania and Radiation Management Corporation of Philadelphia.

*Emergency preparedness; Emergency medical services

0180
Lippin, Paula. "Managing Under the Specter of Three Mile Island." **Administrative Management** 40(June 1979):28-29, 68, 70, 74.

Faced with the possibility of mass evacuation, leaders in the business community near TMI responded with contingency plans "to ensure the continuation of business in the face of potential disaster." Managers from Agway, Inc., Commonwealth National Bank of Harrisburg, Hershey Foods Corp., and Community General Osteopathic

Hospital are interviewed.
*Local business; Evacuation; Emergency planning

0181
MacLeod, Gordon K. "A Role for Public Health in the Nuclear Age." **American Journal of Public Health** 72(March 1982):237-238. (12 refs.) (Editorial)

In his editorial regarding Maxwell's article (*AJPH* 72(March 1982):275-278), MacLeod comments on the health effects of TMI and the implications of the lack of preparedness of public health officials during the accident.
*Public health; Emergency preparedness; Health effects

0182
MacLeod, Gordon K. "Some Public Health Lessons from Three Mile Island: A Case Study in Chaos." **AMBIO/A Journal of the Human Environment** 10(1, 1981):18-23. (9 refs.)

MacLeod, Pennsylvania's Secretary of Health during the TMI crisis, points out the deficiencies and lack of preparedness of Pennsylvania's public health sector. Inadequate resources and misinformation led to chaos. Decisions were made by engineers and physicists instead of doctors. Also discussed is the increased frequency of hypothyroidism downwind of two Pennsylvania reactor sites.
*Public health; Emergency preparedness; Health effects; Hypothyroidism

0183
Maxwell, Christopher. "Hospital Organizational Response to the Nuclear Accident at Three Mile Island: Implications for Future-Oriented Disaster Planning." **American Journal of Public Health** 72(March 1982):275-279. (14 refs.)

The focus is on the emergency response of the four hospitals in Dauphin County. Crash plans successfully reduced hospital census to less than 50 percent within 48 hours of the accident. Problems discussed are: staffing, communications, administrative response, emergency/critical care services, and hospital evacuation and transportation. Recommendations included stockpiling of potassium iodide, development of contingency staffing plans, and planning of mass medical evacuation and transportation systems.
*Emergency preparedness; Hospitals; Health care

0184
Mendes, Robert F. "Develop Nuclear Emergency Plans; Mind Lessons of Pa. Plant Accident." **Business Insurance** 13(May 14, 1979):43, 46. (News)
*Emergency planning; Safety; Lessons learned

0185
Muller, H. Arnold. "Health Care Lessons Learned From the Three Mile Island Incident." **Bulletin of the New York Academy of Medicine** 59(December 1983):994-997.

Muller, Pennsylvania's Secretary of Health and formerly with the Hershey Medical Center (TMI's medical facility), indicates that the most important lesson learned from TMI is to be prepared for a nuclear accident. As part of a panel presentation, he talks about problems encountered, such as the potential need for moving patients, triaging large numbers of people, the need for a health physics expert, and difficulties with communication (problems with phone service and recognizing the most reliable source of information).
*Emergency medical services; Emergency planning; Lessons learned; Communications

0186
"Nuclear Emergencies: Coordinating the Federal Response." **EPRI Journal** 5(June 1980):39-43.

Based on the recommendation of the Kemeny Commission, the Federal Emergency Management Agency (FEMA) was directed to assume primary responsibility for coordinating off-site planning and "coordinating federal response to emergencies at nuclear power plants." FEMA's role, along with the planning process, is described.

*Federal Emergency Management Agency; Emergency preparedness; Emergency planning; Government

0187
"Six Pennsylvania Hospitals Near Nuclear Site `Fought' Confusion During Crisis." **Hospitals** 53(May 1, 1979):19-20.

This brief commentary draws attention to the problem of unclear lines of authority in a disaster such as TMI with possible need for mass evacuation. Hospitals did not know who had the authority to order evacuation -- the governor, Civil Defense, or the Department of Health. A list of precautionary measures taken by the six hospitals nearest the site of the accident are noted.

*Hospitals; Emergency planning; Disaster planning; Communications

0188
Small, William E. "Mobilizing for Three Mile Island -- The Disaster That Wasn't." **American Pharmacy** 19(June 1979):8-9.

Two days after the TMI accident, the FDA ordered the emergency production of 250,000 30-ml bottles of potassium iodide solution, to be shipped to Harrisburg in a 72-hour period. Although never used, the exercise was not in vain -- much was learned in arranging for the production and transportation of the drug.

*Emergency preparedness; Potassium iodide

0189
Smith, J. Stanley, Jr., and Fisher, James H. "Three Mile Island: The Silent Disaster." **JAMA** 245(April 24, 1981):1656-1659. (4 refs.)

During the TMI incident from March 28 to April 4, 1979, the Dauphin County Office of Emergency Preparedness developed an emergency plan to evacuate the 200,000 residents in a 20-mile radius from the plant, including the coordination of the evacuation of four short-term and fifteen long-term care facilities in Dauphin County. Key problems identified in the planning process were communications, transportation, and manpower resources.

*Emergency planning; Evacuation; Hospitals

0190
Strohl, George R. "Nuclear Threat: Hospitals Need to Know." **Osteopathic Hospitals** 23(July/August 1979):6, 8-9.

The lack of preparedness of hospitals in response to the TMI accident is described along with a discussion on the need for planning for nuclear emergencies.

*Hospitals; Emergency preparedness

0191
"Telcos Were Lifeblood of the Three Mile Island Emergency." **Communications News** 16(June 1979):36. (News)

*Emergency preparedness; Telephones; Communications

0192
Vinsel, Douglas B. "Hospitals Must Plan for Nuclear Accidents." **Hospitals** 54(August 16, 1980):113-118, 121.

York Hospital, located close to TMI, determined that its emergency plan was inadequate to handle a mass emergency as a result of a nuclear accident. A major concern was the need to evacuate the hospital. Vinsel discusses the need for a comprehensive emergency plan to handle nuclear accidents, the changes made at York Hospital as a result of TMI, and York Hospital's evacuation plan.

*Emergency planning; Evacuation; Hospitals

0193
Von Hippel, F. "Available Thyroid Protection." **Science** 204(June 8, 1979):1032. (Letter)

*Potassium iodide; Emergency preparedness

0194
Wald, Niel. "The Three Mile Island Incident in 1979: The State Response." In: **The Medical Basis for Radiation Accident Preparedness**, edited by Karl F. Hubner and Shirley A. Fry, 491-500. New York: Elsevier/North Holland, 1980. (5 refs.)

Wald, a consultant to the Pennsylvania State Health Department during the accident, covers a variety of topics related to emergency preparedness and public health: the sequence of events regarding state notification of the accident, communications problems, radiation dosimetry, the evacuation decision, the question of use of potassium iodide, and emergency planning.

*Emergency preparedness; Pennsylvania; Public health; Evacuation

0195
Weidner, William A.; Miller, Kenneth L.; Latshaw, Robert F.; and Rohrer, G. Victor. "The Impact of a Nuclear Crisis on a Radiology Department." **Radiology** 135(June 1980):717-723. (17 refs.)

Sequential events at Three Mile Island are described from 4 a.m., March 28 to March 31, 1979. A discussion of the radioisotopes which escaped during the accident is presented along with radiation dosages. The emergency radiation plan in effect at the Milton S. Hershey Medical Center at the time of the accident is described. Because of the events at TMI, a re-evaluation of the plan included modifications for larger numbers of contaminated patients and the possibility of evacuating the medical center itself.

*Disaster planning; Radiation dosage; Emergency medical services; Hospitals

# 4

# Health Effects

## GENERAL

0196
"Agencies to Study Effects of Three Mile Island Accident." **National Underwriter (Prop Ed)** 83(July 20, 1979):25-26. (News)
*Health effects; Research and development

0197
"Agencies to Study Effects of Three Mile Island Accident." **National Underwriter (Life Ed)** 83(August 4, 1979):11. (News)
*Health effects; Research and development

0198
Altoff, Robert J.; Sternglass, Ernest J.; and Maruyama, Yosh. "Professional Platform: What Will be the Probable Long-term Health Impact of Three-Mile Island?" **Applied Radiology** 7(July-August 1979):24, 117. (1 ref.)
A hospital physicist, a professor of radiological physics, and the chairman of a radiation medicine department comment on the health impact they perceive that the TMI accident will have. Amount of radiation reportedly released, the accuracy of its measurement, and the significance it may have are discussed.
*Health effects; Radiation dosage

0199
Axelrod, David. "Introduction." **Bulletin of the New York Academy of Medicine** 59(December 1983):865-868. (2 refs.)
The introduction to the Symposium on Health Aspects of Nuclear Power Plant Incidents (April 7 and 8, 1983), this essay discusses the levels of risk posed by an accident such as TMI. These can generally be divided into actual health risks -- such as effect of radiation on health -- and perceived risks -- such as fear of nuclear power resulting in demoralization. Rational development of policies and programs is stressed.
*Health effects; Risk

0200
"B&W Fine Sought; Infant Death Rate Said Unchanged." **Nuclear News** 23(May 1980):29-30. (News)
*Babcock & Wilcox; Health effects; Infant mortality

0201
Baker, Donald G., and Riddervold, Hans O. "Point of View: A Nuclear Accident in Perspective." **Virginia Medical** 107(June 1980):437-439. (6 refs.)
After summarizing the health risks from low level radiation, the authors use 1.5 millirems as the average dosage received to the population in a 50-mile radius of TMI as the basis for calculating the expected number of cancer and genetic effects of the accident. Comparisons are made with the BEIR Report. The authors conclude that

there is "no significantly greater risk of cancer or genetic sequela as a result of the nuclear accident" for the population near TMI as opposed to similar populations in the United States.

*Health effects; Radiation, low level; Population exposure

0202
Campbell, William H. "TMI -- A Catastrophe Prevented." **Pennsylvania Health** 40(1, 1979):20-21.

The Department of Health has begun an intensive census of residents located in the five-mile radius of TMI. The initial study will examine pregnancy outcome; other studies include chromosome abnormalities, incidence of leukemia, incidence of mental disorders, fetal thyroid disorders, and health economics.

*Health effects; Census

0203
Dickson, David. "US to Study Health Effects of Accident." **Nature** 279(June 7, 1979):464. (News)

*Health effects; Research and development; Health surveys

0204
Fabrikant, Jacob I. "Health Effects of the Nuclear Accident at Three Mile Island." **Health Physics** 40(February 1981):151-161. (16 refs.)

The author, a member of the President's Commission on the Accident at Three Mile Island, analyzes the real and potential health effects of the TMI accident in terms of the effects of low level radiation and the psychological and behavioral response of the population. The author presents documented data on what is known about the health effects of low level radiation, much of which results from laboratory experiments. He discusses radiation dosages released at TMI and reaches the conclusion that no increase in cancer, teratogenesis, or mutations will be traced to TMI. The major health effect was behavioral, with the highest levels of demoralization being shown by teenagers, mothers with preschool children, and the workers at the nuclear plant.

*Health effects; Radiation, low level; Radiation dosage; Health effects - psychological

0205
Finston, Roland A. "Feedback: Readers Reply to the July/Aug Professional Platform on Three Mile Island." **Applied Radiology** 7(November-December 1979):206. (Letter)

*Health effects

0206
Goldhaber, Marilyn K.; Staub, Sharon; and Tokuhata, George K. "Goldhaber, et al, Respond." **American Journal of Public Health** 74(May 1984):520. (5 refs.) (Letter)

*Health effects; Abortion, spontaneous; Infant mortality

0207
Goldhaber, Marilyn K.; Staub, Sharon L.; and Tokuhata, George K. "Spontaneous Abortions After the Three Mile Island Nuclear Accident: A Life Table Analysis." **American Journal of Public Health** 73(July 1983):752-759. (20 refs.)

The incidence of spontaneous abortion among women pregnant during the TMI accident was conducted for the population in the five-mile area of the nuclear plant. Using the life table method, it was found that the estimated incidence of spontaneous abortion was 15.1 % (16.1 % for spontaneous abortions and stillbirths), which was comparable to baseline studies of fetal loss.

*Abortion, spontaneous; Health effects; Infant mortality; Pregnancy outcome

0208
Goldhaber, Marilyn K.; Tokuhata, George K.; Digon, Edward; Caldwell, Glyn G.; Stein, Gary F.; Lutz, Guy; and Gur, David. "The Three Mile Island Population Registry." **Public Health Reports** 98(November-December 1983):603-609. (25 refs.)

A population registry of persons residing within five miles of TMI was developed by the Pennsylvania Department of Health in conjunction with the Centers for Disease Control and the Bureau of the Census. The information collected will provide baseline data for future studies on mobility, morbidity and mortality. Plans for analysis include a link to death certificate listings, a link to cancer registry data, and special health surveys.

*Population registry; Health effects; Health surveys

0209
Houts, Peter S.; Hu, Teh Wei; Henderson, Rugh A.; Cleary, Paul D.; and Tokuhata, George. "Utilization of Medical Care Following the Three Mile Island Crisis." **American Journal of Public Health** 74(February 1984):140-142. (12 refs.)

Utilization of primary health care following the TMI crisis was studied through Blue Cross-Blue Shield records, a sample of patients of the Department of Family & Community Medicine at the Hershey Medical Center, interviews with persons living within five miles of TMI, and questionnaires sent to primary care physicians in a 25-mile radius of TMI. Results indicate that there was only a slight increase in utilization of health services following TMI. In the Hershey study, persons with a high use pattern prior to the accident "tended to be more upset during the crisis than low practice utilizers."

*Health effects; Health care utilization; Health care

0210
Johnson, Carl J. "Re: Spontaneous Abortions Following Three Mile Island Accident." **American Journal of Public Health** 74(May 1984):520. (2 refs.) (Letter)

*Abortion, spontaneous; Health effects; Infant mortality

0211
MacLeod, Gordon K. "Independent Investigation of TMI Health Findings in Order." **Pennsylvania Medicine** 84(September 1981):33. (2 refs.)

MacLeod responds to comments by Muller, (*Pennsylvania Medicine* 84(September 1981):32-33), in reference to his paper in *Pennsylvania Medicine* 84(September 1981):25-31.

*Health effects

0212
MacLeod, Gordon K. "Medical Ethics in the Nuclear Age." **Pennsylvania Medicine** 84 (September 1981):25-31. (7 refs.)

MacLeod, Pennsylvania's Secretary of Health at the time of the accident, reports that neonatal hypothyroidism and neonatal death rates increased after the TMI accident. Calling the death rate statistically significant, he cautions that further epidemiological investigation of other possible environmental causes is needed.

*Health effects; Hypothyroidism; Infant mortality

0213
Muller, H. Arnold. "Data on TMI Accident Disputed." **Pennsylvania Medicine** 84(September 1981):32-33. (1 ref.)

Muller disputes information on neonatal hypothyroidism, infant mortality, a radiation health unit, and radiation monitoring as reported by Gordon MacLeod in *Pennsylvania Medicine* 84(September 1981):25-31.

*Health effects

0214
"Population Dose and Impact on Health of Three Mile Island Accident." **Public Health Reports** 94(July-August 1979):387. (1 ref.)

This is a summary, with order information, of the government publication, "Population Dose and Health Impact of the Accident at the Three Mile Island Nuclear Station."
*Health effects; Radiation dosage

0215
Repass, Paul E. "Feedback: Readers Reply to the July/Aug Professional Platform on Three Mile Island." **Applied Radiology** 7 (November-December 1979):43. (Letter)
*Health effects; Radiation dosage

0216
"Scientists Challenge Baby Deaths at Three Mile Island." **New Scientist** 86(April 24, 1980):180. (News)
*Health effects; Infant mortality

0217
Syed, Ibrahim B. "Feedback: Readers Reply to the July/Aug Professional Platform on Three Mile Island." **Applied Radiology** 7 (November-December 1979):43, 206. (Letter)
*Radiation dosage; Health effects; Nuclear industry

0218
"Three Mile Island Studies Coordinated by Health Dept." **Journal of Environmental Health** 42(September/October 1979):94. (News)
*Health effects

0219
Tokuhata, George K. "Impact of TMI Nuclear Accident Upon Pregnancy Outcome, Congenital Hypothyroidism and Infant Mortality." In: **Energy, Environment, and the Economy**, edited by Shyamal K. Majumdar, 135-149. Easton, PA: Pennsylvania Academy of Science, 1981. (16 refs.)
Tokuhata, who coordinated all health-related research activities related to TMI for the Commonwealth of Pennsylvania, reports that "no major radiation effect upon pregnancy outcome is expected." He concludes that clustering of cases of congenital hypothyroidism in Lancaster County during 1979 "were not related to the TMI nuclear accident." Further, it was determined that TMI had no effect on "quarterly or annual variations in fetal mortality."
*Health effects; Pregnancy Outcome; Hypothyroidism; Infant mortality

0220
Tokuhata, George K. "Three Mile Island Nuclear Accident and its Effect on the Surrounding Population." Chapter 25 in **Management of Radioactive Materials and Wastes: Issues and Progress**, edited by S.K. Majumdar and E. Willard Miller, 326-341. Easton, PA: Pennsylvania Academy of Science, 1985. (32 refs.)
Tokuhata reviews the overall studies related to the health effects of TMI on the local population, including psychological health, psycho-behavioral studies, physical health studies, and long-term health studies such as the TMI Population Registry. Also discussed are the radiation dosage and radiation exposure of the local population in relation to health risks.
*Health effects; Health effects - psychological; Population exposure

0221
Torrey, Lee. "Radiation Cloud Over Nuclear Power." **New Scientist** 86(April 24, 1980):197-199.
The author interviewed Professor Ernest J. Sternglass of the University of Pittsburgh School of Medicine. Dr. Sternglass believes the public has been misled with regard to the effects of radiation, and cites research that he has done that shows an increase in infant mortality near Three Mile Island in the four months following

the accident.

*Health effects; Infant mortality; Radiation dosage; Sternglass, Ernest J.

0222

Upton, Arthur C. "Health Impact of the Three Mile Island Accident." **Annals of the New York Academy of Sciences** 365(1981):63-75. (15 refs.)

Upton summarizes the health effects of ionizing radiation, looks at radiation doses from the accident, and evaluates the potential impact of the accident on the risk of cancer, genetic damage, teratogenic effects, and mental stress. Includes discussion.

*Health effects; Health effects - cancer; Genetic effects; Teratogenesis; Health effects - psychological; Stress; Radiation dosage; Radiation, low level

0223

Vider, Manuel. "Feedback: Readers Reply to the July/Aug Professional Platform on Three Mile Island." **Applied Radiology** 7 (November-December 1979):43. (1 ref.) (Letter)

*Health effects - cancer; Health effects; Mutagenesis

0224

Wilson, Robert A. "Feedback: Readers Reply to the July/Aug Professional Platform on Three Mile Island." **Applied Radiology** 7 (November-December 1979):43. (Letter)

*Radiation dosage; Health effects

0225

Witten, Manley. "Three Mile Island: Genetics of Nuclear Debate -- Who are the Losers?" **Lab World** 31(July 1980):14-18.

While government officials claim that there have been no harmful effects from the TMI accident, it may be many years before the health effects can be determined. Potential risks include hypothyroidism, cancer, leukemia, and genetic defects. Laboratory tests and procedures in use in Pennsylvania to detect TMI-related health effects are mentioned.

*Health effects

## CANCER

0226

Abrams, Herbert L. "Radiation Risks from Nuclear Power: Final Round." **New England Journal of Medicine** 303(September 11, 1980):645. (1 ref.) (Letter)

*Radiation dosage; Health effects - cancer

0227

Ad Hoc Population Dose Assessment Group. "Preliminary Dose and Health Impact of the Accident at the Three Mile Island Nuclear Station." **Nuclear Safety** 20(September-October 1979):591-594.

This is a summary of the report (GPO stock number 017-001-00408-1) by the Ad Hoc Population Dose Assessment Group, which "has concluded that the off-site collective dose associated with the radioactive material represents minimal risks of additional health effects to the off-site population, i.e., an increase of 1 cancer death over the 325,000 which would otherwise be expected."

*Radiation dosage; Health effects - cancer; Ad Hoc Population Dose Assessment Group

0228

Ahrens, Richard; Damon, Dwight; Isaac, Erich; von Graevenitz, Alexander; Todorovick, Miro; and Wilson, Richard. "Radiation Risks from Nuclear Power

Exaggerated." **New England Journal of Medicine** 302(May 22, 1980):1205-1206. (7 refs.) (Letter)

*Health effects - cancer; Radiation dosage

0229
Berkheiser, Samuel W. "Breast Cancer Incidence and the TMI Accident." **Pennsylvania Medicine** 90 (April 1987):40-42. (16 refs.)

In an analysis of the overall incidence of breast cancer, including bilateral breast cancer, "this study found few differences in the incidence of breast cancer" for the five years prior to and after the TMI accident.

*Health effects - cancer; Breast cancer

0230
Berkheiser, Samuel W. "Review of Leukemia, Lymphoma, and Myeloma Before and After the TMI Accident." **Pennsylvania Medicine** 89(October 1986):50-52. (16 refs.)

In a review of bone marrow examinations at Polyclinic Medical Center, Harrisburg, PA, and York Hospital, York, PA, no evidence was found of "increased incidence of leukemia, lymphoma, or myeloma" for the five years before and after the TMI accident.

*Health effects - cancer; Leukemia; Lymphoma; Myeloma

0231
Berkheiser, Samuel W. "Thyroid Cancer and TMI." **Pennsylvania Medicine** 88(May 1985):50-51. (17 refs.)

A baseline study of the incidence of thyroid cancer occurring before and after the Three Mile Island incident is presented for the eleven year period 1974 through 1984. The study population consisted of patients at Polyclinic Medical Center in Harrisburg, PA. The study revealed "no appreciable difference in the incidence of thyroid cancer for the five years following the TMI accident."

*Health effects - cancer; Thyroid cancer

0232
Bryant, P.M.; Hallam, J; Hemming, C.R.; and Jones, J.A. "Risks of Local and Global Public Health Effects from the Purge of Krypton-85 at Three Mile Island in 1980." **Health Physics** 43(October 1982):598-601. (7 refs.)

Local, regional and global public radiation exposure and risks associated with the purge of krypton-85 from TMI from June 6 to July 1980 were calculated. The authors compared their estimates with predictions by NRC staff and estimates obtained by EPA staff from environmental monitoring; agreement was close. Risk to the world population of fatal cancer was 1 in 200, and risk was 1 in 5 of a case of non-fatal skin cancer.

*Health effects - cancer; Venting; Krypton; Radiation dosage; Population exposure

0233
Chivian, Eric; Leaf, Alexander; Cope, Oliver: Winters, Thomas; and Geiger, H. Jack. "Nuclear Power." **New England Journal of Medicine** 301(October 4, 1979):790. (10 refs.) (Letter)

*Radiation dosage; Health effects - cancer

0234
"Looking for an Increase in Deaths from Cancer." **New Scientist** 97(March 31, 1983):974-975. (News)

*Health effects - cancer

0235
Maruyama, Y., and Wilson, L.C. "Report from KMA Cancer Committee -- Three Mile Island and Medical Radiation Risk and Benefit Considerations." **Journal of the Kentucky Medical Association** 77(August 1979):401.

The near disaster of Three Mile Island has raised public awareness of and interest in the hazards of radiation. It also makes medical personnel more conscious of the risks and benefits of medical radiation, raising the issue of "unnecessary medical exposure."

*Health effects - cancer; Radiation hazards

0236
Morgan, Karl Z. "Radiation Risks from Nuclear Power: Final Round." **New England Journal of Medicine** 303(September 11, 1980):645. (2 refs.) (Letter)

*Radiation dosage; Health effects - cancer

0237
"Three Mile Island Data Inconclusive, Study Says." **Chemical & Engineering News** 62(September 3, 1984):7-8. (News)

*TMI Public Health Fund; Radiation dosage; Health effects - cancer

0238
Tokuhat[a], George, and Digon, Edward. "TMI 2 No Cause for Cancer." **Nuclear Engineering International** 31(January 1986):20-21. (News)

*Health effects - cancer

0239
Upton, Arthur C. "Radiation Risks from Nuclear Power Exaggerated." **New England Journal of Medicine** 302(May 22, 1980):1205. (3 refs.) (Letter)

*Radiation dosage; Health effects - cancer

0240
Warren, Shields. "Risk of Cancer Subsequent to Low-Dose Radiation." **Journal of Forensic Sciences** 25 (October 1980):721-726. (21 refs.)

A physician discusses the effects of low-dose radiation exposure on subsequent carcinogenesis as it relates to the TMI accident. Noting that some cancers seem to appear with greater frequency among nuclear workers, and that a higher incidence of leukemia has been recorded in Japanese atomic bomb survivors, caution in interpreting these results is urged. The rate at which radiation is received is important and effects of low-dose radiation are difficult to separate from other environmental or natural factors.

*Health effects - cancer; Radiation dosage; Radiation, low level

## PSYCHOLOGICAL

0241
Bartlett, Glen S.; Houts, Peter S.; Byrnes, Linda K.; and Miller, Robert W. "The Near Disaster at Three Mile Island." **International Journal of Mass Emergencies and Disasters** 1(1983):19-42. (refs.)

Using data drawn from a series of telephone surveys of adults living within 50 miles of TMI and a questionnaire survey of high school students within 25 miles of the plant, the reactions of residents to the TMI accident are evaluated. The data show that factors such as younger age, lower parent's education, and living closer to the plant "were all associated with a stronger negative affective response to the accident and with the likelihood of having evacuated the area." A theoretical framework for the sequence of events at TMI is given, and it is concluded that the existing theory of disaster response is applicable to TMI.

*Disaster theory; Health effects - psychological; Affective disorders; Children; Evacuation; Telephone surveys

0242
Baum, Andrew; Fleming, Raymond; and Singer, Jerome E. "Coping with Victimization by Technological Disaster." **Journal of Social Issues** 39(1983):117-138. (41 refs.)

The Three Mile Island accident was the focus of this study on coping with chronic stress as a result of a technological disaster. "Findings suggested that both emotionally-focused coping and self-blame were associated with less stress than were problem-focused coping and denial." Emotional regulation and assumption of responsibility for difficulties were related to each other, suggesting that control-oriented coping "can be effective in reducing distress associated with technological catastrophes."

*Health effects - psychological; Stress, chronic; Coping; Disasters, technological

0243
Baum, Andrew; Fleming, Raymond; and Singer, Jerome E. "Stress at Three Mile Island: Applying Psychological Impact Analysis." **Applied Social Psychology Annual** 3(1982):217-248. (45 refs.)

Baum et al. summarize their continuing investigations of stress among residents living near Three Mile Island. The study began at the time of the accident and has continued through venting and cleanup activities. It has evaluated attitudes toward nuclear power, sources and measures of stress at TMI, analysis of the effects of venting, symptoms of stress, and biochemical measures of stress. The chronic stress symptoms are mostly "subclinical" in nature. Faults in the study are pointed out along with the analysis.

*Health effects - psychological; Stress, chronic; Attitude to nuclear power

0244
Baum, Andrew; Gatchel, Robert J.; and Schaeffer, Marc A. "Emotional, Behavioral, and Psychological Effects of Chronic Stress at Three Mile Island." **Journal of Consulting and Clinical Psychology** 51(August 1983):565-572. (25 refs.)

In a study evaluating the psychophysiological impact of TMI, residents living near the accident were compared with people near a coal-fired power plant and people living more than twenty miles from any power plant. Self-report and behavioral measures of stress, along with urinary catecholamine levels, were used to determine that "residents of the TMI area exhibited more symptoms of stress (self-report, performance, and catecholamine levels) more than 1 year after the nuclear accident than did people living under different circumstances." The intensity of the problem is "subclinical."

*Health effects - psychological; Stress, chronic; Psychophysiological studies

0245
Baum, Andrew; Schaeffer, Marc A.; Lake, C. Raymond; Fleming, Raymond; and Collins, Daniel L. "Psychological and Endocrinological Correlates of Chronic Stress at Three Mile Island." **Perspectives on Behavioral Medicine** 2(1985):201-217. (44 refs.)

Endocrinological response (adrenal activity) in relation to chronic stress in residents living near TMI was studied following the accident. One year after the accident, higher levels of epinephrine, norepinephrine and cortisol were exhibited in TMI-area residents as opposed to a control group, suggesting that the TMI residents had a higher level of stress.

*Health effects - psychological; Stress, chronic; Psychophysiological studies

0246
Bromet, Evelyn, and Dunn, Leslie. "Mental Health of Mothers Nine Months After the Three Mile Island Accident." **Urban & Social Change Review** 14(Summer 1981):12-15. (24 refs.)

The study found "both acute and chronic mental health effects among mothers of small children" living near TMI. Clinical manifestations were found in the initial two-month period, while subclinical symptoms were found as late as nine months after the accident.

*Health effects - psychological; Stress; Stress, chronic; Mothers

0247
Bromet, Evelyn; Schulberg, Herbert C.; and Dunn, Leslie. "In Reply." **Archives of General Psychiatry** 40(September 1983):1037. (2 refs.) (Letter)

*Health effects - psychological; Mental health; Media; News coverage

0248
Bromet, Evelyn; Schulberg, Herbert C.; and Dunn, Leslie. "Reactions of Psychiatric Patients to the Three Mile Island Nuclear Accident." **Archives of General Psychiatry** 39(June 1982):725-730. (26 refs.)

A study of patients in the community mental health system evaluated reactions to the TMI accident. A sample of 151 patients living near TMI was compared with 64 patients from the Shippingport-Beaver Balley area of Pennsylvania, which also has a nuclear plant. No significant differences were found between the mental status (e.g. state of anxiety or depression episodes) of the TMI group and the comparison group. Two risk factors, quality of network support and viewing TMI as dangerous, were found to be associated with mental health.

*Health effects - psychological; Mental health; Social support; Attitude to TMI

0249
Bromet, Evelyn J.; Hough, Linda; and Connell, Melanie. "Mental Health of Children Near the Three Mile Island Reactor." **Journal of Preventive Psychiatry** 2(3-4, 1984):275-301. (29 refs.)

A sample of children from the TMI area and from a comparison site near a similar nuclear plant was used to assess mental health of children near TMI. "Mental status was determined 3 1/2 years after the accident using measures of Social Competence, Behavior Problems, Fearfulness and Self-Esteem." The study found no significant difference between the two groups. Prior research studies in this area are critically reviewed. It is suggested that researchers "take into account the quality of the family environment in designing post-disaster intervention strategies."

*Health effects - psychological; Children; Mental health

0250
Bromet, Evelyn J.; Parkinson, D.; Schulberg, H.; Dunn, L; and Gondek, P. "Mental Health of Residents Near the Three Mile Island Reactor: A Comparative Study of Selected Groups." **Journal of Preventive Psychiatry** 1(3, 1982):225-276.

Unavailable.

*Health effects - psychological; Stress; Mental health

0251
Chisholm, Rupert F.; Kasl, Stanislav; and Eskenazi, Brenda. "The Nature and Predictors of Job Related Tension in a Crisis Situation: Reactions of Nuclear Workers to the Three Mile Island Accident." **Academy of Management Journal** 26(September 1983):385-405. (38 refs.)

Job tension of workers at TMI versus workers at the nearby Peach Bottom plant were examined. There were three tension components: quantitative overload, role conflict, and qualitative overload. Levels of tension were much higher at TMI.

*Nuclear workers; Tension, job-related; Stress; Health effects - psychological; Telephone surveys; Peach Bottom plant

0252
Chisholm, Rupert F., and Kasl, Stanislav V. "The Effects of Work Site, Supervisory Status and Job Function on Nuclear Workers' Responses to the TMI Accident." **Journal of Occupational Behaviour** 3(January 1982):39-62. (29 refs.)

The effects of the TMI accident on workers at the plant was compared with workers at the nearby Peach Bottom plant. Major effects included "perceived

radiation exposure during the incident, evacuation behaviour, uncertainty and conflict regarding the accident, job related attitudes and tension, symptoms experienced during the accident period, and demoralization." Results varied "according to supervisory status and functional work area."
*Plant workers; Stress; Health effects - psychological; Peach Bottom plant

0253
Chisholm, Rupert F.; Kasl, Stanislav V.; Dohrenwend, Bruce P.; Dohenwend, Barbara Snell; Warheit, George J.; Goldsteen, Raymond L.; Goldsteen, Karen; and Martin, John L. "Behavioral and Mental Health Effects of the Three Mile Island Accident on Nuclear Workers: A Preliminary Report." **Annals of the New York Academy of Sciences** 365(1981):134-145. (14 refs.)
Workers at the TMI plant were surveyed by telephone to evalute the psychological effects of the accident; workers at the nearby Peach Bottom plant served as the control. Results were analyzed by degree of upset, demoralization, threat to physical health, uncertainty about occupational future, hostility from the community, and trust in authorities. The basic finding was that "TMI employees reacted differently to the accident than did nuclear workers at the Peach Bottom plant." Therefore, it was concluded that working at the accident site had a major psychological effect on nuclear workers. The study was part of the "Report of the Task Group on Behavioral Effects" of the President's Commission. An appendix includes the survey questions and their results.
*Nuclear workers; Health effects - psychological; Stress; Peach Bottom plant; Telephone surveys

0254
Chisholm, Rupert F.; Kasl, Stanislav V.; and Mueller, Lloyd. "The Effects of Social Support on Nuclear Worker Responses to the Three Mile Island Accident." **Journal of Occupational Behaviour** 7(3, 1986):179-193. (31 refs.)
"This study examines the effects of social support on the reactions of nuclear workers at the Three Mile Island plant and a comparison site to stress during the 1979 crisis." It is concluded that "social support may have positive (main) effects on employee reactions to stress even under severe crisis conditions." Findings are compared between TMI workers and the comparison site, the Peach Bottom nuclear plant.
*Social support; Stress; Nuclear workers; Health effects - psychological; Peach Bottom plant

0255
Cleary, Paul D., and Houts, Peter S. "The Psychological Impact of the Three Mile Island Incident." **Journal of Human Stress** 10(Spring 1984):28-34. (18 refs.)
The psychological impact of TMI is examined based on data from a panel study of persons living within five miles of the accident, and a telephone survey of persons within 55 miles. "Active coping strategies were associated with continued high levels of distress." Additionally, adverse reaction to the accident was associated with living closer to the reactor, younger people, women, and married individuals. The study included analyses related to distress and number of friends, organizational activities, use of psychotropic drugs, and avoidance behavior. The nature of the stressor should be considered when models are developed for coping and response to stressful situations.
*Health effects - psychological; Stress; Coping; Telephone surveys

0256
Collins, Daniel L.; Baum, Andrew; and Singer, Jerome E. "Coping with Chronic Stress at Three Mile Island: Psychological and Biochemical Evidence." **Health Psychology** 2(Spring 1983):149-166. (33 refs.)
This study measured the psychological, biochemical and behavioral aspects of stress during the "protracted stress situation" of TMI and its two-year aftermath. Results were compared to indices of coping from the Ways of Coping Inventory.

Results indicated that "reappraisal-based emotional management appeared to be more effective in reducing psychological and behavioral consequences of stress" than was "use of denial or problem-oriented modes of coping...."

*Health effects - psychological; Stress, chronic; Coping; Psychophysiological studies

0257

Cornely, Paul, and Bromet, Evelyn. "Prevalence of Behavior Problems in Three-Year-Old Children Living Near Three Mile Island: A Comparative Analysis." **Journal of Child Psychology and Psychiatry & Allied Disciplines** 27(July 1986):489-498. (19 refs.)

The Behavior Screening Questionnaire was used to assess behavior problems in three-year-old children living near TMI versus two other sites in Pennsylvania. The prevalence of behavior problems was 11%, with no significant differences among sites.

*Health effects - psychological; Preschool children; Behavior problems; Stress

0258

Davidson, Laura M., and Baum, Andrew. "Chronic Stress and Posttraumatic Stress Disorders." **Journal of Consulting and Clinical Psychology** 54(June 1986):303-308. (27 refs.)

Residents near Three Mile Island and a control group were evaluated for chronic and posttraumatic stress. Residents near TMI exhibited more symptoms of chronic and posttraumatic stress than the control group. The study showed a "strong relationship between self-report, behavioral, and psychological measures of chronic stress and characteristics central to posttraumatic stress."

*Health effects - psychological; Stress, posttraumatic; Stress, chronic

0259

Davidson, Laura M.; Baum, Andrew; and Collins, Daniel L. "Stress and Control-Related Problems at Three Mile Island." **Journal of Applied Social Psychology** 12(September/October 1982):349-359. (17 refs.)

Residents near TMI were compared with residents of Frederick, Maryland, in a study which evaluated the relationship between stress and control-related problems. "Residents at TMI reported more loss of control and performed more poorly on the behavioral task than did comparison subjects." Residents near TMI that reported more loss of control had more stress symptoms than the other group.

*Health effects - psychological; Stress; Loss of control

0260

Davidson, Laura M.; Baum, Andrew; Fleming, India; and Gisriel, Martha M. "Toxic Exposure and Chronic Stress at Three Mile Island." **Advances in Environmental Psychology** 6(1986):35-46. (28 refs.)

Residents within five miles of TMI were studied beginning 15 months after the accident. Data were compared with three control groups. Using psychological, behavioral, and physiological indices of stress, residents of TMI continued to show higher levels of stress than the control groups more than two years after the accident.

*Health effects - psychological; Stress, chronic

0261

Dohrenwend, Bruce P. "Psychological Implications of Nuclear Accidents: The Case of Three Mile Island." **Bulletin of the New York Academy of Medicine** 59(December 1983):1060-1076. (7 refs.)

In a presentation at the Symposium on the Health Aspects of Nuclear Power Plant Incidents, April 7 and 8, 1983 at the New York Academy of Medicine, the results are presented of the survey conducted by the Task Group on Behavioral and Mental Health Effects of the President's Commission. The survey, involving residents in the twenty-mile area of TMI, showed varying levels of distress among the general population and workers at TMI. Demoralization and distrust are a primary focus for the study. Results were most pronounced among residents in the five-mile area.

*Health effects - psychological; Stress

0262

Dohrenwend, Bruce P.; Dohrenwend, Barbara Snell; Warheit, George J.; Bartlett, Glen S.; Goldsteen, Raymond L.; Goldsteen, Karen; and Martin, John L. "Stress in the Community: A Report to the President's Commission on the Accident at Three Mile Island." **Annals of the New York Academy of Sciences** 365(1981):159-174. (7 refs.)

This is an adaptation of the report of the Task Group on Behavioral Effects of the President's Commission. The survey results looked at how upset the population was at the time of the accident, who evacuated, how demoralized the people were, attitude toward living in the TMI area and to nuclear power, trust in authorities, etc. TMI had "substantial immediate psychological effects on the people living in the area," and has had "a lasting impact on the population of the area in terms of their distrust of authorities with respect to nuclear power."

*Health effects - psychological; Stress; Local residents; Stress, chronic

0263

Fabrikant, Jacob I. "The Effects of the Accident at Three Mile Island on the Mental Health and Behavioral Responses of the General Population and Nuclear Workers." **Health Physics** 45(September 1983):579-586. (22 refs.)

Fabrikant summarizes the report of the Behavioral Effects Task Group of the President's Commission on the Accident at Three Mile Island. The study included residents of the TMI area, school age children, and TMI nuclear workers, and investigated such aspects as leaving the area during the accident, demoralization, and attitude toward nuclear power.

*Health effects - psychological; Local residents; Nuclear workers

0264

Fienberg, Stephen E.; Bromet, Evelyn J.; Follmann, Dean; Lambert, Diane; and May, Sherryl M. "Longitudinal Analysis of Categorical Epidemiological Data: A Study of Three Mile Island." **Environmental Health Perspectives** 63(November 1985):241-248. (25 refs.)

The paper is a longitudinal, statistical analysis of the "psychological well-being of the mothers of young children living within 10 miles" of TMI. A series of statistical model techniques are used; inadequacies of the models are discussed, and "a new class of mixture models for logistic response structures" is applied to the TMI data. The evidence is mixed "on the implications of the perception of danger as a risk factor for psychological stress in the aftermath of the TMI accident."

*Health effects - psychological; Stress, chronic; Mothers; Physical threat

0265

Fleming, Raymond; Baum, Andrew; Gisriel, Martha M.; and Gatchel, Robert J. "Mediating Influences of Social Support on Stress at Three Mile Island." **Journal of Human Stress** 8(September 1982):14-22. (33 refs.)

A sample of residents living near TMI was studied using symptom reporting, task performance, and urinary catecholamines; three control groups were used. "More than a year after the accident, living near the damaged reactor was associated with elevations in all indices of stress compared with control levels." These stress indices were mediated by social support.

*Health effects - psychological; Social support; Stress; Psychophysiological studies

0266

Gatchel, Robert J.; Schaeffer, Marc A.; and Baum, Andrew. "A Psychophysiological Field Study of Stress at Three Mile Island." **Psychophysiology** 22(March 1985):175-181. (13 refs.)

Residents near TMI and from a control group near Frederick, MD, participated in a longitudinal study to measure stress through biochemical performance and self-report measures. In measurements taken prior to, during, and after the venting of krypton-85 from TMI, "TMI subjects exhibited higher levels of stress on each mode of this battery." The results suggest that venting "had specific acute effects on chronic problems being experienced by TMI subjects, but no global effect across all measures."

*Health effects - psychological; Stress, chronic; Venting; Psychophysiological studies

0267
Goldsteen, Raymond, and Schorr, John K. "The Long-Term Impact of a Man-made Disaster: An Examination of a Small Town in the Aftermath of the Three Mile Island Nuclear Reactor Accident." **Disasters** 6(1, 1982):50-59. (15 refs.)

A sample of residents in Newberry Township and Goldsboro was interviewed by telephone in September 1979, with a reinterview of a subsample in March 1980; in-person interviews were conducted with community leaders. Interviews dealt with residents' perception of danger and risk, and their fear and attitudes regarding nuclear power. Findings were that, after the TMI accident, "there are constant or increasing levels of distress reported by community residents."

*Health effects - psychological; Stress, chronic; Attitude to nuclear power; Local residents; Telephone surveys

0268
Goldsteen, Raymond L.; Schorr, John; and Goldsteen, Karen S. "What's the Matter with Those People: Rethinking TMI." **International Journal of Mass Emergencies and Disasters** 2(November 1984):369-387. (26 refs.)

This study evaluated the psychosocial consequences of the TMI accident on "the adult population of a community situated almost entirely within five miles of the reactor." A telephone survey was conducted in October-November 1979 and October-November 1980. Findings indicated that residents were stressed at the time of both interviews, and that "perceived threat to physical health is more highly associated with distress than personal or demographic characteristics." Indications of distress were demoralization, attitude toward nuclear power, and attitude toward continuing to live in the TMI area.

*Health effects - psychological; Stress, chronic; Psychosocial effects; Physical threat; Telephone surveys

0269
Hartsough, Don M., and Savitsky, Jeffrey C. "Three Mile Island; Psychology and Environmental Policy at a Crossroads." **American Psychologist** 39(October 1984):1113-1122. (48 refs.)

This article reviews five of the major research studies on psychological stress following the TMI accident, and discusses the legal background of psychological damage in case law that preceded the court decision which will require that the psychological effects of TMI be evaluated under the National Environmental Policy Act (*PANE v. U.S. NRC*). TMI has raised new issues for psychologists in relation to environmental controversies.

*Environmental policy; Health effects - psychological; Stress; Legal aspects; People Against Nuclear Energy v. U.S. Nuclear Regulatory Commission

0270
Holden, Constance. "Psychiatrists Study Three-Mile Trauma." **Science** 208(April 4, 1980):37. (News)

*Health effects - psychological; Research and development

0271
Houts, Peter S., and Goldhaber, Marilyn K. "Psychological and Social Effects on the Population Surrounding Three Mile Island After the Nuclear Accident on March 28,

1979." In: **Energy, Environment and the Economy**, edited by Shyamal K. Majumdar, 151-164. Easton, PA: Pennsylvania Academy of Science, 1981. (14 refs.)

This paper reviews "studies dealing with the decision to evacuate, the economic impact of the crisis, the effect of the crisis on mobility of persons living near the plant, attitudes toward TMI following the crisis and psychological stress experienced by people in the vicinity of the facility." The studies are consistent in their findings that the major impact was felt during the weeks immediately after the accident and that the "long term effects were of a low magnitude."

*Health effects - psychological; Stress

0272

Kasl, Stanislav V.; Chisholm, Rupert F.; and Eskenazi, Brenda. "The Impact of the Accident at the Three Mile Island on the Behavior and Well-Being of Nuclear Workers. Part I: Perceptions and Evaluations, Behavioral Responses, and Work-Related Attitudes and Feelings." **American Journal of Public Health** 71(May 1981):472-483.

Telephone interviews were conducted six months after the accident with TMI workers and workers at the Peach Bottom plant. "TMI workers experienced more uncertainty and conflict at the time of the accident." Additionally, they had lower job satisfaction and greater uncertainty about their job futures. The study was conducted by the Public Health and Safety Task Force of the President's Commission on the Accident at Three Mile Island.

*Health effects - psychological; Nuclear workers; Stress; Telephone surveys; Peach Bottom plant

0273

Kasl, Stanislav V.; Chisholm, Rupert F.; and Eskenazi, Brenda. "The Impact of the Accident at the Three Mile Island on the Behavior and Well-Being of Nuclear Workers. Part II: Job Tension, Psychological Symptoms, and Indices of Distress." **American Journal of Public Health** 71(May 1981):484-495. (39 refs.)

As compared with workers at the Peach Bottom plant, TMI workers experienced greater job tension, lower self-esteem, and at the time of the accident, "more periods of anger, extreme worry and extreme upset, and more psychophysiological symptoms." Some symptoms persisted six months later.

*Health effects - psychological; Nuclear workers; Stress; Peach Bottom plant; Psychophysiological studies

0274

Mendelson, George. "No Accident." **Archives of General Psychiatry** 40(September 1983):1036-1037. (1 ref.) (Letter)

*Health effects - psychological; Mental health; Media; News coverage

0275

Mileti, Dennis S.; Hartsough, Donald M.; Madson, Patti; and Hufnagel, Rick. "The Three Mile Island Incident: A Study of Behavioral Indicators of Human Stress." **International Journal of Mass Emergencies and Disasters** 2(March 1984):89-113. (35 refs.)

In this study, unobstrusive behavioral indicators (cardiovascular deaths, psychiatric admission, suicide, automobile accidents, alcohol sales, and crime) were used to measure stress for the population as a whole near Three Mile Island before, during and after the accident. The results reflected a mild level of stress.

*Health effects - psychological; Stress; Behavioral indicators

0276

"Nuclear Hitches Still Plague Americans." **New Scientist** 88(November 6, 1980):351. (News)

*Health effects - psychological

0277
Parkinson, David K., and Bromet, Evelyn J. "Correlates of Mental Health in Nuclear and Coal-Fired Power Plant Workers." **Scandinavian Journal of Work, Environment and Health** 9(August 1983):341-345. (8 refs.)

The mental health of nuclear workers at TMI was compared with that of workers from another nuclear plant and workers from two coal-fired power plants. Workers at the coal-fired plants "perceived less stress but more problems with workplace exposures than the nuclear plant workers." The overall results suggested workers did not experience long-term psychological difficulties.

*Health effects - psychological; Stress; Nuclear workers; Tension, job-related; Radiation exposure, occupational

0278
Schaeffer, Marc A., and Baum, Andrew. "Adrenal Cortical Response to Stress at Three Mile Island." **Psychosomatic Medicine** 46(May/June 1984):227-237. (39 refs.)

This study looked at "the relationship between biochemical, psychologic, and behavioral components of chronic stress" associated with persons living near TMI. TMI residents had "higher levels of urinary cortisol" as well as catecholamines, self report or stress symptoms and decrements in task performance, as compared with control subjects. Urinary cortisol levels were higher in males than females. There was no significant relationship between urinary cortisol and coping style. Stress response was persistent, but subclinical in intensity.

*Health effects - psychological; Stress, chronic; Coping; Psychophysiological studies

0279
Schwebel, Milton, and Schwebel, Bernice. "Children's Reactions to the Threat of Nuclear Plant Accidents." **American Journal of Orthopsychiatry** 51(April 1981):260-270. (11 refs.)

In the wake of the TMI accident, questionnaires and interviews were conducted of the children in elementary and secondary schools from ten school districts in New Jersey. Students indicated that they believed there would be a serious nuclear accident (70%), and 85% said that they cared about such an accident. Overall, 50% of the students believed there would be serious physical consequences, 10% believed there would be psychological consequences, and 32% believed they would be able to escape such an accident. Only 34% believed that nuclear plants should be closed. Results were compared with a similar survey about nuclear power conducted in the 1960s.

*Health effects - psychological; Children; Attitude to nuclear power

0280
Schwebel, Milton, and Schwebel, Bernice. "Children's Reactions to the Threat of Nuclear Plant Accidents." **Annual Progress in Child Psychiatry and Child Development** (1982):371-383. (11 refs.)

Children in a New Jersey school district were surveyed following the accident at TMI. Reprinted from the *American Journal of Orthopsychiatry* 51(2, 1981):260-270.

*Health effects - psychological; Stress; Children; Attitude to nuclear power

0281
Shivers, Lynne. "`Three Mile': The Accident Isn't Over." **Christianity and Crisis** 39(November 12, 1979):274-275, 284-286.

As a result of TMI, local residents have experienced anxiety and stress. Shivers discusses the topic informally, including interviews with local residents.

*Health effects - psychological; Stress; Local residents; Anxiety

0282
Shore, James H., editor. "The Three Mile Island Disaster: A Search for High-Risk Groups." Chapter 1 in **Disaster Stress Studies: New Methods and Findings**, 2-19. Washington, DC: American Psychiatric Press, 1986. (34 refs.)

In a study sponsored by the National Institute of Mental Health, three high risk groups (mothers of preschool children, workers at the nuclear plant, and persons within ten miles of TMI who had recently received psychiatric treatment) were studied over a three-year period. The group most affected by the TMI accident was mothers with preschool children.

*Health effects - psychological; Stress

0283
Solomon, Zahava. "Stress, Social Support and Affective Disorders in Mothers of Pre-School Children -- A Test of the Stress-Buffering Effect of Social Support." **Social Psychiatry** 20(2, 1985):100-105. (25 refs.)

Women with pre-school children living near TMI were compared to women in a control group from Beaver County, PA in a study which evaluated mental status, stress, and social support. "Results indicated that women with less adequate social support were more at risk for psychiatric disorder following the Three Mile Island nuclear accident." Symptoms of major depression and subclinical symptoms of depression were more prevalent among the TMI mothers.

*Health effects - psychological; Stress; Social support; Affective disorders; Mothers

0284
"Two Miles from Three Mile Island: Stress and Nuclear Crisis." **Behavioral Medicine** 6(May 1979):26-29.

Local residents' reactions to the TMI accident are illustrated, primarily through the practice of Joseph Leaser, M.D., a family physician located within two miles of the plant. Patients utilized physicians in the area for reassurance regarding health effects, safety, and the decision to evacuate. Stress was a major psychological effect. Also discussed is the evacuation center at the Hersheypark Arena.

*Health effects - psychological; Stress; Physicians; Evacuation; Health care

0285
Van Der Smissen, E. "Feedback: Readers Reply to the July/Aug Professional Platform on Three Mile Island." **Applied Radiology** 7(November-December 1979):43. (Letter)

*Health effects - psychological

0286
Wert, Barbara J. "Stress Due to Nuclear Accident: A Survey of an Employee Population." **Occupational Health Nursing** 27(September 1979):16, 19-24.

A survey was conducted of Pennsylvania Blue Shield employees in Central Pennsylvania to evaluate stress following the TMI accident. Among the results were that the evacuee rate was highest for employees living closest to TMI, that the educational level of evacuees was slightly higher than the level of those who stayed, and that employees who evacuated had a higher number of stress symptoms than those who stayed.

*Health effects - psychological; Stress; Evacuation; Health surveys

0287
"Woke Up This Morning with Those Post-TMI Blues." **New Scientist** 95(September 2, 1982):607. (News)

*Health effects - psychological

# 5

# Radioactive Releases and the Environment

0288
Berger, C.D. "The Effectiveness of a Whole Body Counter During and After an Accident Situation at Nuclear Facilities." **Health Physics** 40(May 1981):685-692. (9 refs.)

Whole body counters to provide rapid assessment of body burdens of radionuclides are part of emergency preparedness plans for nuclear facilities. During the TMI accident, whole body counters were used for this purpose, but improper acquisition of data and analysis "resulted in questionable assignments of body burden." The problem areas were "background subtraction techniques, personnel entrance procedure, counter location, and electronics stability." Recommendations are made to improve the reliability of results.

*Whole body counters; Body burden; Radiation dosage; Nuclear workers

0289
Beyea, Jan. "Three Mile Island -- Six Years Later." **Journal of Nuclear Medicine** 26(November 1985):1345-1346. (2 refs.) (Letter)

*Radiation dosage; Population exposure

0290
Carter, J.A. et al. "Isotope-Dilution Resin Bead Mass Spectrometry -- An Ultra Trace Technique for Measuring Nuclides in Three Mile Island Water." **International Journal of Environmental and Analytical Chemistry** 8(4, 1980):241-248.

Unavailable.

*Radionuclides; Spectrometry

0291
Culliton, Barbara J., and Waterfall, Wallace K. "Person-rems and the Future." **British Medical Journal** 2(August 11, 1979):375-376.

After reviewing two reports regarding the biological consequences of low-level radiation, the authors discuss radiation at Three Mile Island.

*Radiation dosage; Radiation, low level; Population exposure; Health effects

0292
Fabrikant, Jacob I. "Report of the Public Health and Safety Task Force: Public Health and Safety Summary." **Progress in Nuclear Energy** 6(1-3, 1980):327-359. (6 refs.)

Reprinted from the U.S. government document.

*Radiation dosage; Health effects; Public health

0293
Fainberg, Anthony. "More on Three Mile Island." **Bulletin of the Atomic Scientists** 35(November 1979):65. (Letter)

*Radiation dosage

0294
Field, R. William; Field, Elizabeth; Zegers, David A.; and Steucek, Guy L. "Iodine-131 in Thyroids of the Meadow Vole (Microtus Pennsylvanicus) in the Vicinity of Three Mile Island Nuclear Generating Plant." **Health Physics** 41(August 1981):297-301. (26 refs.)

Following the accident, meadow voles (Microtus pennsylvanicus) were trapped in three locations near TMI. "Thyroids of voles caught 1.9 km from the reactor (Site III) contained significantly higher amounts of 131 I than those of voles caught further away." The authors propose that the vole be used to monitor I-131 contamination of ecosystems.

*Voles; Iodine radioisotopes; Ecological contamination; Ecosystems; Radiation dosage; Wildlife

0295
Field, R. William; Zegers, Davis A.; Steucek, Guy L.; and Field, Elizabeth A. "Regarding 131 I in Meadow Vole Thyroids." **Health Physics** 44(February 1983):177-180. (12 refs.) (Letter)

In an extensive letter, the authors respond to Kirk's commentary (*Health Physics* 44(February 1983):175-177) about his article originally published in *Health Physics* 41(August 1981):297-301. Discussed are iodine 131 in vole thyroid samples and overall radiation dosage to the vole thyroid.

*Iodine radioisotopes; Voles; Radiation dosage; Wildlife

0296
Gerusky, Thomas M. "Three Mile Island: Assessment of Radiation Exposures and Environmental Contamination." **Annals of the New York Academy of Sciences** 365(1981):54-62. (6 refs.)

Information about the radiation releases during the TMI accident are summarized, and estimates of radiation exposure and environmental contamination are given. The author, a member of the Bureau of Radiation Protection of the Penna. Department of Environmental Resources, was involved in the emergency activities surrounding the accident. Includes discussion.

*Radiation dosage; Population exposure; Environmental effects

0297
Gold, R.; Roberts, J.H.; Ruddy, F.H.; Preston, C.C.; Milroy, W.N.; Rao, S.V.; Greenborg, J.; and Fricke, V.R. "Neutron Dosimetry with Solid-State Track Recorders in the Three-Mile Island Unit-2 Reactor Cavity." **Nuclear Tracks and Radiation Measurements** 10(3, 1985):447-459. (14 refs.)

Solid-state track recorder (SSTR) neutron dosimetry used in the TMI-2 reactor cavity reveals "that at least 2 tonnes of fuel, which is roughly 4 fuel assemblies, is lying at the bottom of the vessel." Neutron streaming upward from the reactor cavity region "dominates the observed neutron intensity."

*Dosimetry, neutron; Reactor core analysis; Instrument function

0298
Gold, Raymond; Ruddy, Frank H.; Roberts, James H.; Preston, Christopher C.; Ulseth, James A.; McElroy, William N.; Leitz, Fred J.; Hayward, Benjamin R.; and Schmittroth, Frank A. "Application of Solid State Track Recorder Neutron Dosimetry for Three Mile Island Unit 2 Reactor Recovery." **Nuclear Tracks** 7(1/2, 1983):13-30. (15 refs.)

"Application of neutron dosimetry for assessment of fuel distribution throughout the Three Mile Island-2 (TMI-2) reactor core region and the primary coolant system is advanced." A detailed TMI-2 scoping experiment using solid state track recorder neutron dosimetry is described.

*Dosimetry, neutron; Reactor core; Cooling system

0299
Gricar, Barbara Gray, and Baratta, Anthony J. "Bridging the Information Gap at Three Mile Island: Radiation Monitoring by Citizens." **Journal of Applied Behavioral Science** 19(February 1983):35-49. (34 refs.)

The Citizen Radiation Monitoring Program was designed to reverse the effects of the erosion of public confidence by involving citizens in twelve communities in monitoring, interpreting and publicizing radiation levels during the purge of radioactive krypton one year after the accident. The program was designed "to provide citizens with information about radiation exposure and to rebuild public confidence in information supplied by government agencies."

*Citizen monitoring; Radiation monitoring; Education; Information exchange

0300
Gricar, Barbara Gray, and Baratta, Anthony J. "Reply to Professor Key." **Journal of Applied Behavioral Science** 19(February 1983):51-53. (1 ref.)

Gricar and Baratta reply to Key's comments (*Journal of Applied Behavioral Science* 19(February 1983):49-51), regarding their article in that same journal issue, pp. 35-49.

*Citizen monitoring; Radiation monitoring; Education; Information exchange

0301
"Harrisburg Leak `the Worst Accident in Nuclear History'." **Nature** 278(April 5, 1979):497. (News)

*Reactor accidents; Risk; Radionuclides; Radioactive releases; Radiation, low level

0302
Ibser, H.W. "Three Mile Island." **Science** 204(May 25, 1979):795. (Letter)

*Radiation dosage

0303
Jester, W.A.; Baratta, A.J.; and Granlund, R.W. "Evaluation of Radiation Monitor Effectiveness for the Detection of 85 Kr." **Health Physics** 42(June 1982):827-832. (6 refs.)

Twelve radiation detector systems were evaluated for 85 Kr sensitivity in preparation for the purging of TMI Unit-2. The 85 Kr test room of the Pennsylvania State University's Breazeale Nuclear Reactor was used to calibrate the instruments. As a result of the test, the Environmental Protection Agency "chose to use the Eberline PAC-4G Gas Flow Proportional Counters with a Model AC21B B Probe on their mobile monitoring vans."

*Cleanup; Venting; Krypton; PSU Noble Gas Monitor; Citizen monitoring

0304
Jester, William A., and Baratta, Anthony J. "Monitoring Krypton-85 During Three Mile Island Unit 2 Purging." **Nuclear Technology** 56(March 1982):478-483. (6 refs.)

During the period from June 29 to July 11, 1980, the Penn State noble gas monitor was used to monitor the krypton-85 purged from TMI-2 primary containment. The Penn State results agreed "with the PAC-4G monitor used in the collection van, the results using the EPA system, and results obtained through the TMI area community monitoring program."

*Radiation monitoring; PSU Noble Gas Monitor; Venting; Krypton

0305
Key, William. "Comments." **Journal of Applied Behavioral Science** 19(February 1983):49-51.

Key comments on the article by Gricar and Baratta (*Journal of Applied Behavioral Science*, v.19, February 1983, pp.35-49).

*Citizen monitoring; Radiation monitoring; Education; Information exchange

0306
Kirk, William P. "131 I in Thyroids in Meadow Voles Near Three Mile Island Nuclear Generating Plant." **Health Physics** 44(February 1983):175-177. (4 refs.) (Letter)

This is an extensive letter that comments on a paper by Field et al. in *Health Physics* 41(August 1981):297-301.

*Voles; Iodine radioisotopes; Ecological contamination; Ecosystems

0307
Kunz, Charles; Wahlen, Martin; Peterson, Kendall R.; and Rodriguez, Daniel J. "Krypton-85 Purge at Three Mile Island: A Comparison of Measured and Calculated Surface Air Concentrations." **Health Physics** 49(September 1985):522-526. (6 refs.)

Krypton-85 was purged from TMI Unit-2 between June 28 and July 11, 1980. Measurements from seven downwind locations resulted in a whole-body dose of less than 1% of the skin dose at distances of 36 to 63 km downwind.

*Venting; Krypton; Radiation dosage; Whole body count

0308
McLean, A.S. "Radiation Doses After Three Mile Island." **Lancet** 1(May 12, 1979):1039. (Letter)

*Radiation dosage; Population exposure

0309
Marshall, Eliot. "The Radiation Studies Begin." **Science** 204(April 20, 1979):281. (News)

*Radiation dosage; Radionuclides; Radioactive releases

0310
Maxon, Harry R. "Fallout at Three Mile Island." **Annals of Internal Medicine** 91(September 1979):486-487. (4 refs.)

Maxon summarizes and comments on the radiation exposure estimates released by the Ad Hoc Population Dose Assessment Group. The maximal dose for any one individual was reported to a man located 1.1 miles NNE of TMI, who probably received 37 mrem, with a maximum possible dose of 180 mrem.

*Radiation dosage; Ad Hoc Population Dose Assessment Group

0311
Morse, Roger A.; Van Campen, Darrell R.; Gutenmann, Walter H.; Lisk, Donald J.; and Collison, Clarence. "Analysis of Radioactivity in Honeys Produced Near Three-Mile Island Nuclear Power Plant." **Nutrition Reports International** 22(September 1980):319-321. (2 refs.)

Honey samples collected near TMI were assayed for radioactivity. "None of the samples exhibited gamma radiation which was significantly different from that of the background."

*Honey; Radioactivity; Contamination; Environmental effects

0312
Nelson, Mark, and Rosenstein, Marvin. "Radiation Doses After Three Mile Island Accident." **Lancet** 2(July 21, 1979):145. (refs.) (Letter)

*Radiation dosage; Population exposure

0313
"New Dosimeter Developed for TMI Clean-up." **Nuclear Engineering International** 27(October 1982):32. (News)

*Dosimeters; Cleanup

0314
Norman, Colin. "Assessing the Effects of a Nuclear Accident." **Science** 228(April 5,

1985):31-33. (News)
*Radionuclides; Radioactive releases; Radiation dosage

0315
Olsen, C.R.; Larsen, I.L.; Cutshall, N.H.; Donoghue, J.F.; Bricker, O.P.; and Simpson, H.J. "Reactor-Released Radionuclides in Susquehanna River Sediments." **Nature** 294 (November 19, 1981):242-245. (16 refs.)
The release history and radionuclide concentration data for sediment samples of 137 Cs, 134 Cs, 60 Co, and 58 Co in the lower Susquehanna River are given. These and other radionuclides were introduced by the Three Mile Island and Peach Bottom reactors. No evidence of a ruptured fuel element was found.
*Radionuclides; Radioactive releases; Ecology; Susquehanna River; Peach Bottom plant; Water contamination, radioactive

0316
Pasciak, Walt; Branagan, Edward F., Jr.; Congel, Frank J.; and Fairobent, James E. "A Method for Calculating Doses to the Population from 133 Xe Releases During the Three Mile Island Accident." **Health Physics** 40(April 1981):457-465. (6 refs.)
This article gives a detailed description of a method to calculate off-site xenon releases from TMI. The method uses "the topological time-averaged meteorological dispersion factors derived from meteorological data obtained during the accident" in addition to thermoluminescent dosimeters located around the site. This method estimated that the population located within 50 miles of TMI received about 3400 person-rems.
*Radionuclides; Radioactive releases; Xenon radioisotopes; Dosimeters, thermoluminescent; Radiation dosage; Population exposure

0317
Perham, Christine. "EPA's Role at Three Mile Island." **EPA Journal** 6(October 1980):18-20, 29.
The involvement of the Environmental Protection Agency at Three Mile Island is detailed, including installations of monitoring devices and establishment of water sampling stations.
*Environmental Protection Agency; Radiation monitoring; Water sampling

0318
Plato, Phillip A.; Hudson, Charles G.; Katzman, David; and Sha, Richard. "Calibration of Environmental and Personal Dosimeters by Submersion in and Distant Exposure to 133 Xe." **Health Physics** 38(April 1980):523-528. (3 refs.)
Panasonic personal dosimeters and environmental dosimeters were calibrated to measure the level of 133 Xenon, both in and near the source of the radionuclide. The dosimeters for this study were calibrated for use in and around TMI.
*Xenon radioisotopes; Dosimeters; Radiation dosage

0319
"Radiation Doses After Three Mile Island." **Lancet** 2(July 21, 1979):161. (1 ref.) (News)
*Radiation dosage; Population exposure; Health effects

0320
Reyburn, Jerry H. "Three Mile Island -- A Forester's Personal Experience." **Journal of Forestry** 78(April 1980):186, 242. (Letter)
*Ecology; Radiation dosage

0321
Riley, R.J.; Zanzonico, P.B.; Masterson, M.E.; St. Germain, J.M.; and Laughlin, J.S. "Evaluation of the Response to Xenon-133 Radiations by Thermoluminescent Dosimeters Used During the Accident at Three Mile Island." **Health Physics** 42(March 1982):329-334. (7 refs.)

The accuracy and sensitivity of three types of thermoluminescent dosimeters used during the TMI accident were evaluated. The dosimeters, supplied by Teledyne Isotopes and Radiation Management Corporation, all over-responded to xenon-133 radiation.

*Xenon radioisotopes; Dosimeters, thermoluminescent; Radiation dosage

0322
Ruddy, F.H.; Roberts, J.H.; Gold, R.; and Preston, C.C. "Applications of Solid State Track Recorder Neutron Dosimetry for Fuel Debris Location in the Three Mile Island Unit 2 Makeup and Purification Demineralizers." **Nuclear Tracks and Radiation Measurements** 8(1-4, 1984):473-481. (4 refs.)

Neutron dosimetry is being used at TMI to locate fuel debris and to aid in the recovery effort. This report describes the measurements and calibration in the TMI-2 demineralizer A Cubicle, from which "the amount of fuel present in TMI-2 Demineralizer A was estimated" at 1.7 kg, which "corresponds to a total neutron source of about 500 neutrons/sec."

*Dosimetry, neutron; Fuel debris; Cleanup

0323
Shuping, Ralph. "Use of Photographic Film to Estimate Exposure Near TMI." **Health Physics** 41(July 1981):195-199. (3 refs.)

Photographic film purchased from shops near Three Mile Island was analyzed to estimate radiation exposure levels from the accident. Film from Maryland was used as a control. No difference was observed in net density levels of the films from the two locations.

*Radiation dosage; Film, photographic; Population exposure

0324
Solon, Leonard R., and Sidel, Victor W. "Risks of Nuclear Power." **Annals of Internal Medicine** 92(January 1980):125-126. (3 refs.) (Letter)

*Radiation dosage; BEIR Committee; Risk

0325
Toke, L.F.; Carson, B.H.; Baker, G.G.; McBride, M.H.; Plato, P.A.; and Miklos, J.A. "Performance Testing of the Environmental TLD System for the Three Mile Island Nuclear Station." **Health Physics** 46(May 1984):1013-1020. (5 refs.)

Using appropriate performance standards, the environmental thermoluminescent dosimeter (TLD) system in use at TMI (Panasonic UD-801), was tested for "TLD uniformity, reproducibility, energy dependence and directional dependence." The four qualifying tests were satisfactorily completed and were within the guidelines for environmental TLDs."

*Dosimeters, thermoluminescent

0326
Wahlen, Martin; Kunz, Charles O.; Matuszek, John M.; Mahoney, William E.; and Thompson, Roger C. "Radioactive Plume from the Three Mile Island Accident: Xenon-133 in Air at a Distance of 375 Kilometers." **Science** 207(February 8, 1980):639-640. (5 refs.)

The presence of a radioactive plume containing xenon-133 was measured in Albany, New York from March 29 to April 2, 1979. Xenon-133 was measured in ambient air via two techniques, with recordings "as high as 3900 picocuries per cubic meter." The dose received from TMI amounted to ".004 percent of the annual dose from natural sources."

*Radionuclides; Radioactive releases; Xenon radioisotopes; Albany, NY

# 6

# Accident Investigations/Commissions

## PRESIDENT'S COMMISSION

0327
"A-Power Licensing Clamp Tighter After TMI Report." **Engineering News-Record** 203(November 8, 1979):11-12. (News)
*President's Commission; Licensing

0328
Dickson, D. "End Nuclear Complacency -- 3 Mile Island Report." **Nature** 282(November 8, 1979):120-121.
This is a summary of the President's Commission on Three Mile Island -- The Kemeny Report. The overall finding "has resulted in a damning indictment of official attitudes toward the safety of nuclear power." Human error caused a series of minor failures to turn into a major disaster. The report found fault with the "competence, background and training of both operators and management." The report comments on health effects, costs, the seriousness of the accident, worker health and safety, and the structure and shortcomings of the NRC.
*President's Commission; Recommendations; Safety

0329
Dynes, Russell R. "The Presidential Commission on the Accident at Three Mile Island." **American Behavioral Scientist** 26(May/June 1983):607-621. (12 refs.)
Dynes, a member of the President's Commission on the Accident at Three Mile Island, analyzes the Commission in the context of other presidential commissions; in particular, he looks at the role of social scientists in the TMI Commission.
*President's Commission; Social scientists

0330
English, Robert E. "Report of the Technical Assessment Task Force from Volume II: Chemistry." **Progress in Nuclear Energy** 6(1-3, 1980):91-115. (34 refs.)
Reprinted from the U.S. government document.
*Zirconium; Fission products; Hydrogen bubble; Technical assessment

0331
Forscher, Frederick, and Bland, William M., Jr. "Report of the Technical Assessment Task Force from Volume IV: Pilot-Operated Relief Valve Design and Performance." **Progress in Nuclear Energy** 6(1-3, 1980):239-325. (28 refs.)
Reprinted from the U.S. government document.
*Valves; Equipment function; Technical assessment

0332
Graham, John. "Fireworks at TMI Commission Session Finale." **Nuclear News** 22(Mid-September 1979):23B. (News)
*President's Commission; Nuclear Regulatory Commission; Licensing

0333
Graham, John. "Kemeny Report." **Nuclear News** 22 (December 1979):26-28. (News)
*President's Commission; Recommendations

0334
Graham, John. "Three Mile Island; President's Commission Confronts the `Truth'." **Nuclear News** 22(July 1979):26-30. (News)
*President's Commission; Recommendations

0335
Graham, John. "TMI/Kemeny Commission: Recollections of a Day in March." **Nuclear News** 22(September 1979):30-32. (News)
*President's Commission; Evacuation

0336
Green, Harold P. "On the Kemeny Commission." **Bulletin of the Atomic Scientists** 36(March 1980):46-48. (Editorial)
Green urges implementation of the recommendations of the Kemeny Commission. The focus is on people-related problems and restructuring of the NRC.
*President's Commission; Recommendations

0337
Hayes, William C. "Looking Back in Anger." **Electrical World** 192(December 1, 1979):3. (Editorial)
In an essay in reaction to the Kemeny report, Hayes describes a "sense of betrayal by our government, as represented by the Nuclear Regulatory Commission."
*President's Commission; Nuclear Regulatory Commission

0338
Jorgenson, Barbara. "Credibility and the President's TMI Commission." **Public Relations Journal** 36(March 1980):36-38.
Credibility is discussed by the public information director of the President's Commission. She covers such areas as the internal organization of the commission, staff problems, dealing with the NRC, and media leaks.
*President's Commission; Credibility; Public relations; Media

0339
Kemeny, John G. "Saving American Democracy: The Lessons of Three Mile Island." **Technology Review** 82(June/July 1980):64-75.
Kemeny, Chairman of the President's Commission on the Accident at Three Mile Island, gives insight into the inner workings of the Commission, relating analysis and decisions of the Commission to democracy and the political system. Jeffersonian democracy is no longer adequate to handle the complex issues of modern technology. Findings of the Commission are discussed, in particular the three levels of people problems: human error, management problems, and the Nuclear Regulatory Commission.
*President's Commission; Democratic theory; Politics

0340
"The Kemeny Commission's Recommendations." **New Scientist** 84(November 8, 1979):428. (News)
*President's Commission; Recommendations

0341
"Kemeny Report Further Shows Nation's Nuclear Future." **Power** 123(December 1979):12. (News)
*President's Commission; Recommendations

0342
Kenward, Michael. "Critical Reactions." **New Scientist** 84(November 1, 1979):330. (Editorial)
*President's Commission; Recommendations

0343
Lanouette, William J. "The Kemeny Commission Report." **Bulletin of the Atomic Scientists** 36(January 1980):20-31.
The major recommendations of the Kemeny Commission report are discussed along with its implication for the nuclear industry.
*President's Commission; Recommendations; Nuclear industry

0344
Lanouette, William J. "Kemeny Panel Will Draw Blueprint for Future of Nuclear Power." **National Journal** 11(August 18, 1979):1366-1369.
The President's Commission on the Accident at Three Mile Island was put together by the Carter Administration in less than one week, and was given six months to complete the investigation and put together a report. Some "inside" background on the investigation and some of the discoveries are discussed. One example of information discovered was a Babcock & Wilcox Company internal memo which diagnosed a similar loss of coolant accident in 1978.
*President's Commission

0345
"Lessons of Three Mile Island." **Economist** 273(November 3, 1979):31-32. (News)
*President's Commission; Recommendations

0346
Lewis, Carolyn. "A Reporter Feels the Heat." **Columbia Journalism Review** 18(January 1980):34-37.
Lewis, a member of the President's Commission on the Accident at Three Mile Island, provides an inside view of the workings of the Commission and its failure to recommend a moratorium on new nuclear plants. From the start, the Commission is characterized as being overly concerned with public relations and the media.
*President's Commission; Public Relations; Media

0347
Lombardo, Thomas G. "TMI: An Insider's Viewpoint." **IEEE Spectrum** 17(May 1980):52-55. (3 refs.)
Lombardo interviews Patrick E. Haggerty, a member of the President's Commission on the Accident at TMI. Haggerty discusses TMI, the nuclear industry, the NRC, and the IEEE.
*President's Commission; Haggerty, Patrick E.; Nuclear industry

0348
Marrett, Cora Bagley. "The Accident at Three Mile Island and the Problem of Uncertainty." **Annals of the New York Academy of Sciences** 365(1981):280-291. (19 refs.)
Marrett, a member of the President's Commission, focuses on the lessons that the commissioners learned from TMI. Particularly, she chronicles the communications difficulties and commissioners' "search for clarity and certainty in the midst of incomplete information." She discusses "problems surrounding the relationship between the office of the governor and the utility," and briefly comments on evacuation advisories.
*President's Commission; Communications; Lessons learned

0349
Marshall, Eliot. "Carter Backs `Spirit of Kemeny Report'." **Science** 206(December 21, 1979):1380. (News)

*President's Commission; Recommendations; Carter administration; Nuclear Regulatory Commission

0350
Marshall, Eliot. "Kemeny Report: Abolish the NRC." **Science** 206(November 16, 1979):796-798. (News)
*President's Commission; Recommendations

0351
Marshall, Eliot. "A Review to Review the Review." **Science** 206(November 16, 1979):797. (News)
*President's Commission; Carter administration

0352
Miller, William H. "Decisions This Fall to Sway Nuclear Fate." **Industry Week** 203(October 1, 1979):18-19, 22. (News)
*President's Commission; Nuclear industry; Public opinion

0353
Pelham, Ann. "Kemeny Commission Cites Nuclear Risks ... Sees Need for Fundamental Changes." **Congressional Quarterly Weekly Report** 37(November 24, 1979):2666-2667.

The major recommendations of the Kemeny Commission are summarized. No moratorium was recommended and the NRC was sharply criticized. The report "made it clear that the future of nuclear power is a political question."
*President's Commission; Recommendations; Politics; Nuclear industry

0354
Peterson, Russell W. "Lessons of Three Mile Island." **Technology in Society** 2(3, 1980):295-302.

Peterson, a member of the President's Commission on Three Mile Island, indicates that the primary lesson to be learned from TMI is the fear of a nuclear catastrophe. He warns that "the public must not be lulled into complacency by the promotional efforts of the nuclear industry." He also indicates that the environment is more important than the energy problem.
*President's Commission; Lessons learned; Nuclear industry; Public relations

0355
"President Responds to Kemeny Report." **Nuclear News** 23(January 1980):42-43. (News)
*President's Commission; Carter administration

0356
"Presidential Commission Issues TMI Accident Report." **Public Utilities Fortnightly** 104(November 22, 1979):33. (News)
*President's Commission; Recommendations

0357
"President's Investigation." **Science** 204(May 11, 1979):595. (News)
*President's Commission

0358
"Recommendations of the Kemeny Commission." **Nuclear News** 22(December 1979):29-32. (News)
*President's Commission; Recommendations

0359
"Report of the President's Commission on the Accident at Three Mile Island." **Nuclear Safety** 21(March-April 1980):234-244.

The Preface and the Commission Recommendations of the report are printed here.

*President's Commission; Recommendations

0360
"Staff Reports to the President's Commission on the Accident at Three Mile Island; Selected Reports from: The Technical Assessment Task Force, The Public Health and Safety Task Force, The Office of the Chief Counsel." **Progress in Nuclear Energy** 6(1-3, 1980):1-436.

Reprints of the President's Commission Report.

*President's Commission; Recommendations

0361
"Three Mile Island Panel Issues Report." **Chemical & Engineering News** 57(November 5, 1979):7. (News)

*President's Commission; Recommendations

0362
"Three Mile Island Panel Sets Hearings." **National Underwriter (Prop Ed)** 83(May 18, 1979):2. (News)

*President's Commission; Liability insurance

0363
Torrey, Lee. "The Accident at Three Mile Island." **New Scientist** 84(November 8, 1979):424-427. (News)

*President's Commission; Recommendations; Kemeny, John G.; Cleanup

0364
"Two Commissions: Battle in the Making." **Nuclear News** 22(October 1979):22-23. (News)

*President's Commission; Nuclear Regulatory Commission

**OTHER INVESTIGATIONS**

0365
"Accident Investigations Gear Up as New Violation Disclosed." **Engineering News-Record** 202(April 19, 1979):11. (News)

*Accident investigations

0366
"Accident that Should Not Have Happened." **Nuclear Engineering International** 24(October 1979):34. (News)

*Accident investigations; President's Commission; Accident - causes

0367
"Aftermath of the Three Mile Island Accident." **Public Utilities Fortnightly** 103(May 10, 1979):35-36. (News)

*Accident investigations; Regulation; Congress

0368
Christiansen, Donald. "Where Do We Go From Here?" **IEEE Spectrum** 16(November 1979):110-111. (Editorial)

While reports from most investigations on TMI are still outstanding, Christiansen has compiled a list of recommendations that have emerged thus far. These observations relate to design factors, operating factors and failure analysis.

*Accident investigations; Recommendations; Lessons learned

0369
Cookson, Clive. "Three Mile Island in a Sea of Inquiries." **New Scientist** 82(May 17, 1979):525. (News)
*Accident investigations

0370
Cottrell, William B. "Developments Pertaining to the Three Mile Island Accident." **Nuclear Safety** 20(September-October 1979):613-623. (24 refs.)
This article discusses developments related to TMI that have occurred over the past several months, including information on the following: Lessons Learned Task Force, Advisory Committee on Reactor Safeguards (ACRS) Reports on TMI, Congressional investigations, and Interim Report by Metropolitan Edison, a NRC Report on Babcock & Wilcox feedwater transients, a nuclear program review by the Tennessee Valley Authority, radioactive sampling, and liability insurance.
*Accident investigations; Recommendations; Lessons learned

0371
"Deaf Ears Heard Harrisburg Warning." **New Scientist** 82(April 26, 1979):245. (News)
*Accident investigations

0372
Greenberg, Dan. "Harrisburg Fallout." **New Scientist** 82(April 19, 1979):170. (Editorial)
*Accident investigations; Nuclear industry

0373
Hatcher, James E. "The Three Mile Island Nuclear Accident: GAO's Role." **GAO Review** 15(Winter 1980):26-28.
Hatcher summarizes events surrounding the accident, in particular, the response of the various government agencies to the accident, the confusion resulting from lack of planning and coordination, and the various commission reports and congressional investigations. GAO's role stems from congressional interest; the agency is involved in following the work of the many groups investigating TMI.
*Accident investigations; General Accounting Office; Government; Congress

0374
Holland, J.R., and Wechsler, M.S. "Engineers' Committee on Three Mile Island." **Journal of Metals** 32(January 1980):34, 45. (News)
*Accident investigations; Engineers

0375
"House Panel Finds Little Danger at Three Mile Island Plant." **Public Utilities Fortnightly** 105(April 24, 1980):56. (News)
*Accident investigations; House Science and Technology Committee; Safety

0376
Kenward, Michael. "More Lessons from Three Mile Island." **New Scientist** 85(January 17, 1980):146. (News)
*Accident investigations; Recommendations; Lessons learned

0377
Marshall, Eliot. "Assessing the Damage at TMI." **Science** 204(May 11, 1979):594-596. (News)
Marshall discusses the various investigations and inquiry commissions that have been initiated in response to the TMI accident. The damage which must be assessed ranges from the actual plant damage and clean-up requirements to the financial market both of GPU Nuclear in particular and the nuclear industry in general.
*Accident investigations; Cleanup; Costs; Nuclear industry; Financing

0378
"Mayor of Harrisburg Misled Inquiry." **Nuclear Engineering International** 30(March 1985):8. (News)
*Accident investigations; Health effects; Reed, Stephen

0379
"NRC Criticized Again for Three Mile Island." **Chemical & Engineering News** 58(February 4, 1980):8. (News)
*Rogovin report; Recommendations

0380
"NRC Halts Plant Licensing Pending Accident Review." **Engineering News-Record** 202(May 31, 1979):9. (News)
*Accident investigations; Nuclear Regulatory Commission; Licensing

0381
"NRC Management Hit Hard by its own TMI Study Team." **Engineering News-Record** 204(January 31, 1980):14. (News)
*Rogovin commission; Recommendations; Nuclear Regulatory Commission

0382
"NRC Publishes TMI Tragedy of Errors." **Nuclear Engineering International** 24(September 1979):10. (News)
*Accident investigations; Operators; Human error; Radiation dosage

0383
Okrent, David, and Moeller, Dade W. "Implications for Reactor Safety of the Accident at Three Mile Island, Unit 2." **Annual Review of Energy** 6(1981):43-88. (90 refs.)

A brief description and chronology of the accident are given, followed by a summary of the results, recommendations and actions of the various studies and broad implications of the accident for nuclear power in the U.S.
*Accident investigations; Nuclear industry; Lessons learned; Accident sequence

0384a
"Report Estimates a TMI-Style Nuclear Accident Every 13.5 Years." **Industrial Research & Development** 24(October 1982):56-57. (News)
*Oak Ridge National Laboratory Report; Risk

0384b
"Reporting to the Nation on TMI." **EPRI Journal** 5(May 1980):36. (News)
*Accident investigations; Nuclear industry

0385
"The Rogovin Report on Three Mile Island 2." **Nuclear Safety** 21(May-June 1980):389-393.

The "Foreward" and "Summary" of the Rogovin report are printed here.
*Rogovin report; Recommendations

0386
"Seasoned Investigator Picked to Lead TMI Accident Inquiry." **Engineering News-Record** 202(June 21, 1979):47. (News)
*Rogovin, Mitchell; Accident investigations

0387
"Senate Investigators Hit NRC in Three Mile Island Probe." **Engineering News-Record** 203(October 11, 1979):12. (News)

*Accident investigations; Senate Subcommittee on Nuclear Regulation; Nuclear Regulatory Commission

0388
"Senate Releases Report on Three Mile Island Accident." **Public Utilities Fortnightly** 106(July 31, 1980):36. (News)
*Hart report; Recommendations

0389
Sheridan, Peter J. "After Three Mile Island: A Momentous Decision is Impending." **Occupational Hazards** 41(October 1979):68-73.
TMI marks the end of an era. Summarized here are reports by a task force of the House Subcommittee on Energy and the Environment, and by the NRC. Published prior to the Kemeny report, the task force report warned against design error, equipment malfunctions and faulty construction, procedural errors, and operator errors. The NRC also found "35 technical violations in the course of its investigation... ."
*Accident investigations; House Subcommittee on Energy and the Environment; Recommendations

0390
"Study Said Safety Gear Might Fail at A-Plant." **Engineering News-Record** 202(April 26, 1979):14. (News)
*Accident investigations

0391
"Summary of TMI-2 Lessons Learned Task Force Report." **Nuclear Safety** 20(November-December 1979):735-741. (7 refs.)
The TMI-2 Lessons Learned Task Force, established by Harold R. Denton, which made recommendations in the areas of design and plant operations, along with implementation of short-term recommendations, is summarized.
*Lessons learned; Nuclear reactor design; Nuclear power plants

0392
"Three Mile Island; the Latest Report." **Nuclear News** 23(August 1980):38, 41. (News)
*Hart report; Senate Committee on Environment & Public Works; Senate Subcommittee on Nuclear Regulation

# 7

# Nuclear Industry: Safety, Occupational, and Financial Issues

## NUCLEAR INDUSTRY

0393
"A-Power Trials Continue in Wake of TMI Accident." **Engineering News-Record** 203(July 26, 1979):10. (News)
*Nuclear industry; Plant cancellations

0394
"Accident Clouds Future for U.S. Nuclear Power." **Chemical & Engineering News** 57(April 9, 1979):8. (News)
*Nuclear industry

0395
"Accident Triggers Pitched Battle over Nuclear Future." **Engineering News-Record** 202(April 5, 1979):10-15. (News)
*Nuclear industry

0396
Aquilante, Daniel. "Human Factors Problems at Three Mile Island." **Industrial Design** 26(September 1979):17. (News)
*Human error; Control room design

0397
"Assessment: The Impact and Influence of TMI." **EPRI Journal** 5(June 1980):24-33.
In the third of four articles, the overall impact of TMI on the nuclear industry and the regulatory environment is discussed. Most technical questions have been answered, but "the economic implications are still unfolding." Radioactivity releases and health effects are also discussed.
*Nuclear industry; Regulatory environment; Economic effects

0398
"At Harrisburg." **America** 140(April 14, 1979):297-298. (Editorial)
TMI forced the nation to rediscover the dangers of atomic power. There are always costs for technological developments.
*Nuclear industry; Risk

0399
Burchill, William E. "Impact of TMI on Combustion Engineering Technical Activities." **Progress in Nuclear Energy** 10(3, 1982):267-283. (26 refs.)
This paper describes the impact of the TMI accident on the technical activities of Combustion Engineering, the designer and manufacturer of nuclear steam supply systems (NSSSs). Studies of plant transient characteristics that have been completed since the accident are highlighted. Changes in the vendor/utility relationship as a result of the accident are covered, along with safety functions, emergency procedure guidelines, training and new instrumentation.

*Combustion Engineering; Nuclear steam supply system; Transients; Lessons learned

0400
Coffman, F., and Feinroth, H. "Three Mile Island -- the Challenge, the Saga, the Outlook." **Nuclear Energy** 21(December 1982):365-369.

As a result of the TMI accident, the nuclear industry has been challenged to minimize the probability of recurrence of an accident as serious as TMI. The industry has also been challenged to explain TMI to the public. Lessons have been learned from the clean-up (venting the krypton and cleaning up the water in the reactor building). Decontaminating the building and defueling the reactor are also discussed. The nuclear industry still faces a major challenge.

*Nuclear industry; Cleanup; Lessons learned

0401
Cohen, David. "After Three Mile Island." **New Scientist** 87(September 25, 1980):948. (Editorial)

*Nuclear industry; Lessons learned

0402
David, Peter. "Disaster by Many Small Cuts." **Nature** 306(December 15, 1983):631. (News)

*Nuclear industry; Cleanup

0403
Edwards, Jack. "Lessons from Three Mile Island." **New Scientist** 97(January 13, 1983):79-84.

The accident at TMI is described, and the implications are analyzed in terms of the positive effects on the safety of nuclear power plants. The implications of this loss of coolant accident (LOCA), and the resultant changes in procedures and safety regulations are discussed and applied to the nuclear industry in Great Britain.

*Nuclear industry; Lessons learned; Great Britain

0404
Eisenhut, Darrell G. "TMI Plus 5: Nuclear Power on the Ropes, Part II: NRC as Referee." **IEEE Spectrum** 21(April 1984):33-39.

Eisenhut, the NRC's licensing director, "reports on the key regulations stemming from the Three Mile Island accident. Over 130 items must be improved at more than 65 sectors," as part of the implementation of Nureg-0737, "Clarification of TMI Action Plan Requirements."

*Nuclear industry; Nuclear Regulatory Commission; Regulation; TMI - anniversary (5)

0405
Ember, Louis R. "Nuclear Safety in Doubt a Year After Accident." **Chemical & Engineering News** 58(April 7, 1980):22-24.

The effects of TMI have been far-ranging on the nuclear industry and on its owner, GPU. Discussed here are the Kemeny Commission's report, a report by Ralph Nader's Critical Mass Energy Project, reactions of the Union of Concerned Scientists and the Sierra Club, effects on the NRC, and political and economic effects on the industry.

*Nuclear industry; TMI - anniversary (1); Accident investigations

0406
Feinroth, H., and Coffman, F.E. "Discussion: Three Mile Island -- The Challenge, the Saga, the Outlook." **Nuclear Energy** 22(February 1983):53-55.

This is a discussion of a paper published in *Nuclear Energy* 21(December 1982):365-369.

*Nuclear industry

0407
Feld, Bernard T. "Three Mile Island." **Bulletin of the Atomic Scientists** 35(May 1979):6. (Editorial)
The TMI accident is viewed in relation to the nuclear industry.
*Nuclear industry

0408
Fischetti, Mark A. "TMI Plus 5: Nuclear Power on the Ropes." **IEEE Spectrum** 21(April 1984):26-27.
In this introduction to a five-part special report, the economic woes of the nuclear industry, five years after the events triggered by TMI, is described. No new nuclear plants have been ordered in the U.S. since TMI, numerous plant orders have been cancelled, and construction plans for nuclear plants altered.
*Nuclear industry; Economic effects; TMI - anniversary (5)

0409
Fischetti, Mark A. "TMI Plus 5: Nuclear Power on the Ropes, Part III: Band-aids and Better." **IEEE Spectrum** 21(April 1984):39-43.
Primarily because of improving computer technology, worldwide changes have improved five vital areas in the nuclear industry: "hardware, control-room instrumentation, operator training, reactor maintenance, and communications within industry." Nuclear plants around the world are used to illustrate these changes.
*Nuclear industry; Lessons learned; International reaction; TMI - anniversary (5)

0410
Fischetti, Mark A. "TMI Plus 5: Nuclear Power on the Ropes, Part V: But Challenges Remain." **IEEE Spectrum** 21(April 1984):49-55. (refs.)
This final section discusses issues such as "overregulation and deliberate licensing procedures, educating the public, and forming cooperative nuclear facilities." The cleanup of TMI remains a challenge to the entire nuclear industry.
*Nuclear industry; Regulation; TMI - anniversary (5)

0411
Greenberg, Dan. "Receding Impact of TMI." **New Scientist** 85(March 27, 1980):1027-1028. (Editorial)
*Nuclear industry

0412
"Have Three Mile Island's Lessons Been Learned?" **Economist** 274(February 2, 1980):79-80. (News)
*Nuclear industry; Lessons learned

0413
Hayes, William C. "A Nuclear Accident is a National Event." **Electrical World** 192(November 15, 1979):3. (Editorial)
Hayes comments on nuclear power in general, safety concerns, and the effects of the TMI accident on the nuclear industry.
*Nuclear industry

0414
Hayes, William C. "TMI Anniversary Marks a Fruitful Year." **Electrical World** 193(April 1, 1980):3. (Editorial)
Hayes describes a rebirth of the nuclear industry as a result of TMI. Instead of dying, the industry has faced and dealt with three basic issues: "generic safety, the quality of operator training, and the preparedness of utilities to meet such crises as the TMI-type accident."
*Nuclear industry; TMI - anniversary (1); Safety

0415
Hayes, William C. "TMI: Broadcasting the Worst Kind of Signals." **Electrical World** 198(February 1984):3. (Editorial)

Hayes comments that "as long as TMI-2 remains in its present unsecured state, its presence will negatively effect every nuclear utility in this country."

*Nuclear industry; Cleanup

0416
Holdren, John P. "Environmental Liabilities of Nuclear Power." **Bulletin of the Atomic Scientists** 36(January 1980):31-32. (Editorial)

Holdren's discussion centers around the risk factors of nuclear power. Three Mile Island must be evaluated in the context of the full range of environmental liabilities.

*Nuclear industry; Environmental liabilities; Risk

0417
Horgan, John. "TMI Plus 5: Nuclear Power on the Ropes, Part IV: Learning from Elsewhere." **IEEE Spectrum** 21(April 1984):43-49.

"Incidents at sites other than TMI have expanded the nuclear knowledge base and may help contribute to the solutions of new and old problems." Incidents at plants in the U.S. and Canada include: ruptures, leaks, and corrosion of steam generator tubes; pipe cracks in reactors; mistakes by operators; and poor quality control of construction.

*Safety; Nuclear industry; Reactor accidents; TMI - anniversary (5)

0418
Howard, Robert B. "Of Catastrophes, Nuclear and Otherwise." **Postgraduate Medicine** 66(July 1979):27-28.

A physician comments on the "near disaster proportions at Three Mile Island" along with several other disasters within a month's time involving modern technology. In accepting the risk of nuclear technology, we must insist that industry and government minimize risks and "that they tell us the truth."

*Risk; Nuclear power

0419
"Industry Response to Three Mile Island." **EPRI Journal** 4(November 1979):33-36.

In forums such as congressional committees, industry representatives described a three-part approach to apply lessons learned from TMI. The self-help approach addresses "the consequences of TMI on a technical design level, at the utility operations level, and on a financial risk-sharing basis." Two new institutions have been established, the Nuclear Safety Analysis Center at EPRI and the Institute of Nuclear Power Operations.

*Nuclear industry; Industry response; Lessons learned

0420
Iwler, Louis. "Nuclear Industry Mulls Lessons of TMI." **Electrical World** 193(January 1, 1980):18-19. (News)

*Nuclear industry; Lessons learned; President's Commission

0421
Iwler, Louis. "Three Mile Island Serious but not Fatal to Nuclear." **Electrical World** 191(May 15, 1979):23-24. (News)

*Nuclear industry; Safety; Regulation

0422
Jacobson, N.H. "Edison Electric Institute; Focus on TMI-2." **Nuclear News** 22(May 1979):46, 48-50. (News)

*Nuclear industry; Edison Electric Institute

0423
Kenward, Michael. "... And the Nuclear Backlash." **New Scientist** 85(January 17, 1980):146-147. (News)
*Nuclear industry

0424
Kenward, Michael. "Nuclear Bubble Bursts." **New Scientist** 82(April 5, 1979):2. (Editorial)
*Nuclear industry

0425
Lauren, T.C. "Lessons of Three Mile Island." **Optical Spectra** 13(5, 1979):6.
Unavailable.
*Lessons learned

0426
Lewis, Floyd W. "Nuclear Regulation After Three Mile Island." **Progress in Nuclear Energy** 7(2, 1981):103-109.
Lewis looks at the NRC, its role in nuclear regulation, and the future of nuclear energy in the U.S. in light of the Three Mile Island accident. He views the legacy of TMI as a mandate for reform of the NRC. As a result of the accident, the industry has been challenged to "create forward-looking action."
*Regulation; Nuclear Regulatory Commission; Nuclear industry; Lessons learned

0427
McCormick, N.J. "Changes in the Nuclear Power Industry After TMI." **Progress in Nuclear Energy** 10(3, 1982):245-248. (9 refs.)
In this introduction to a special issue of *Progress in Nuclear Energy*, the Three Mile Island accident is viewed as being beneficial to the nuclear power industry and the general public. The accident "has led to institutional changes and to changes in reactor instrumentation, operation training, and emergency preparedness."
*Nuclear industry; Institutional response; Lessons learned

0428
Mauro, Charles. "Three Mile Island: A Human Factors Problem." **Industrial Design** 26(September 1979):44.
*Human error; Control room design

0429
Miller, William H. "What Next for Nuclear Power?" **Industry Week** 201(April 16, 1979):17-19. (News)
*Nuclear industry

0430
"More Doubts About Nuclear Power." **Chemical Week** 124(April 11, 1979):22. (News)
*Nuclear industry; Public opinion; Safety

0431
Negin, C.A. "Regulatory Turbulence in the Wake of Three Mile Island." **Nuclear Engineering International** 24(October 1979):41-43.
"The entire complex of design, manufacture, operation, and regulation" have been placed on trial because of Three Mile Island. Recommendations of the Lessons Learned Task Force are discussed. The cost of TMI to the nuclear industry will be high in terms of new safety requirements and changes in the regulatory and licensing system.
*Nuclear industry; Regulation; Lessons learned

0432
"Nuclear Power: Can We Live With It?" **Technology Review** 81(June/July 1979):32-47.

This is "a panel discussion on nuclear risk, the lessons of Three Mile Island, and the future of nuclear power." Participants in the panel were Norman C. Rasmussen, Joel Yellin, Daniel J. Kleitman, Richard B. Stewart, Steven J. Marcus, and Leonard A. Phillips.

*Nuclear power; Lessons learned; Risk; Energy policy

0433
Olds, F.C. "TMI Update." **Power Engineering** 84(July 1980):28. (News)

*Nuclear industry; Public opinion

0434
Palladino, Nunzio J. "TMI Plus Five; Challenges for the Future: A Nuclear Regulator's Perspective." **Nuclear News** 27(April 1984):59-62.

Palladino looks at the future of the nuclear industry ten to twenty years after TMI. Discussed in the five-year context are: public confidence, industry actions as a result of TMI, and quality assurance.

*Nuclear industry; TMI - anniversary (5)

0435
Payne, Jon. "Three Mile Island: A One-Way Door." **Nuclear News** 23(March 1980):33. (Editorial)

*Nuclear industry

0436
Payne, Jon. "Three Mile Island: Many Lessons." **Nuclear News** 22(May 1979):31. (Editorial)

Payne's editorial covers initial reaction to the accident.

*Nuclear industry; Lessons learned; Public opinion

0437
Rawlins, David H., ed. "Impact of the Accident at Three Mile Island on an NSSS Vendor -- A Westinghouse Perspective." **Progress in Nuclear Energy** 10(3, 1982):285-297. (8 refs.)

Rawlins discusses the response of Westinghouse "to the demands of the highly intensive post-TMI environment." One lesson of TMI was the need to view the nuclear power plant as "an integration of components, systems, operations and people." Among the topics discussed are the impact of TMI on: licensing and Westinghouse initiatives in regulatory reform; design and analysis; plant operating procedures; and training.

*Westinghouse; Nuclear plant construction; Lessons learned; Nuclear industry

0438
Reed, Cordell. "Impact of Post-TMI Regulations on a Nuclear- Oriented Utility." **Progress in Nuclear Energy** 10(3, 1982):259-266.

The TMI accident, and the resulting regulations, had a great impact on the operations of the Commonwealth Edison Company, including management, control room operation, training, plant modifications, monitoring, emergency planning, and risk assessment. "In the long run [TMI] will have an overall positive effect on the entire nuclear industry."

*Commonwealth Edison Company; Lessons learned; Nuclear power plants; Nuclear industry

0439
Rotblat, Joseph. "Harrisburg: A Lesson." **New Scientist** 82(May 10, 1979):426. (Editorial)

*Nuclear industry; Lessons learned

0440
Rubinstein, Ellis. "Special Issue; Three Mile Island and the Future of Nuclear Power." **IEEE Spectrum** 16(November 1979):30-111.

This is a special issue that deals with the effects of the Three Mile Island accident on the nuclear industry. The introduction is on pp. 30-31. Articles in this issue directly related to TMI are listed separately.

*Overview; Nuclear industry

0441
Sugarman, Robert. "Nuclear Power and the Public Risk." **IEEE Spectrum** 16(November 1979):59-79.

In evaluating the public risk of nuclear power, Sugarman discusses: human factors engineering, in particular human error as related to TMI; safety systems, with emphasis on TMI; and accident mitigation.

*Nuclear industry; Risk; Human error; Safety

0442
Szalay, Robert A. "The Reaction of the Nuclear Industry to the Three Mile Island Accident." **Annals of the New York Academy of Sciences** 365(1981):222-229. (10 refs.)

TMI has shown that "responsibility for the safety of nuclear power plants belongs to their operators and managers." The industry responded with immediate help to stabilize the plant. Combined industry effort has created the Nuclear Safety Analysis Center, to address safety concerns; the Institute of Nuclear Power Operations, to improve operations and training; and the Nuclear Electric Insurance Limited, for financial protection due to plant outages. Additionally, a Policy Committee was formed within the Atomic Industrial Forum "to establish an official industry liaison with NRC."

*Nuclear industry; Safety; Industry response

0443
"Three Mile Island: Nuclear Power in the Balance." **Conservation News** 44(May 15, 1979):2+.

Unavailable.

*Nuclear industry

0444
"TMI: A Turning Point in Nuclear Power -- But Which Way?" **Mechanical Engineering** 102(February 1980):49-50. (News)

*Nuclear industry; Safety

0445
"TMI Brings Lessons, Challenges." **Nuclear News** 23(Mid-March 1980):23-29. (News)

*Nuclear industry; Lessons learned; TMI - anniversary (1)

0446
Udall, Morris K. "The Future of Nuclear Generation -- A Post-TMI Perspective." **Public Utilities Fortnightly** 105(June 19, 1980):81-82. (Editorial)

TMI has caused the nuclear power industry to be more safety conscious. Udall views the role of Congress as establishing regulatory agencies and overseeing them, while agencies such as the NRC must be responsible for the technical decisions.

*Nuclear industry; Safety; Regulation

0447
Udall, Morris K. "A Legislator's View of the Nuclear Option." **Progress in Nuclear Energy** 7(2, 1981):93-102.

Udall describes his view of the future of nuclear power in the United States, and how his "view of the safety issue was changed by Three Mile Island and the inquiries that followed." He discusses his experiences as related to the accident, including inadequate reporting, operator error, and safety issues.

*Nuclear industry; Safety; Politics

0448
Weinberg, A.M. "Can We Fix Nuclear Energy?" **Annals of Nuclear Energy** 6(1979):473-482.

In evaluating the future of nuclear energy, Weinberg makes constant reference to the impact and lessons of Three Mile Island.

*Nuclear industry; Energy policy

0449
Weinberg, Alvin M. "Can We Fix Nuclear Energy?" **Across the Board** 16(August 1979):20-31.

In a general discussion on the nuclear power industry, TMI is used as a "benchmark," with constant reference being made to the accident, both "before" and "after."

*Nuclear energy; Energy policy

0450
Weinberg, Alvin M. "Three Mile Island in Perspective: Keynote Address." **Annals of the New York Academy of Sciences** 365(1981):1-12.

"Three Mile Island has demonstrated that the central problem of nuclear energy is...reactor safety." Weinberg and others at the Institute for Energy Analysis at Oak Ridge propose four requirements for a nuclear energy system: technical fixes, professionalization of the nuclear cadre, restructuring of the nuclear utility industry, and a rational siting policy. The TMI accident affected the industry as a whole. The utilities and government have responded, but these responses may not be sufficient. Nuclear energy is necessary in the long term, and TMI may have been the "salvation" of the industry. Includes discussion.

*Nuclear industry; Safety

0451
"Will TMI Kill Nuclear Power?" **Electrical World** 191(June 1, 1979):11. (News)

*Nuclear industry

## SAFETY/OCCUPATIONAL ISSUES

0452
Ahrons, Richard; Damon, Dwight; Isaac, Erich; von Oss, Carel; von Graevenitz, Alexander; and Wilson, Richard. "Nuclear Power Defended." **New England Journal of Medicine** 301(July 5, 1979):52. (Letter)

*Nuclear reactor design

0453
"B&W Designers Respond to TMI-2 Incident." **Design News** 36(February 18, 1980):14-15. (News)

*Babcock & Wilcox; Nuclear reactor design

0454
Breen, Robert J. "Defense-in-Depth Approach to Safety in Light of the Three Mile Island Accident." **Nuclear Safety** 22(September-October 1981):561-569. (10 refs.)

The analysis and follow-up to the TMI accident have furnished many reactor safety lessons, which have strengthened the "defense-in-depth approach to reactor safety." The three-level concept "(Level 1: prevent; Level 2: detect and mitigate;

Level 3: contain and protect)" is used to analyze the TMI accident; the lessons learned are discussed in the context of the industry as a whole.

*Safety; Nuclear industry; Lessons learned

0455

Burns, Robert D., III. "Report of the Technical Assessment Task Force from Volume II: WASH 1400 -- Reactor Safety Study." **Progress in Nuclear Energy** 6(1-3, 1980):117-140. (38 refs.)

Reprinted from the U.S. government publication.

*Reactor Safety Study (WASH 1400); Technical assessment; Safety

0456

Butler, Paul. "Lessons Learned from TMI." **Engineer** 248 Supplement(June 28, 1979):S9, S11. (Editorial)

The TMI accident will lead to improved nuclear reactor safety.

*Safety; Nuclear industry; Lessons learned

0457

Clark, Philip R. "TMI Plus Five; Looking Beyond the Lessons: A Utility Manager's Perspective." **Nuclear News** 27(April 1984):63-68.

Clark, President and Chief Executive Officer of GPU Nuclear, looks primarily at safety and risk as related to the nuclear industry. He focuses on the organization of GPU Nuclear, the resources of the company, and quality assurance activities.

*Safety; GPU Nuclear; Management; Quality assurance; TMI - anniversary (5)

0458

Corcoran, W.R.; Finnicum, D.J.; Hubard, F.R., III; Musick, C.R.; and Walzer, P.F. "The Operator's Role and Safety Functions." **Combustion** 51(June 1980):28-39. (16 refs.)

Roles for the operator in nuclear safety are outlined, along with how "safety functions can be used to reduce economic loses and increase safety margins." The operator's three roles in plant safety are to "set up the plant to respond properly to adverse events," to "operate the plan so as to minimize the frequency and severity of adverse events," and to "assist in mitigating the consequences of adverse events."

*Operators; Nuclear workers; Safety

0459

Deutsch, Robert W. "Plant Safety and Public Risk in Nuclear Energy." **Public Utilities Fortnightly** 106(November 20, 1980):18-21.

The future of nuclear power, public risk from nuclear energy and public attitudes toward new technologies are discussed in light of the experience of the Three Mile Island accident.

*Safety; Risk; Nuclear industry; Public opinion

0460

"Directions: Making Nuclear Plants Safer." **EPRI Journal** 5(June 1980):34-38.

The TMI accident is viewed from positive actions that have resulted. These include an emphasis on safety and accident prevention. Preparation for similar emergencies is an important outcome. This is the fourth and final article in a series on TMI.

*Safety; Accident prevention; Nuclear industry; Lessons learned

0461

Douglas, John. "California Report Pinpoints Hazards in Layout of Three Mile Island Control Room." **Nature** 279(June 7, 1979):464. (News)

*Control room design

0462
Egan, Joseph R. "To Err is Human Factors." **Technology Review** 85(February/March 1982):23-29. (15 refs.)

As part of a general discussion on the sources of human error in nuclear power plants, the human factors aspects of the TMI accident is discussed. Despite the NRC's pursuit of human factors, primarily in response to TMI, the "social factor" of working relationships has been largely ignored.

*Human factors; Operators; Human error

0463
Embrey, N.S. "B&W Responds to TMI-2." **Electrical World** 194(September 1, 1980):34-36.

The author discusses improvements that have been made to Babcock & Wilcox equipment since TMI. The four top priority areas of concern are: man/machine interface, data collection and analysis, improved supplier interface, and operator training and qualifications.

*Babcock & Wilcox; Nuclear reactor design

0464
Eytchison, Ronald M. "Report of the Technical Assessment Task Force from Volume III: Selection, Training, Qualification and Licensing of Three Mile Island Reactor Operating Personnel." **Progress in Nuclear Energy** 6(1-3, 1980):141-238. (158 notes)

Reprinted from the U.S. government document.

*Nuclear workers; Training, occupational; Technical assessment

0465
Feuer, Dale. "Training at Three Mile Island: Six Years Later." **Training** 22(March 1985):26-40.

Following recommendations of the NRC, operator training has emerged as a major area of concern for the owners of TMI. Training is being revamped, along with a focus on integrity and management competence.

*Training, occupational; Nuclear workers; Operators; Management

0466
Hagen, E.W., and Mays, G.T. "Human Factors Engineering in the U.S. Nuclear Arena." **Nuclear Safety** 22(May-June 1981):337-346. (36 refs.)

Human factors engineering in the U.S. nuclear power industry is chronicled. The focus is on human factors at the TMI accident, where at least three types of human errors were involved: operator error, operator training/procedures, and instrumentation display. The causes of human error at TMI are overviewed; reactions and changes in human factors engineering initiated or recommended by the NRC, the industry and other services, as a result of TMI, are discussed.

*Human factors; Human error; Operators; Training, occupational

0467
Hanauer, Stephen H. "The Human Factor in Nuclear Power Plant Safety: Progress Since Three Mile Island." **Progress in Nuclear Energy** 10(3, 1982):299-347. (145 refs.)

Hanauer reviews the "post-Three Mile Island assessments of human factor considerations in nuclear power plant safety." The paper discusses: analyses of human factors in the TMI accident, including recommendations of the Kemeny Commission; human capabilities and limitations, including human error and reliability; the human factors improvements recommended by the NRC Action Plan; operators, operator error, and training; management and administration; procedures in nuclear power plants; and control room design and man-machine interface. TMI is viewed as the event which triggered an emphasis on people versus hardware, and examples from TMI are used throughout the paper.

*Human factors; Human error; Operators; Safety; Lessons learned; President's Commission; Recommendations; Control room design

0468
Hanes, Lewis F.; O'Brien, John F.; and DiSalvo, Raymond. "Control-Room Design: Lessons from TMI." **IEEE Spectrum** 19(June 1982):46-52.

Changes and new developments in control room design, implemented because of the TMI accident, are discussed. "Improved instrumentation plus computer-based decision aids under development should help minimize the potential consequences of accidents."

*Control room design; Decision making, computer assisted; Lessons learned

0469
"The Impact of the Three-Mile Island Accident on Nuclear Safety Research." **OECD Observer** no. 107(November 1980):48-50.

A report, "Nuclear Safety Research in the OECD Area: The Response to the Three Mile Island Accident," describes four major safety research programs as a result of TMI: projects related to reactor systems and their operation, the man/system interface, evaluating the residual risk of a major accident, and reducing possible consequences of an accident.

*Safety; Research and development; Risk

0470
Kennedy, Richard T. "Nuclear Power in the Post-TMI Era." **Public Utilities Fortnightly** 105(June 19, 1980):82, 84, 85. (Editorial)

TMI has had a major impact on the industry's view on safety. It remains to be seen whether the accident has "signed the death warrant of the nuclear industry" or "signaled the rebirth" of a safer industry.

*Nuclear industry; Safety

0471
Kerr, William. "Whatever Happened to Reactor Safety." **Bulletin of the New York Academy of Medicine** 59(December 1983):907-911. (4 refs.)

Experience at Browns Ferry and Three Mile Island have improved safety at nuclear plants, particularly in the areas of fire protection, operator training and reactor design. Much remains to be done in the area of the quality of "design, construction, and operation of nuclear power plants."

*Safety; Operators; Nuclear reactor design

0472
Kletz, T.A. "Three Mile Island: Lessons for the HPI." **Hydrocarbon Processing** 61(June 1982):187-188, 190-192. (18 refs.)

The article looks at the accident in three phases, discusses the mechanical faults and operator errors in dealing with the accident, and comes up with nine steps that the hydrocarbon processing industry can use to prevent disasters. The author suggests that rules may not be enough, and that the nuclear industry must accept more responsibility for safety.

*Safety; Accident prevention; Lessons learned; Accident sequence

0473
Lamarsh, John R. "Safety Considerations in the Design and Operation of Light Water Nuclear Power Plants." **Annals of the New York Academy of Sciences** 365(1981):13-19.

As part of a conference on TMI, Lamarsh explains the operation of light water nuclear reactors, the safety features of multiple barrier design, and what occurs to fission products in a reactor accident. Includes discussion.

*Light water reactors; Safety

0474
Lee, Van M. "Siren System Designed for Three Mile Island." **Electrical World** 198(February 1984):64-66. (News)
*Safety; Siren system

0475
Lihach, Nadine. "Valve Test Program: Response to TMI." **EPRI Journal** 8(May 1983):14-17.
Following TMI, EPRI was asked to test the safety and relief valves that safeguard PWR primary systems from overpressure. Results from three facilities showed that "performance was satisfactory for relief values under all conditions, and performance was generally satisfactory for safety valves." Results are available to the entire industry.
*Valves; Equipment function; Safety

0476
Livingston, W.L. "The Legacy of TMI-2 for the Nuclear Power Plant Operations System." **Combustion** 52(November 1980):10-18. (3 refs.)
The action plan developed by the Nuclear Regulatory Commission as a result of TMI was NUREG 0660 (May 1980). The impact of NUREG 0660 on the operation of nuclear power plants is discussed.
*NUREG-0660; Nuclear power plants; Safety

0477
Livingston, William L. "The Impact of Three Mile Island on Process Control Computer Technology." **Control Engineering** 27(May 1980):86-91.
Since Three Mile Island, the NRC has reversed its position and encourages "the application of process computers for an important role in control room operations...." This article describes the design of "man-computer and man-control system interfaces."
*Computers; Control room design; Human factors

0478
Livingston, William L. "TMI Lessons-Learned Demand Full Systems Approach to Control Design." **Control Engineering** 27(October 1980):93-95.
The full systems approach to control room design is emphasized. Control room design is discussed with respect to the task action plan in NUREG-0660.
*Control room design; Lessons learned; NUREG-0660

0479
Long, R.L.; Gaines, D.P.; and Knief, R.A. "Nuclear Personnel Training After TMI-2: The GPUN Response." **Progress in Nuclear Energy** 10(3, 1982):349-361. (22 refs.)
"This paper briefly describes the significance of the TMI-2 accident to nuclear personnel training [and] then describes in some detail the response of GPUN [GPU Nuclear] in the upgrading and expansion of training for all personnel involved in the operations, maintenance, technical and administrative support of the GPUN power plants." Included are GPUN training and education organization, GPUN training facilities, operator training and requalification programs, technician training, general employee and radiation worker training, and training/educational development programs.
*Nuclear workers; Operators; Training, occupational; GPU Nuclear

0480
Mynatt, F.R. "Nuclear Reactor Safety Research Since Three Mile Island." **Science** 216(April 9, 1982):131-135. (21 refs.)
Three aspects of the Three Mile Island accident have affected nuclear reactor safety research and development. The small release of radioactive iodine "has led to a new emphasis on the physical chemistry of fission product behavior in accidents." The severely damaged nuclear core has opened up research in a new area, that of

"degraded core accidents," while the role of the operators in the accident has created a new emphasis on human reliability. Research in all areas will help phase out present reactors and contribute to a new generation of reactors.
*Safety; Research and development; Nuclear reactor design

0481
"Nuclear Accident Will Redirect Safety Efforts." **Chemical & Engineering News** 58(April 14, 1980):5-6. (News)
*House Subcommittee on Energy Research & Production; Safety

0482
"Nuclear Heat Stress Program Appraised." **Public Utilities Fortnightly** 113(January 5, 1984):62-63. (News)
*Nuclear workers; Heat stress

0483
Olds, F.C. "Post TMI Plant Designs." **Power Engineering** 84(August 1980):55-62.
While the article's intended focus is on changes in plant design which have resulted from the TMI accident, it also includes a description and interpretation of the sequence of events during the accident, including what is labeled "the phantom accident," which occurred on Friday, and the communication problems.
*Nuclear reactor design; Accident sequence; Communications

0484
Phung, Doan L. "Light Water Reactor Safety Before and After the Three Mile Island Accident." **Nuclear Science and Engineering** 90(August 1985):509-520. (39 refs.)
The author first reviews light water reactor safety through the TMI accident, including reactor design criteria, operational problems, and the Rasmussen study (WASH-1400). The accident at TMI is described briefly, followed by the repercussions of the accident and a more detailed look at safety concerns by the NRC and the industry since the accident. Because of TMI, nuclear plants are safer than before the accident.
*Safety; Nuclear industry; Lessons learned; Light water reactors

0485
Phung, Doan L. "Technical Note: LWR Safety After TMI." **Nuclear Safety** 25(May-June 1984):317-323.
This is an adaptation of the Executive Summary of a report sponsored by the Andrew Mellon Foundation, *Assessment of LWR Safety Since TMI*, by Doan L. Phung. Specific questions addressed include what has the NRC done since TMI, what has the industry done since TMI, and how much safer are light-water reactors (LWRs) after TMI?
*Safety; Light water reactors; Nuclear Regulatory Commission; Nuclear industry

0486
Pigford, T.H. "Management of Nuclear Safety; A Review of TMI After Two Years." **Nuclear News** 24(March 1981):41-48.
Pigford reviews lessons learned from TMI. The fourteen lessons center on safety, planning, and training.
*Safety; Lessons learned; TMI - anniversary (2)

0487
"Post-TMI Safety Fixes Confound Utilities." **Engineering News-Record** 207(September 17, 1981):53-54. (News)
*Safety

0488
Rasmussen, Norman C. "A Review of the Reactor Safety Study." In: **The Medical Basis for Radiation Accident Preparedness**, edited by Karl F. Hubner and Shirley A. Fry, 519-531. New York: Elsevier/North Holland, 1980.

Rasmussen reviews "the methods used by the Reactor Safety Study (WASH 1400) in assessing the risk to the public from nuclear power plant operation." He then relates the study to the accident at TMI and how well the study "identified specific events at TMI as a possible source of trouble."

*Reactor Safety Study (WASH 1400); Risk; Probability; Safety

0489
Roth, Jack. "Three Mile Island." **IEEE Spectrum** 17(February 1980):16. (Letter)
*Safety

0490
Shelton, R.E. "Emergencies and Rationality: The Case of Three Mile Island." **International Journal of Mass Emergencies and Disasters** 2(March 1984):41-60. (37 refs.)

Rationality theory is applied to the accident at Three Mile Island. The actions of the operators at TMI are explained in terms of practical rationality. They were, in fact, reacting in an appropriate way to correct the malfunctions that they believed had occurred; the failure in proper diagnosis can be traced back to poor training. The practical rationality concept can aid in understanding the organizational behavior of emergencies.

*Nuclear workers; Rationality theory; Organizational behavior

0491
Sheridan, Thomas B. "Human Error in Nuclear Power Plants." **Technology Review** 82(February 1980:22-33. (9 refs.)

Three Mile Island demonstrated what can occur due to human operator error in a nuclear power plant. Since TMI, the Nuclear Regulatory Commission has put new emphasis on "human reliability in nuclear plants to determine how error contributes to overall risks -- and how to improve overall safety."

*Human error; Nuclear workers; Operators

0492
Simon, Louis P. "What's Really Wrong with the Nuclear Power Industry: An Operator's Viewpoint." **Public Utilities Fortnightly** 108(November 19, 1981):26-30.

The Kemeny report failed to address the issue of a lack of standardization in nuclear plant design. According to Simon, such standardization may have "mitigated against some of the factors which have been officially determined as contributing to the occurrence of the Three Mile Island unit 2 reactor accident."

*Nuclear reactor design; Operators; Nuclear industry

0493
Stadie, Klaus. "Improving Nuclear Safety: A Post Harrisburg View." **OECD Observer** 99(1979):18-22.

The TMI accident has had a major impact on safety concerns of the nuclear industry both in the U.S. and worldwide.

*Safety; Nuclear industry; Lessons learned; International reaction

0494
"TMI Cleanup Crews Face Increased Radiation Risk." **Engineering News-Record** 212(January 26, 1984):13. (News)
*Cleanup; Nuclear workers; Risk; Radiation exposure, occupational

0495
"TMI Heat Stress Control a `Model for the Industry'." **Nuclear Engineering**

**International** 29(April 1984):48. (News)
*Nuclear workers; Heat stress

0496
"TMI-2; Adding to Safety Knowledge." **Nuclear News** 23(March 1980):56-57. (News)
*Safety; Cleanup

0497
Vesperman, Gary C. "Three Mile Island." **IEEE Spectrum** 17(February 1980):14. (Letter)
*Safety

0498
Zebroski, E.L., and Leverett, M.C. "The Nuclear Safety Analysis Center." **Nuclear Safety** 22(January-February 1981):1-9. (5 refs.)
The Nuclear Safety Analysis Center (NSAC), set up immediately after the TMI accident by the Electric Power Research Institute, is "charged with analyzing the accident, deriving lessons to be learned from it, and performing other safety-related functions." NSAC's work discussed here includes a brief analysis of the TMI accident, including lessons learned and health effects; a response to regulatory issues; and information on NSAC's generic safety studies.
*Nuclear Safety Analysis Center; Industry response; Safety; Regulation

0499
Zebroski, Edward L. "Scope and Functions of the U.S. Nuclear Safety Analysis Center." **Progress in Nuclear Energy** 7(2, 1981):111-118.
The Nuclear Safety Analysis Center (NSAC) was the one of the utility industry's responses to the TMI accident. It's "initial mission was to carry out a detailed technical analysis of the events that occurred at Three Mile Island" and to interpret lessons to be learned.
*Nuclear Safety Analysis Center; Nuclear industry; Industry response

## ECONOMICS

0500
Abbotts, John. "Nuclear Power After Three Mile Island." **Business and Society Review** 37(Spring 1980-81):18-24.
The economic fallout from TMI has not been limited to its owners, but has extended to the nuclear industry as a whole. The accident has caused confusion among industry and government experts.
*Economic effects; Nuclear industry; Financing; General Public Utilities; Government

0501
"Atomic Reactions." **Economist** 271(May 5, 1979):28, 45. (News)
*Economic effects; Nuclear industry; Activism

0502
Barrett, W. Brian; Heuson, Andrea J.; and Kolb, Robert W. "The Effect of Three Mile Island on Utility Bond Risk Premia: A Note." **Journal of Finance** 41(March 1986):255-261. (23 refs.)
The TMI accident caused "an increase in the risk premia attached to all electric utility bonds." Firms with nuclear plants had a larger increase in risk premia, and firms using Babcock & Wilcox reactors also suffered a larger risk premia than other firms.
*Stocks and bonds; Risk

0503
Blake, E. Michael. "GPU Faces Cash Crunch as Tax Time Nears." **Nuclear News** 24(March 1981):36. (News)
*General Public Utilities; Financing

0504
Blake, E. Michael. "GPU Gets Rate Hike, New Loan; Pays Taxes." **Nuclear News** 24(May 1981):34. (News)
*General Public Utilities; Financing; Rate hearings

0505
Bowen, Robert M.; Castanias, Richard P.; and Daley, Lane A. "Intra-Industry Effects of the Accident at Three Mile Island." **Journal of Financial and Quantitative Analysis** 18(March 1983):87-111. (28 refs.)
Bowen examines the "*ex post* returns for privately owned electric utilities (other than GPU)." Daily return data for a sample of utilities were analyzed statistically. It is suggested that the pricing of the equity of utilities has extensive interdependencies, as evidenced with TMI causing a "significant, downward price shift related to a utility's commitment to nuclear power and/or BW plants."
*Stock market; Utilities; Equity; Nuclear industry; Financing

0506
Brooks, Leroy D., and D'Souza, Rudolph E. "Electric Utility Returns and Risk in the Light of Three Mile Island." **Public Utilities Fortnightly** 110(November 11, 1982):26-32. (4 refs.)
An examination of "abnormal returns and risk measures on nuclear-dependent utilities and nondependent utilities" before, at the time of, and following the TMI accident, shows that "investors associate a decline in future profitability or increased risk with nuclear-associated utilities." Since TMI, objective measures "indicate a clear reduction in risk, relative to both the market and to nonnuclear-dependent utilities"; it is almost impossible to determine "the past and current ex ante risk of nuclear-dependent utilities."
*Stocks and bonds; Risk; Financial investments; Nuclear industry; Financing

0507
Chen, Carl R. "The Structural Stability of the Market Model After the Three Mile Island Accident." **Journal of Economics and Business** 36(February 1984):133-140. (17 refs.)
In a study of alpha and beta in the market model following the Three Mile Island accident, data indicate that, for 70 utilities divided into two portfolios (nuclear and non-nuclear), there was no change in the non-nuclear portfolio, and alpha fell and beta rose briefly for the nuclear portfolio.
*Stock market; Utilities; Economic effects

0508
Davidson, Wallace N., III, and Chandy, P.R. "A Note on the Effect of the Accident at Three Mile Island on the Nuclear Construction Industry." **Akron Business and Economic Review** 17(Spring 1986):23-27. (14 refs.)
Since TMI, no new nuclear plants have been ordered and 75 have been cancelled. The authors prove the hypothesis that "TMI caused an immediate downward shift in the stock returns" for the four major firms involved in the nuclear construction industry. However, the results did not show an increased risk level of these firms.
*Nuclear industry; Nuclear plant construction; Stock market; Risk

0509
"Economic Shock Wave Follows A-Plant Accident, Clouding Utility Plans." **Engineering News-Record** 202(April 12, 1979):36-38. (News)
*Economic effects; Nuclear industry

0510
Evans, Nigel. "Hidden Costs of the Accident at Three Mile Island." **Energy** 7(September 1982):723-730. (9 refs.)

While some costs of the TMI accident are evident, for example, cleanup costs and the cost of writing off an essentially new reactor, the accident has another cost which can be identified as the cost of replacement generation costs which are due to a drop in the performance of nuclear reactors worldwide. These costs worldwide are estimated to be approximately the same as the costs of the TMI cleanup operation.

*Costs; Cleanup; Nuclear industry; Replacement power costs

0511
"Fallout from Harrisburg." **Economist** 271(May 12, 1979):121. (News)

*General Public Utilities; Financing

0512
Fisher, M.J. "NRC Probe to Determine if TMI Activated Price-Anderson." **National Underwriter (Prop Ed)** 83(July 27, 1979):2, 32. (News)

*Price-Anderson Act; Extraordinary nuclear occurrence

0513
Gamble, Hays B., and Downing, Roger H. "Effects of Nuclear Power Plants on Residential Property Values." **Journal of Regional Science** 22(November 1982):457-478. (18 refs.)

Two studies are reported: one examines residential property values around four nuclear power plants before March 1979, and "the second examines the effects of property values in the vicinity of the TMI plant following that accident." In the first study, regression analyses found "no significant positive or negative effects on property values" near four nuclear plants in Northeast U.S. Using the same methodology, there was no evidence of adverse effects on residential property values near TMI after the March 1979 accident. A sharp decline of sales within ten miles of the plant occurred, but returned to near normal conditions within four to eight weeks of the accident.

*Residential property values; Real estate

0514
"GPU Chief Outlines Key Recovery Steps." **Public Utilities Fortnightly** 107(June 4, 1981):38. (News)

*General Public Utilities; Financing

0515
"GPU Rate Hike Denied; Credit Ceiling Cut." **Nuclear News** 23(October 1980):32. (News)

*General Public Utilities; Financing; Rate hearings

0516
"GPU Reports Financial Impact of TMI." **Electrical World** 191(May 1, 1979):11. (News)

*General Public Utilities; Financing

0517
Graham, John G. "Financing GPU ... After the Three Mile Island Accident." **Journal of Commercial Bank Lending** 62(May 1980):45-55.

Graham, Treasurer of General Public Utilities Corporation, reviews GPU's financial condition just prior to the TMI accident, and the financial and emotional situation just after the accident. Financial problems included the cost of replacement power, short-term debt, and reactions of the regulators. Actions taken by GPU included an increase in bank lines, a deep cut in expenditures to husband cash resources, pursuit of permanent financing, and negotiations for a revolving credit agreement. Discussed also are the implications of rate-setting commissions, and some lessons learned from the accident.

*General Public Utilities; Financing; Lessons learned

0518

Hill, Joanne, and Schneeweis, Thomas. "The Effect of Three Mile Island on Electric Utility Stock Prices: A Note." **Journal of Finances** 38(September 1983):1285-1292. (9 refs.)

The effect of TMI on the stock returns of public utility firms is examined. The analysis was conducted using risk-adjusted returns over monthly time periods surrounding TMI for GPU and for a sample of electric utilities; additionally, the sample was divided into nuclear and non-nuclear utilities. There was a clear effect on GPU's returns, but the results "did not indicate a statistically significant TMI effect on the utility sample as a whole." The impact was less on non-nuclear firms than on nuclear- based utilities.

*Stock market; Utilities; Economic effects; Nuclear industry; Financing

0519

Hu, Teh-Wei, and Slaysman, Kenneth S. "Health-Related Economic Costs of the Three Mile Island Accident." **Socio-Economic Planning Sciences** 18(3, 1984):183-193. (5 refs.)

The authors calculate that the total value of health-related economic costs of the TMI accident for residents in the five-mile radius over a ten-month period was $178,419. The largest cost items were work days lost and physician visits.

*Economic effects; Health care; Local residents; Evacuation

0520

"Impact of TMI." **Nuclear News** 23(May 1980):58. (News)

*Liability insurance

0521

"Insurer Paid Claims to Some 7,500 Following Pa. Nuclear Plant Accident." **National Underwriter (Prop Ed)** 83(April 20, 1979):17. (News)

*Insurance claims; Evacuation

0522

Jaffe, Alfred I. "Affects of Three Mile Island Crisis." **National Underwriter (Prop Ed)** 83(April 27, 1979):35, 37-38.

Jaffe comments on how the TMI accident will affect the insurance industry, in particular the insurance of the nuclear industry.

*Liability insurance; Insurance industry

0523

Kettler, D.J. "Nuclear Versus Fossil Cost Post TMI." **Combustion** 52(January 1981):25-32.

The costs of operating a nuclear unit versus a coal-fired unit are compared; included are "estimates of capital investment, fuel, and total bus bar energy cost... ." The potential impact of TMI on the base nuclear plant estimate is evaluated.

*Costs; Nuclear power

0524

Komanoff, Charles. "Capital Calamity; The Economic Fall-Out from Harrisburg." **Ecologist** 9(September 1979):209-210.

The accident at TMI will have economic consequences for the entire nuclear industry. The economic fallout includes an escalation of capital costs for nuclear power plants. Reprinted from *Critical Mass Journal*.

*Economic effects; Nuclear industry

0525
Lyons, Lois J. "American Nuclear Insurers Look at Lessons from Three Mile Island." **National Underwriter (Prop Ed)** 86(March 26, 1982):3, 48-49. (News)
*Insurance; Financing; Nuclear industry; Lessons learned

0526
Lyons, Lois J. "Insurers Still Guessing at Damages at Nuke Site." **National Underwriter (Prop Ed)** 83(April 13, 1979):1, 79. (News)
*Property damage; Liability insurance

0527
McLeod, Douglas. "TMI Rebate Settlement Might Not be Insured." **Business Insurance** 17(January 31, 1983):1, 78. (News)
*Liability insurance; Babcock & Wilcox; Court cases

0528
"More Insurance Claims Filed for TMI Accident." **National Underwriter (Prop Ed)** 86(November 19, 1982):23. (News)
*Liability insurance; Class action suits

0529
Nelson, Jon P. "Three Mile Island and Residential Property Values: Empirical Analysis and Policy Implications." **Land Economics** 57(August 1981):363-372. (21 refs.)

This paper examines residential property values for houses situated within five miles of TMI. Using a statistical analysis based on the hedonic price model and the interactive dummy variable method, it was determined for houses within four miles of the plant "that the accident at TMI caused neither an absolute decline in prices nor a slower appreciation rate for housing sales recorded during May-December 1979." The analysis, when extended to a five-mile area, found no statistically significant difference between prices in that area and two control areas.

*Residential property values; Economic effects; Real estate

0530
"NRC Determines Reactor Accident Not Extraordinary." **Public Utilities Fortnightly** 105(May 22, 1980):34. (News)
*Liability insurance; Extraordinary nuclear occurrence; Insurance

0531
"NRC to offer Ruling on 3-Mile Island Nuclear Accident." **National Underwriter (Prop Ed)** 83(August 17, 1979):47. (News)
*Price-Anderson Act; Extraordinary nuclear occurrence

0532
"Nuclear Insurance: How Much is Enough (and from Whom)?" **Technology Review** 81(June/July 1979):73. (News)
*Price-Anderson Act

0533
Rodda, William H. "The Impact of Three Mile Island. Part 1." **National Underwriter (Prop Ed)** 84(May 9, 1980):39, 45.

Three Mile Island's impact on the insurance industry relates to the liability limitations of the Price-Anderson Act. Possible changes in nuclear liability include raising the limit and government takeover of nuclear insurance operations.

*Price-Anderson Act; Liability insurance; Insurance industry

0534
Rodda, William H. "The Impact of Three Mile Island. Part 2." **National Underwriter (Prop Ed)** 84(May 16, 1980):27-29.

The TMI accident raises the question of whether this is an "insurable hazard in which the insurance business should participate." Changes in the handling of liability insurance are discussed.

*Liability insurance; Insurance industry

0535
Rodda, William H. "Insurance Implications of Nuclear Power Plant Incidents." **Professional Safety** 25(August 1980):40-43.

The effects of the accident at Three Mile Island on the insurance business are discussed. The changes will go "beyond the effects on the nuclear insurance pools and the current limitations on liability for nuclear power plant incidents." The sources of the liability fund, part of the Price-Anderson Act, are detailed.

*Insurance industry; Liability insurance; Price-Anderson Act

0536
Roy, Kalyan K., and Walz, Daniel T. "Three Mile Island and Utility Returns: Further Evidence." **Akron Business and Economic Review** 18(Spring 1987):75-81. (10 refs.)

In a long-term study of fifty-three public utilities from January 1963 through December 1982, it was found that the TMI accident had "no differential effect on the systematic risk of nuclear utilities as compared with non-nuclear electric utilities."

*Economics; Stock market; Utilities; Equity; Risk; Economic effects

0537
Simon, LeRoy J. "Managing Nuclear Risks." **Risk Management** 28(October 1981):12-18.

Simon discusses risk management of nuclear facilities following the TMI accident. Details concerning the insurance aspects of TMI are given in a one-page section (p. 17), including property loss, liability claims and class action suits.

*Risk management; Insurance; Liability insurance; Class action suits

0538
Studness, Charles M. "Utility Capital Spending in the Aftermath of Three Mile Island." **Public Utilities Fortnightly** 105(March 13, 1980):36-37.

The TMI accident will result in increased capital spending by utilities. The effects of the accident include: increases in the allowance for funds used during construction, NRC design changes, increased costs from inflation due to a lengthened construction timetable, an accelerated construction schedule for non-nuclear units, and cancellation of nuclear units planned for the late 1980s or 1990s.

*Capital expenses; Nuclear plant construction; Utilities; Financing

0539
"Three Mile Island Accident Not an ENO, NRC Staff Believes." **Nuclear News** 23(February 1980):35-36. (News)

*Extraordinary nuclear occurrence; Nuclear Regulatory Commission

0540
"TMI Gets Temporary Relief." **Engineering News-Record** 206(May 7, 1981):11-12. (News)

*Metropolitan Edison; Financing; Pennsylvania Public Utilities Commission; Cleanup

0541
"TMI Lesson: Woefully Inadequate Insurance." **National Underwriter (Prop Ed)** 86(July 9, 1982):10, 14-15. (News)

*Liability; Insurance; Cleanup; Financing; Shanaman, Susan M.

0542
"TMI's Cost Impact Estimated." **Electrical World** 195(October 1981):59. (News)
*Economic effects; Costs; Insurance; Legislation

0543
"$20 Million in Claims Paid to Gen. Public Utilities Over TMI." **National Underwriter (Prop Ed)** 83(September 28, 1979):50. (News)
*Liability insurance; General Public Utilities; Financing

0544
Weese, Samuel H. "Another Look at `Nuclear Liability After Three Mile Island.'" **Journal of Risk and Insurance** 50(June 1983):323-327. (11 refs.)
Weese responds to Wood's views on nuclear liability and TMI in *Journal of Risk and Insurance* 48(September 1981):450-464.
*Liability; Insurance; Risk

0545
"Who Pays the Price?" **Economist** 271(April 7, 1979):111. (News)
*Stocks and bonds

0546
Wood, William C. "Nuclear Liability After Three Mile Island." **Journal of Risk and Insurance** 48(September 1981):450-464. (14 refs.)
Liability and insurance for nuclear plants are analyzed in light of the Three Mile Island accident. One conclusion is that "private insurers' implicit estimates of the risks are more accurate than those of the government." Existing liability and insurance arrangements, based on the assumption that nuclear risk is overestimated, should be reconsidered.
*Liability; Insurance; Risk

0547
Wood, William C. "`Nuclear Liability After Three Mile Island' -- Author's Reply." **Journal of Risk and Liability** 50(June 1983):328-329. (3 refs.)
Wood comments on Weese (*Journal of Risk and Insurance* 50(June 1983):323-327), which was a commentary on Wood's earlier article in *Journal of Risk and Insurance* 48(September 1981):450-464.
*Liability; Insurance; Risk

0548
Zimmerman, Martin B. "The Valuation of Nuclear Power in the Post-Three Mile Island Era." **Energy Journal** 4(April 1983):15-29. (7 refs.)
The author reviews the chronology of events at Three Mile Island, and presents a model that evaluates the impact of the accident on stock market prices of utilities as a measure of the valuation of nuclear power. The conclusion is that "what changed was the expectation of getting new plants licensed and not the expected costs of operation."
*Stock market; Nuclear industry; Accident sequence

# 8

# Media and Communications

0549
Baer, John H. "Conflicting News Clouds Nuclear Plant Mishap." **Editor & Publisher** 112(April 7, 1979):10, 17.
*News coverage; Reporters; Communications

0550
Baer, John H. "95 Reporters Tour Three Mile Nuke Plant." **Editor & Publisher** 114(April 18, 1981):44. (News)
*Media; News coverage; Reporters; Tours

0551
Brandon, George. "News Media `Exonerated' for Problems in TMI Stories." **Editor & Publisher** 114(July 18, 1981):20. (News)
*Media; News coverage

0552
Burnham, David. "The Press and Nuclear Energy." **Annals of the New York Academy of Sciences** 365(1981):107-109. (8 refs.)
Burnham's comments on the government and the nuclear industry misleading the press and the public, as illustrated by Three Mile Island.
*Press coverage; Nuclear industry; Public relations

0553
Burns, Christopher. "Three Mile Island: The Information Meltdown." **Information Management Review** 1(Summer 1985):19- 25.
Unavailable.
*Communications; Information flow

0554
Christiansen, Donald. "TMI and the Press." **IEEE Spectrum** 16(November 1979):92-95.
IEEE members were surveyed for their opinion regarding press and media coverage of TMI. The results showed that 68 percent felt the coverage had been counterproductive. A table giving coverage of newspapers, news magazines, major TV networks and radio shows that the print media rated the highest regarding thoroughness of coverage and accuracy. Eighteen respondents to the IEEE questionnaire are quoted on their opinions of TMI press coverage.
*Public relations; Press coverage; Media

0555
Consoli, John. "Judge Rejects Prior Restraint on Shopper." **Editor & Publisher** 113(February 9, 1980):11, 47. (News)
*Newspapers; News coverage; Security; Court cases

0556
Eisenbud, Merril, moderator. "The Role of the Media: Panel Discussion." **Annals of the New York Academy of Sciences** 365(1981):123-133.

Panel members were David Burnham, Henry Myers, William B. Murray, David M. Rubin, Arthur Tamplin, and Richard Vollmer.

*Media; News coverage

0557
Farrell, Thomas B., and Goodnight, G. Thomas. "Accidental Rhetoric: The Root Metaphors of Three Mile Island." **Communication Monographs** 48(December 1981):271-300. (94 refs.)

This essay contends that TMI presented a crisis in which the inadequacies of accidental rhetoric point to "the failure of technical reason itself to offer communication practices capable of mastering the problems of our age." A detailed reconstruction of the discourse at TMI is followed by the conclusion that "the limits of technical communicative discourse are severe, recurrent, and perhaps irreparable."

*Accidental rhetoric; Rhetoric; Communications; Public relations

0558
Fenichel, Michael, and Dan, Peter. "Heads from *Post* and *Times* on Three Mile Island." **Journalism Quarterly** 57(Summer 1980):338-339, 368.

Two major newspapers, the New York *Times* and the New York *Post* were analyzed for bias in the reporting of the Three Mile Island accident. In examining the headlines and sub-headlines, the *Post* was found to use more non-factual information; the *Times* also used a higher proportion of referents to the event and to official reactions than did the *Post*.

*Newspapers; News coverage; Journalism

0559
Friedman, Sharon M. "Blueprint for Breakdown: Three Mile Island and the Media Before the Accident." **Journal of Communication** 31(Spring 1981):116-128. (17 refs.)

The results of the Public's Right to Information Task Force of the President's Commission concluded that "the public's right to know had been `ill-served' by Metropolitan Edison, the Nuclear Regulatory Commission, and most of the other institutions directly involved." The blame for the communications breakdown falls to both the utility and the media. An analysis of media coverage prior to the accident indicated that the utility directed its information at convincing citizens that TMI was a good neighbor, while the media coverage was "flaccid" and did not seek out negative issues because their staffs "perceived public disinterest in nuclear power plants."

*Media; Press coverage; Communications; Right to know

0560
Friendly, Fred; Diamond, Stuart; Eisenbud, Merril; and Lester, Martha. "Panel II Discussion: Session III Public Information Needs." **Bulletin of the New York Academy of Medicine** 59(December 1983):1087-1094.

This is a panel discussion regarding information needs and the flow of information that surrounded the TMI accident, and whether the information flow has improved in the four years following the accident.

*Right to know; Information flow

0561
Hyde, Richard C. "Three Mile Island: PR's Balaklava?" **Public Relations Journal** 35(June 1979):10-14.

Metropolitan Edison Company was faced with public relations problems and the need for crisis communications. "Hill and Knowlton was retained by Met-Ed to develop a program for employee communications and community and governmental relations." The firm was faced with a unique situation of dealing with citizens' fear, anger and mistrust, and with handling national and worldwide communications.

*Public relations; Communications; Metropolitan Edison; Hill and Knowlton; Credibility

0562
Lester, Martha S. "Public Information During a Nuclear Power Plant Accident: Lessons Learned from Three Mile Island." **Bulletin of the New York Academy of Medicine** 59(December 1983):1080-1086. (4 refs.)

At the time of the accident, the author was the assignment editor at WHP-AM-FM-TV (Harrisburg's CBS affiliate), which is the area's designated emergency broadcast station. She chronicles the confusion and mishandling of public information that occurred during the TMI accident. The need for a co-ordinated information effort is emphasized, with the governor's office playing a major role. A resource for rumor control was cited as an important public information tool. Evacuation planning and health resource organization have been updated since the accident.

*Press releases; Right to know; Public relations; Disaster planning

0563
Marcus, S. "Technology and the Press at Three Mile Island." **Technology Review** 81(June/July 1979):72. (News)

*Press coverage; Communications

0564
Mazur, Allan. "The Journalists and Technology: Reporting About Love Canal and Three Mile Island." **Minerva** 22(Spring 1984):45-66. (62 refs.)

The immense publicity given to Love Canal and TMI "gave these events a symbolic significance which transcended their technical malfunctions." The attention from the press, radio and television was high because these events were "unique and exceedingly severe." This paper covers the pattern of attention, national news organs and national news media, and the "shaping" of each event. Journalistic bias is discussed in relation to Love Canal and TMI.

*Journalism; News coverage; Media; Love Canal

0565
Murray, William B. "Shared Responsibilities of the Utility and the Media in Crisis Situations." **Annals of the New York Academy of Sciences** 365(1981):116-120.

In cases such as TMI, the utility must retain a spokesmanship role, but it should be shared with spokesmen appointed to provide information in certain subject areas. The lessons learned from the communication problems at TMI will benefit the public and the nuclear industry.

*Communications; Public relations; Media; News coverage

0566
Myers, Henry. "Representing the Congress." **Annals of the New York Academy of Sciences** 365(1981):114-115. (2 refs.)

Myers comments that inaccuracies in the press resulted from inadequate and incorrect information provided by the NRC and the utility. He indicated that "those who had the information did not provide it to the press."

*Communications; Press coverage

0567
Nimmo, Dan. "TV Network News Coverage of Three Mile Island: Reporting Disasters as Technological Fables." **International Journal of Mass Emergencies and Disasters** 2(March 1984):115- 145. (42 refs.)

The coverage of the TMI accident by nightly network news "raised questions about the nature of TV news as well as the capacity of the three major networks to inform viewers during disasters." Network news emphasizes "story-telling," and in covering TMI, "CBS, ABC, and NBC respectively and consistently construct[ed]

rhetorical visions of reassurance, threat, and primal assurances."
*Network news; Television; News coverage; Media

0568
Nimmo, Dan, and Combs, James E. "Fantasies and Melodramas in Television Network News: The Case of Three Mile Island." **Western Journal of Speech Communication** 46(Winter 1982):45-55. (26 refs.)

Tapes from network news coverage of TMI for one month following the accident were examined to see if the three networks presented different melodramas and thus provided viewers with different fantasies. Coverage on TMI was placed into one of four categories: popular sensationalist stories, informative stories, feature stories, and didactic stories. Network news coverage was balanced among the first three categories, with didactive stories being in the minority. Among the three networks, "ABC emerges as the network of popular sensationalist news."
*Network news; News coverage; Television

0569
Nimmo, Dan, and Combs, James E. "`The Horror Tonight:' Network Television News and Three Mile Island." **Journal of Broadcasting** 25(Summer 1981):289-293. (2 refs.)

An examination of the reporting styles of network news coverage of the TMI accident shows varied presentations. ABC was "sensationalist," CBS was reassuring and informative, while NBC was "didactic and ambiguous."
*Network news; Television; Media; News coverage

0570
Northart, Leo J. "Credibility, Confidence and Three Mile Island." **Public Relations Journal** 35(November 1979):50, 58.

In light of the credibility gap following TMI, crisis communications planning has become a major concern. At the Northeast District Conference of the Public Relations Society of America, "The Lessons of Three Mile Island" were discussed.
*Communications; Public Relations; Credibility; Lessons learned

0571
Northart, Leo J. "Three Mile Island: The Lessons for Public Relations." **Public Relations Journal** 35(June 1979):8-9.

In an editorial, Northart looks at the damaged credibility of public relations as a result of TMI.
*Public Relations; Lessons learned

0572
Nussbaum, Rudi H. "Responsible Journalism." **Bulletin of the Atomic Scientists** 35(September 1979):63-64. (Letter)
*Media; News coverage; Journalism

0573
Palmer, Lane. "The Scientific Method: A Writer's Best Friend." **Humanist** 42(March/April 1982):29-34, 61.

Palmer concludes that TMI caused "a melt-down of public confidence in nuclear energy." He feels that mistakes made in reporting the accident were as serious as the accident itself. He then proceeds to discuss reporting of scientific information and the need for commitment to truth.
*Journalism; Media; Public opinion

0574
Radolf, Andrew. "An Expensive Rebuttal." **Editor & Publisher** 118(March 30, 1985):13.
*General Public Utilities; Advertising; Philadelphia Inquirer; Public relations; Cleanup

0575
"Reporter Revealed Defects at Nuke Plant on March 13." **Editor & Publisher** 112(April 7, 1979):11, 28; photo credits on p. 4. (News)
*Newspapers; News coverage; Safety; Nuclear reactor design

0576
Rubin, David M. "What the President's Commission Learned About the Media." **Annals of the New York Academy of Sciences** 365(1981):95-106. (20 refs.)
Rubin, a member of the Task Force on the Public's Right to Know of the President's Commission, discusses the news coverage of the accident. Reports singled out for coverage are the CBS Evening News of Friday, March 30 and the Associated Press story on Saturday, March 31 on the hydrogen bubble. The public has an unrealistic expectation of media performance. Most of the errors in media coverage were traceable to inaccurate sources of information, although there are cases of reporter error. A variety of factors, from ill-informed sources and communication problems to lack of scientific background contributed to the overall quality of coverage.
*Media; News coverage; Network news

0577
Sandman, Peter. "Press Coverage of Three Mile Island." **PE - Professional Engineering** 51(September 1981):24-25.
Sandman claims that the press missed the main story at TMI, which was the expert's inability to guarantee our safety, and the confusion and lack of preparation. Lack of technical expertise and failure to ask technical questions also played a part in problems with coverage.
*Media; News coverage; Newspapers

0578
Sandman, Peter M, and Paden, Mary. "At Three Mile Island." **Columbia Journalism Review** 18(July/August 1979):43-58.
"Two journalists who went to cover the press found fear, confusion, and some very good work." This story of press coverage of TMI provides an inside view of the problems facing reporters: lack of knowledge, conflicting stories, deliberate misinformation, and general confusion. One focus is the events surrounding the hydrogen bubble, while another is the intensive coverage of the *Philadelphia Inquirer.* Newspaper coverage is emphasized, although television is included.
*Media; Press coverage; Journalism; Philadelphia Inquirer

0579
Soxman, Kenneth A. "TMI, Tet and the Media." **Bulletin of the Atomic Scientists** 36(March 1980):64. (Letter)
*Media; Public opinion; Nuclear industry

0580
Starr, Chauncey. "The Three Mile Island Nuclear Accident: The Other Lesson." **Annals of the New York Academy of Sciences** 365(1981):292-310. (7 refs.)
Starr diverges from his original topic of technical and operating issues to discuss what he terms as the most important lesson, that "the creation of an informed public opinion is beyond the ability of the utility industry alone." It requires good communication and informed media. An Appendix contains the speech Starr originally intended to present, which documents the major programs that have been initiated by the utilities in response to TMI.
*Public opinion; Communications; Media; Nuclear industry; Institutional response

0581
Stephens, Mitchell, and Edison, Nadyne G. "News Media Coverage of Issues During the Accident at Three-Mile Island." **Journalism Quarterly** 59(Summer 1982):199-204, 259.

Ten news organizations (ABC, CBS, NBC, the New York *Times*, the Washington *Post*, the Los Angeles *Times*, the Associated Press, United Press International, the Philadelphia *Inquirer*, and the Harrisburg *Patriot/Evening News*) were analyzed based on content of the first week of news coverage. Health and safety issues, and nuclear energy issues were classified as "reassuring" versus "alarming," or positive versus negative. Coverage was predominantly reassuring or positive; the major exception was the quality of information being released about the accident, which was predominantly negative.

*News coverage; Newspapers; Media; Network news; Journalism

0582
Stillman, Peter G. "Three Mile Island: A Case of Disinformation." **Democracy** 2(4, 1982):66-78. (11 refs.)

In an essay, Stillman describes the sequence of events at TMI and shows how the accident was a result of disinformation -- either unavailable, unclear, distorted or ambiguous information. The unavailability of clear information caused poor decisions which compounded the problems at TMI. Disinformation was also a problem in dealing with the public and for the Kemeny Commission.

*Information exchange; Disinformation; Accident - causes; Accident sequence

0583
Susskind, Jacob L. "Using Pennsylvania's Three Mile Island Accident as a Case Study to Analyze Newspaper Coverage: A Diary of Events and Suggestions for Teaching Strategies." **Social Studies Journal** 12(September 1983):51-60.

The author followed front page headlines on four newspapers (the Harrisburg *Patriot*, the Philadelphia *Inquirer*, the Washington *Post*, and the New York *Times*) for two weeks following the TMI accident. His analysis was that the press provided responsible coverage. He suggests using this same technique to teach students how to evaluate and analyze current events as reported in newspapers.

*Newspapers; News coverage; Teaching

0584
Tamplin, Arthur. "Representing Public Interest Groups." **Annals of the New York Academy of Sciences** 365(1981):121-122.

Tamplin feels that the role of the media in a nuclear accident such as TMI is essential, and that the media was responsible in its reporting of TMI.

*Media; News coverage

0585
"Three Mile Island: As Confusing as it was Dangerous." **Broadcasting** 96(April 9, 1979):61-62. (News).

*Network news; Media; Television

0586
Vollmer, Richard. "Representing the Nuclear Regulatory Commission." **Annals of the New York Academy of Sciences** 365(1981):110-113.

Vollmer comments on the lack of communication at TMI, and passes on some thoughts from Harold Denton as to how this could have been corrected. Examples are given of difficulties in communicating with the press.

*Nuclear Regulatory Commission; Communications; Press coverage

0587
Williamson, Lenora. "Pa. Governor Praises 3 Mile Coverage." **Editor & Publisher** 112(April 28, 1979):73. (News)

*News coverage; Newspapers; Thornburgh, Richard

# 9

# Cleanup

## GENERAL

0588
Abbotts, John. "Who Pays for Radioactive Rubbish?" **Bulletin of the Atomic Scientists** 40(August/September 1984):24-28. (26 refs.)

As part of a discussion on costs of paying for decommissioning a nuclear plant or disposing of radioactive wastes, the case of Three Mile Island is discussed.

*Waste disposal; Cleanup; Costs

0589
Adam, John A. "TMI Plus 5: Nuclear Power on the Ropes, Part I: A Slow Comeback." **IEEE Spectrum** 21(April 1984):27-33.

Five years after the accident, the cleanup is well behind schedule. Of the four major categories, only decontamination of the containment building and the assessment of the damaged reactor vessel's interior have been completed. Yet to begin are the disassembly of the reactor and defueling. Milestones in the cleanup of Unit 2 are given, and step-by-step technical aspects of the cleanup are discussed.

*Cleanup; Decontamination; TMI - anniversary (5)

0590
"Aid Program Proposed for TMI Cleanup Effort." **Engineering News-Record** 206(February 26, 1981):17-18. (News)

*Cleanup; Financing; Reagan administration

0591
Ardron, K.H., and Cain, D.G. "Core Temperature Transient in the Early Phase of Core Uncovering at TMI-2." **Progress in Nuclear Energy** 7(3, 1981):197-228. (70 refs.)

The paper describes a best-estimate analysis of the initial core boil-down and heat-up transient at Three Mile Island Unit (2) on 28 March 1979. The analysis is directed at understanding the progression of core damage. Results indicate that "observed containment hydrogen levels were due to Zircaloy/stainless steel corrosion"; there was potential for fuel liquifaction; "a substantial fission product release was obtained from fuel over-heating"; and "steam temperatures in the upper plenum ... were generally below 900 C (1650 F)."

*Reactor core analysis; Reactor core uncovery; Transients; Zircaloy

0592
Arnold, Robert C. "Three Mile Island Unit 2: Decontamination Program." **Annals of the New York Academy of Sciences** 365(1981):76-85.

Arnold summarizes progress to date on the decontamination program at TMI. He discusses plant stabilization, decontamination of the auxiliary and fuel buildings, and processing of contaminated water. Future activities include the containment recovery program and the reactor core removal. Figures for the volume of water processed, surface area decontaminated, and exposure histories of workers are given.

A critical path schedule outlines the recovery program. Institutional issues in the recovery are public communications, regulatory interaction, and finances.
*Cleanup; Decontamination; Institutional issues

0593
"August Date for TMI 2 Head Left." **Nuclear Engineering International** 29(May 1984):9. (News)
*Reactor head lift; Cleanup; Financing; Research and development

0594
"B&W Develops System for Shipping TMI-2 Core." **Electrical World** 199(July 1985):102. (News)
*Reactor core; Waste shipment; Fuel canisters; Babcock & Wilcox

0595
Baggoura, Benaissa, and Martin, William R. "Transient Analysis of the Three Mile Island Unit 2 Pressurizer System." **Nuclear Technology** 62(August 1983):159-171. (27 refs.)
This paper describes "details of a numerical model to simulate the dynamic behavior of the TMI-2 pressurizer system." The model includes a fine mesh description of the surge line and a detailed representation of the pressurizer vessel"; it can be used to simulate other pressurized water reactors. Results of two verification tests for the model were in agreement with data from TMI-2.
*Computer models; Computer simulation; Transients; Pressurized water reactors

0596
Baston, V.F.; Hofstetter, K.J.; and Malinauskas, A.P. "Circulation Within the Primary System at TMI-2 with Lowered Coolant Level and at Atmospheric Conditions." **Nuclear Technology** 69(June 1985):308-318. (9 refs.)
As part of an engineering assessment of the Three Mile Island reactor coolant system (RCS) to ascertain the extent of coolant mixing for defueling "four forces fluid movements through the RCS were performed in a feed-and-bleed mode." "Analysis of radionuclide, temperature and dissolved oxygen data ... confirm mixing in the primary system under forced coolant flow conditions with only minimal mixing occurring during static periods."
*Decontamination; Cooling system; Defueling

0597
Bauer, Paul S. "Recovering Three Mile Island; An Interim Progress Report." **Power** 125(July 1981):46-48.
Progress in the cleanup of TMI over the past two years is reviewed.
*Cleanup

0598
Blake, E. Michael. "Executive Conference on TMI-2 Cleanup." **Nuclear News** 28(December 1985):96, 98, 103-105. (News)
*Cleanup; Lessons learned

0599
Buron, H.M., and Bixby, W.W. "TMI-2 Technical Information and Examination Program: Objectives and Current Activities." In: **Three Mile Island Cleanup: Experiences, Waste Disposal, and Environmental Impact**, edited by Lester J. King and James H. Opelka, 16-21. New York: American Institute of Chemical Engineers, 1982. (AIChE Symposium Series, no. 213, v. 78, 1982) (3 refs.)
This paper summarizes objectives and activities of the Technical Information and Examination Program, which is an agreement among the General Public Utilities Corporation, the Electric Power Research Institute, the Nuclear Regulatory Commission, and the Department of Energy "to acquire data and information from the TMI-2 recovery that will enable industry, utilities, and the NRC to improve safety

and operations" of nuclear power plants and prevent future accidents. The group is known as the GEND group.

*GEND; Research and development

0600
Burton, H.M., and Freemerman, R.L. "Reactor Disassembly Activities at Three Mile Island Unit Two." **Progress in Nuclear Energy** 17(2, 1986):141-174. (14 refs.)

This paper discusses the "physical aspects of characterizing the post-accident condition of TMI-2 for planning a safe reactor disassembly. Characterization activities discussed include visual examination, detailed video inspections, radiation measurements, core topography, and core debris grab samples, while reactor disassembly activities discussed include removal of the reactor vessel head assembly, the plenum assembly removal, defueling, and transfer, storage and shipping of fuel canisters.

*Reactor head assembly; Reactor core analysis; Reactor head removal; Defueling

0601
"... But Who Should Foot the Bill?" **Nuclear Engineering International** 27(November 1982):9-10. (News)

*Cleanup; Financing

0602
Butler, Paul. "Three Mile Island: No Sign of Melting Core." **Engineer** 248(May 17, 1979):16-17. (News)

*Reactor core analysis; Fuel melting

0603
"Camera Shows No Fuel Meltdown at TMI Unit." **Chemical & Engineering News** 60(August 2, 1982):5-6. (News)

*Reactor core; Fuel melting; Videocamera use

0604
Carter, Luther J. "Debate over Waste Imperils 3-Mile Cleanup." **Science** 210(October 10, 1980):166-168, 170. (News)

The timetable of the TMI cleanup is threatened by political hangups of radioactive waste disposal. Each phase of the cleanup has occurred in a "highly sensitive political and regulatory environment," but progress has been made in filtering the water and venting the krypton gas. Management problems of the cleanup are described along with some future cleanup activities.

*Cleanup; Politics; Regulatory environment; Waste disposal; Management

0605
"Challenge to EPICOR-II Dropped; CEQ in EIS. **Nuclear News** 23(January 1980):44-46. (News)

*EPICOR; Decontamination

0606
Chu, T.L., and Apostolakis, G.E. "Assessment of the Uncertainties Associated with the Core Uncovery Time in TMI-Type Accidents." **Reliability Engineering** 8(1984):23-56. (30 refs.)

"The core uncovery time in TMI-type accidents, i.e., small loss of coolant accidents due to stuck open power-operated relief valves, is calculated." Many of the valves used for the model are based on the TMI accident.

*Reactor core uncovery; Models

0607
"Cleanup Debate Continues." **Nuclear News** 24(May 1981):97-98. (News)

*Cleanup; Financing; Congress; Hearings

0608
"Cleanup Job at Three Mile Island Gets New Overseer." **Public Utilities Fortnightly** 105(April 10, 1980):33-34. (News)
*Snyder, Bernard; Cleanup

0609
"Cleanup Work Blasted at Three Mile Island." **Engineering News-Record** 211(September 22, 1983):22-23. (News)
*Cleanup; Polar crane; Safety

0610
Collins, E.D.; Campbell, D.O.; King, L.J.; and Knauer, J.B. "Water Decontamination Process Improvement Tests and Considerations." In: **Three Mile Island Cleanup: Experiences, Waste Disposal, and Environmental Impact**, edited by Lester J. King and James H. Opelka, 9-15. New York: American Institute of Chemical Engineers, 1982. (AIChE Symposium Series, no. 213, v. 78, 1982) (4 refs.)

Testing and process improvement for decontaminating high-activity-level water at TMI-2 with the Submerged Demineralizer System is discussed. The process improvement tests were made "to increase the amount of cesium and strontium loaded on the zeolite sorbent, to reduce the cesium and strontium concentrations in the effluent water, and to remove the anionic contaminants from the water."
*Cleanup; Decontamination; Submerged Demineralizer System

0611a
"Complaints Set Back Head Lift at TMI-2." **Nuclear Engineering International** 28(June 1983):8. (News)
*Reactor head lift; Parks, Richard D.; Whistleblowing; GPU Nuclear; Management

0611b
"Computer Aids TMI Cleanup." **Engineering News-Record** 214(April 18, 1985):27.
*Cleanup; Computer models

0612
"Containment Entered; Cleanup Cost Revised." **Nuclear News** 23(September 1980):29-30. (News)
*Containment entry; Cleanup; Costs

0613
"Containment Entry Plan OK; Suit Check Needed." **Nuclear News** 23(June 1980):43-44. (News)
*Cleanup; Containment entry

0614
"Contract Signed to Remove Damaged Core at TMI Unit 2." **Public Utilities Fortnightly** 112(July 21, 1983):53. (News)
*Cleanup; Defueling; Reactor core; Westinghouse

0615
"Cooldown at TMI-2." **Combustion** 51(December 1979):26-29.
Reprinted from *Perspective*, Summer-Fall 1979 issue.
*Cooldown; Pullman Power Products

0616
"Costs Still Climbing at Three Mile Island." **Science** 205(August 3, 1979):475. (News)
*Cleanup; Costs; General Public Utilities; Financing

0617
D'Ambrosia, J.T. "Three Mile Island Waste Management: A DOE Perspective." In: **Three Mile Island Cleanup: Experiences, Waste Disposal, and Environmental Impact**, edited by Lester J. King and James H. Opelka, 27-32. New York: American Institute of Chemical Engineers, 1982. (AIChE Symposium Series, no. 213, v. 78, 1982)

Waste management research and development activities, applicable to TMI-2 cleanup, are being conducted by the Department of Energy. A description and discussion of some of these activities is presented, including: Epicor II resin liners, high integrity container, and Submerged Demineralizer System wastes.

*Waste handling; Research and development; Department of Energy

0618
David, Peter. "Three Mile Island; Damage to Core Revealed." **Nature** 305(September 15, 1983);173. (News)

*Cleanup; Reactor core; Restart

0619
David, Peter. "Three Mile Island; Still a Problem After Five Years." **Nature** 308(March 22, 1984):308. (News)

*Cleanup; Restart; General Public Utilities; Financing; TMI - anniversary (5)

0620
"Defueling: GPUN Confident, NRC Staff Concerned." **Nuclear News** 30(April 1987):26-27. (News)

*Defueling; Cleanup

0621
Demmitt, T.F., and Devine, J.C. "First Experiences with Cleanup Operations at Three Mile Island." **Atomwirtschaft-Atomtechnik** 29(February 1984):93-99.

Unavailable.

*Cleanup

0622
Devine, Jack C. "Dealing with the Liquid Waste from the TMI-2 Accident." **Nuclear Engineering International** 29(September 1984):29-32.

The decisions and actions of the TMI-2 Waste Management Group charted the future course of the TMI recovery program. The processing of the contaminated water in the auxiliary buildings via the Epicor II system, the use of the Submerged Demineralizer System (SDS) in the reactor building, waste management of the water in the reactor coolant system, and waste disposal are all described.

*Cleanup; Decontamination; EPICOR; Waste Management Group; Submerged Demineralizer System

0623
DeVine, John C. "A Progress Report: Cleaning up TMI." **IEEE Spectrum** 18(March 1981):44-49. (3 refs.)

Progress in the cleanup of TMI 2 over the past two years is summarized, including: decontaminating the radioactive water, venting krypton 85, and entering the containment building. Waste disposal has been indefinitely postponed.

*Cleanup; Decontamination; Venting; Krypton; Containment entry

0624
Ember, Lois. "DOE Will Take Three Mile Island High-Level Waste." **Chemical & Engineering News** 59(June 22, 1981):23-24. (News)

*Waste disposal

0625
"Ertel Lobbies for Bill." **Nuclear News** 24(October 1981):106. (News)
*Ertel, Allen; Cleanup; Financing; Legislation

0626
"Experts Seek Answers as Delicate Cool Down Operation Gets Under Way." **Engineering News-Record** 202(April 12, 1979):38-39. (News)
*Cooldown

0627
"Filter System at TMI Installed in Record Time." **Power Engineering** 83(October 1979):126. (News)
*Filters

0628
"Final Cleanup EIS Sees Minimal Health Effects." **Nuclear News** 24(April 1981):51-52. (News)
*Cleanup

0629
"First Removed Unit 2 Fuel Ready to Travel." **Nuclear News** 29(August 1986):38. (News)
*Reactor core; Waste shipment; Fuel debris; Cleanup

0630
Freemerman, R.L., and Rider, R.L. "Preparing for the Reactor Head Lift." **Nuclear Engineering International** 29(June 1984):23-26.

The progress made toward and removal of the reactor vessel head is described. Included are data on the core; information on the refurbishment of the polar crane; and preparation for the reactor pressure vessel head removal, including accessing the vessel and removing the plenum assembly.

*Reactor head removal; Cleanup

0631
Freemerman, R.L., and Rider, R.L. "Taking a Look Inside TMI 2." **Nuclear Engineering International** 27(October 1982):27-32.

The authors discuss work to date on the decontamination and cleanup of Three Mile Island Unit 2, but focus on the pre-head lift examination. The Pre-Head Lift Examination program is described, consisting of a quick look at the reactor vessel, followed by the two major activities: the control rod drive mechanism (CRDM) uncoupling and core mapping, and upper vessel region inspection and sampling.

*Pre-Head Lift Examination Program; Reactor head removal; Cleanup

0632
Freemerman, R.L., and Rider, R.L. "What Was Found Inside TMI 2." **Nuclear Engineering International** 28(April 1983):19-22.

The authors follow up on their article in *Nuclear Engineering International*, October 1982, by describing what was found inside TMI 2 during the "quick look" last summer. They describe the camera inspection, the probe of the rubble bed, the control rod drive mechanism uncoupling, and the data evaluation.

*Cleanup; Containment entry; Reactor core analysis; Videocamera use

0633
"Further Remote Vehicle Use Planned for TMI-2." **Nuclear News** 28(October 1985):64. (News)
*Remote vehicles; Cleanup

0634
"Gov. Thornburgh Proposes Cleanup Cost Sharing." **Nuclear News** 24(August 1981):32-33. (News)

*Thornburgh, Richard; Cleanup; Financing; Submerged Demineralizer System

0635
Harvey, Joan; Piccioni, Richard; and Pisello, Daniel. "TMI: How Many Will the Clean Up Kill?" **Ecologist** 11(June 1981):138-142. (9 refs.)

The authors present figures that indicate that the NRC, Metropolitan Edison and the EPA "carefully underestimate the real damage" and the radiation dosage that will occur during venting in the cleanup of TMI. They claim that the proposed clean-up at TMI "is potentially lethal to a large percentage" of the U.S. population.

*Venting; Radiation dosage; Cleanup

0636
Hayes, William C. "TMI: Charitable Cause -- or a Laboratory?" **Electrical World** 195(February 1981):3. (Editorial)

Hayes suggests that the nuclear industry may miss an opportunity to gain valuable experience should it fail to become involved in financing the TMI cleanup. The aim should be recommissioning, not decommissioning.

*Cleanup; Financing; Research and development

0637
Hayner, George O.; Caldwell, Colin S.; and Eidam, G.R. "Examination of Core Debris Samples from the Three-Mile Island Unit 2 Reactor." **Materials Performance** 24(December 1985):23-29. (12 refs.)

Extensive laboratory analysis of samples from the core of TMI-2 revealed that the "debris was composed primarily of granular ceramic fragments, resolidified metallic, and `lava rock-like' materials." All of the elements found in the debris "were present in the original core materials." The detailed analysis included: radionuclide distribution, elemental examination, immersion density, and a microstructural examination. Values and/or percentages are provided for the analyses.

*Cleanup; Reactor core analysis; Fuel debris

0638
Hofstetter, K.J., and Hitz, C.G. "The Use of the Submerged Demineralizer System at Three Mile Island." **Separation Science and Technology** 18(14-15, 1983):1747-1764. (12 refs.)

The Submerged Demineralizer System (SDS) was designed by GPU Nuclear Corporation, in consultation with the Department of Energy, to process contaminated water at TMI. The system is described, and processing summaries and performance evaluations presented. Over 1.5 million gallons of water have been processed and approximately 340,000 curies of fission product radionuclides have been immobilized.

*Cleanup; Decontamination; Submerged Demineralizer System

0639
Hofstetter, Kenneth J.; Hitz, C.G.; Baston, V.F.; and Malinauskas, Anthony P. "Radionuclide Analyses Taken During Primary Coolant Decontamination at Three Mile Island Indicate General Circulation." **Nuclear Technology** 63(December 1983):461-469. (2 refs.)

Using a feed-and-bleed process to decontaminate the primary reactor coolant system at Three Mile Island, information has been gathered on radionuclide concentration for use in future defueling operations. Analysis of radiocesium samples "indicates general circulation within the primary system, including the reactor vessel and both steam generators." "A standard dilution model ... predicts the radiocesium decontamination rates."

*Cesium radioisotopes; Decontamination; Cooling system

0640
"Hopes Rise for Three Mile Island Clean-up." **Nuclear Engineering International** 26(July 1981):13-14. (1 ref.) (News)
*General Public Utilities; Cleanup; Financing; Environmental effects

0641
"House Unit Examines TMI Cleanup Plans." **Public Utilities Fortnightly** 108(October 22, 1981):35-36. (News)
*House Energy Conservation and Power Subcommittee; Cleanup; Legislation; Financing

0642
"How Robots are Helping at TMI 2." **Nuclear Engineering International** 30(April 1985):27. (News)
*Remote vehicles; Cleanup

0643
Ireland, John R.; Wehner, Thomas R.; and Kirchner, Walter L. "Thermal-Hydraulic and Core-Damage Analyses of the TMI-2 Accident." **Nuclear Safety** 22(September-October 1981):583-593. (24 refs.)
Thermal-hydraulic and core-damage analyses were performed with the Transient Reactor Analysis Code (TRAC) for the accident at TMI-2. The estimates indicated "that the reactor fuel did not reach the melting temperature, that a substantial amount of fuel-rod cladding was severely oxidized, and that some cladding melting and relocation occurred."
*Reactor core analysis; Reactor core damage; Thermal hydraulics; Fuel melting; Fuel rods

0644
Iwler, Louis. "NARUC Group May Air TMI with Reagan." **Electrical World** 195(February 1981):21-22. (News)
*Cleanup; Financing; Shanaman, Susan M.; Reagan administration

0645
Iwler, Louis. "TMI Cleanup is Imperative, Says NRC Staff." **Electrical World** 195(May 1981):25. (News)
*Cleanup; Nuclear workers; Reactor core

0646
"Japan Joins in TMI R and D." **Nuclear Engineering International** 29(February 1984):3-4. (News)
*Cleanup; Research and development; Japan; Reactor core analysis; GPU Nuclear; Management

0647
Joyce, Christopher. "Molten to the Core." **New Scientist** 105(March 28, 1985):16-19.
Progress made over the past six years in cleaning up TMI is discussed along with future cleanup activities. Much is still unknown about the condition of the core.
*Cleanup; Reactor core; TMI - anniversary (6)

0648
Joyce, Christopher. "Nuclear Business Asks Government to Pay for TMI." **New Scientist** 89(January 1, 1981):5. (News)
*Cleanup; Financing; Government

0649
Kalman, G., and Weller, R. "Progress in the Recovery Operations at Three Mile Island Unit 2." **Nuclear Safety** 25(January-February 1984):88-113. (80 refs.)

This article reviews the progress made in the recovery operations at Three Mile Island Unit 2, including "radioactive waste management, decontamination, maintenance of safe shutdown, and efforts toward defueling." Future work, including the difficult task of defueling, is described, and the recovery operation is evaluated.
*Cleanup; Waste handling; Decontamination

0650
Kalman, G., and Weller, R. "Progress in the Recovery Operations at Three Mile Island Unit 2: An Update." **Nuclear Safety** 27(January-March 1986):68-84. (46 refs.)
This article updates information provided in *Nuclear Safety* 25 (January-February 1984), including principal recovery activities such as "efforts in preparation for defueling, decontamination and dose reduction, radioactive waste management, and associated environmental impacts." The article also discusses occupational exposure. Future work is described, including defueling and decontamination of the reactor building basement.
*Cleanup; Decontamination; Waste handling; Radiation exposure, occupational

0651
Kanapilly, George M.; Stanley, John A.; Newton, George J.; Wong, Brian A.; and DeNee, Phillip B. "Characterization of an Aerosol Sample from the Auxiliary Building of the Three Mile Island Reactor." **Health Physics** 45(November 1983):981-989. (8 refs.)
An aerosol filter sample was collected by an air sampler in the auxiliary building of TMI. Analysis for radioisotopic composition indicated that the primary isotopes were Sr-89, Sr-90, Cs-134 and Cs-137. The radionuclides were in a soluble form.
*Cleanup; Reactor auxiliary building; Radionuclides; Radiation analysis; Filters

0652
Komarneni, Sridhar, and Roy, Rustum. "Alternative Radwaste Solidification Route for Three Mile Island Wastes." **Journal of the American Ceramic Society** 65(December 1982):C-198. (10 refs.)
An alternative being considered for handling waste at TMI is radionuclide absorption. This study investigated a "direct sintering option for the fixation of radionuclides, especially 137 Cs, on zeolite ion exchangers." The on-site sintering of zeolites may be a cost-effective option for waste solidification.
*Cleanup; Waste disposal; Zeolites

0653
Kuhns, William G. "Cost Sharing in the TMI-2 Cleanup is in the Self-Interest of All Parties; Otherwise, It Doesn't Make Much Sense." **Electrical World** 195(December 1981):56, 135.
Kuhns, Chairman and Chief Executive Officer of GPU, discusses problems related to the TMI cleanup, including communications, financing, and effects on the nuclear industry.
*Cleanup

0654
"Lack of Funds Delays TMI Cleanup." **Nuclear Engineering International** 28(December 1983):10. (News)
*Cleanup; Financing; General Public Utilities

0655
Lanouette, William J. "One Year Later, the Dangers Persist at Three Mile Island Nuclear Plant." **National Journal** 12(May 3, 1980):730-732.
While the owners of TMI wait for permission to vent krypton-85 gas from the disabled reactor, they warn that delays in disposing of the gas could be more dangerous than venting. Both the NRC and Metropolitan Edison have lost public credibility because of the incident.
*Venting; Krypton; Cleanup; Public opinion; TMI - anniversary (1)

0656
Lefkowitz, Sheldon, and Parlee, Kenneth A. "Polar Crane Guidelines Evolve from TMI Damage Assessment." **Power Engineering** 88(May 1984):40-43.

A 500-ton polar crane, rendered inoperable as a result of the accident at TMI, "has strategic importance in ongoing recovery activities, particularly in completing reactor head removal." Damage to the crane is described along with its refurbishment. The experience gained here "has produced recommendations for everything from insuring crane replacement parts availability to the positioning of an emergency manlift installation."

*Polar crane; Safety; Lessons learned

0657
Lenckus, Dave. "Robots Will Have Big Role in Loss Control." **Business Insurance** 19(September 30, 1985):24, 26, 28. (News)

*Cleanup; Remote vehicles

0658
Lewin, David I. "Becalmed off Three Mile Island." **Mechanical Engineering** 106(July 1984):50.

*Cleanup

0659
Long, R.L.; Crimmins, T.M.; and Lowe, W.W. "Three Mile Island Accident Technical Support." **Nuclear Technology** 54(August 1981):155-173.

This paper describes the buildup of the technical support staff that was required to respond to the TMI-2 accident. Under the leadership of GPU personnel, technical support grew from ten to twenty to nearly 2000 by mid-April. Problems identified, analyzed and solved during initial recovery operations included "estimating core damage, removing decay heat from the core using forced and natural circulation, and elimination of the hydrogen bubble." The paper is divided into three parts: illustrations of the problems that had to be solved; the technical support resources that had to be built up and managed; and a brief history of several problems that were solved. An appendix gives the "TMI Recovery Planning Matrix."

*Cleanup; Technical support; Cooldown; Reactor core; Hydrogen bubble

0660
"Long Haul at Three Mile Island." **Nuclear Engineering International** 25(December 1980):7. (News)

*Cleanup; Environmental effects; General Public Utilities; Financing

0661
Marshall, Eliot. "Problems Continue at Three Mile Island." **Science** 213(September 18, 1981):1344-1345. (1 ref.) (News)

*Cleanup; Costs; General Public Utilities; Financing; General Accounting Office

0662
Marshall, Eliot. "Reagan Pledges Support for TMI Cleanup." **Science** 214(November 6, 1981):636. (News)

*Cleanup; Financing; Reagan administration

0663
Marshall, Eliot. "Thornburgh Gets a Hearing." **Science** 214(October 9, 1981):162. (News)

*Cleanup; Financing; Congress; Thornburgh, Richard

0664
Masters, Richard. "Good Progress Being Made in Three Mile Island Recovery." **Nuclear Engineering International** 27(February 1982):21-25.

Masters discusses the progress in the recovery of Three Mile Island that was reported at the TMI-2 seminar at the ANS Winter Meeting in San Francisco, December 1981. Included are information on sources of funding, the recovery objectives, the building condition, sources of radioactivity, the polar crane, core damage, and fuel removal.

*Cleanup

0665
Meininger, Richard D.; Cannon, Collins P.; and Smyth, George J. "Data Acquisition at TMI-2 Leads to New Preventive Maintenance Program." **Power Engineering** 89(June 1985):50-52.

Sophisticated surveillance techniques, used to monitor equipment at TMI, "are leading to a method of equipment maintenance based on a philosophy of management by anticipation rather than reaction." These surveillance techniques could be implemented at other nuclear plants.

*Equipment monitoring; Equipment function; Surveillance techniques

0666
Milligan, John W. "Reagan, Utility Association Support TMI Cleanup Plan." **Business Insurance** 15(October 19, 1981):2. (News)

*Cleanup; Financing; Edison Electric Institute; Reagan administration

0667
Milligan, John W. "TMI Cleanup Debts Hurt Other Electric Utilities." **Business Insurance** 15(September 28, 1981):2, 4. (News)

*Cleanup; Costs

0668
"Mini TV Camera Tapes TMI-2 Reactor-Core Damage." **Electrical World** 197(January 1983):136-137. (News)

*Reactor core damage; Videocamera use; Cleanup

0669
"More Alarms at Three Mile Island." **New Scientist** 87(July 3, 1980):4. (News)

*Venting; Krypton

0670
Morrell, M.P. "Three Mile Island Unit 2 (TMI-2): Reactor Building Venting Experience." In: **Three Mile Island Cleanup: Experiences, Waste Disposal, and Environmental Impact**, edited by Lester J. King and James H. Opelka, 1-5. New York: American Institute of Chemical Engineers, 1982. (AIChE Symposium Series no. 123, v. 78, 1982)

Radioactive gases, mostly krypton 85, were vented from the TMI-2 reactor building between June 28 and July 11, 1980. This paper describes the preparations for venting, the venting process, and monitoring results. About 44,000 curies of Kr-85 and one curie of Tritium were vented. The maximum integrated doses "calculated for any given sector off-site were 4.5 mrem skin dose and 0.045 mrem whole body dose."

*Venting; Krypton

0671
Newman, Marie Mastin. "TMI-2: Making Strides in Research and Recovery." **Electric Perspectives** (Fall 1982):9-15.

The overall recovery effort at TMI is described. The driving force in the recovery is GEND, which is an acronym that identifies four utility and government organizations that are cooperating in the effort.

*Cleanup; Research and development; GEND

0672
Norman, Colin. "Resins Raise Problems for TMI Cleanup." **Science** 212(June 12, 1981):1252. (News)
*Cleanup; EPICOR; Resins; Decontamination

0673
"NRC, Pa., Positions Leave GPU Caught in Middle." **Nuclear News** 23(November 1980):46. (News)
*Cleanup; General Public Utilities; Financing

0674
"NRC Policy Statement: Cleanup Should Speed Up." **Nuclear News** 24(June 1981):45-46. (News)
*Cleanup; Nuclear Regulatory Commission

0675
"NRC Proposes to Okay Evaporative Release." **Nuclear News** 30(February 1987):32-33. (News)
*Cleanup; Decontamination; Water contamination, radioactive

0676
"Nuclear Reactor Core was Reduced to Rubble." **New Scientist** 95(July 29, 1982):285. (News)
*Reactor core analysis

0677
Ocken, Howard. "Decontamination Technology for TMI-2 Cleanup." **EPRI Journal** 10(November 1985):66-67.
The TMI accident created unique decontamination problems, requiring new mechanical and chemical decontamination techniques. Research and conceptual design of these methods are described.
*Cleanup; Decontamination; Research and development

0678
"On-Site System Helps Decontaminate Three Mile Island." **Computer** 16(May 1983):104. (News)
*Computers; Decontamination

0679
"One-Third of Unit 2 Core Loaded into Canisters." **Nuclear News** 30(July 1987):22. (News)
*Reactor core; Reactor core analysis; Waste shipment

0680
"ORNL Analyzes Three Mile Island Reactor Water." **Analytical Chemistry** 51(August 1979):951A. (News)
*Coolant; Water pollutants, radioactive; Cleanup

0681
Patti, F.J.; Dam, A.S.; and Brolin, E.C. "Securing Three Mile Island: The Initial Recovery Programme." **Nuclear Engineering International** 24(November 1979):60-63.
Staff from Burns and Roe, architect-engineers of Unit 2, worked on recovery efforts for the two months following the accident. Modifications were made to secure and maintain "the plant in a safe condition," and to allow the complete recovery effort to begin. Described are modifications for temperature control, pressure control, radioactivity control, liquid cleanup, and fuel pool waste storage, among others.
*Cleanup; Burns and Roe; Engineering

0682
Plattner, Andy. "Congressional Help Urged for Three Mile Island Plant Nuclear Accident Cleanup." **Congressional Quarterly Weekly Report** 39(March 7, 1981):403-405.

GPU's financial problems related to the cleanup of TMI continue to escalate. Congressional help is being urged by Pennsylvania's U.S. House delegation. Proposed legislation is discussed along with the administration's position.

*General Public Utilities; Financing; Cleanup; Costs; Legislation

0683
Plattner, Andy. "Three Mile Island, Three Years Later: Congress Considering Ways to Help Pay for Cleanup of Troubled Nuclear Plant." **Congressional Quarterly Weekly Report** 40(March 20, 1982):610-611.

Plattner follows the efforts of Pennsylvania legislators to obtain federal help to finance the cleanup of TMI. Political aspects of the legislation, along with opposition to the plan, are given.

*Cleanup; Financing; Legislation; Congress; Politics

0684
"Polar Crane at Center of TMI Safety Review." **Engineering News-Record** 211(July 28, 1983):15-16. (News)

*Cleanup; Polar crane; Safety

0685
"Post-Defueling Plan for Unit 1 Presented." **Nuclear News** 30(January 1987):29. (News)

*Defueling

0686
"Preparatory Work Begun for Unit 2 Defueling." **Nuclear News** 28(December 1985):26. (News)

*Defueling

0687
"The Price Tag on Three Mile Island." **New Scientist** 83(July 26, 1979):268. (News)

*Cleanup; Costs

0688
"Project Agreement Set for TMI Recovery." **Engineering News-Record** 204(April 3, 1980):75. (News)

*Cleanup; Labor

0689
Pwu, L.C., and Broughton, T.G. "RETRAN Simulation of Three Mile Island Unit 2 Startup Overfeed Incident." **Nuclear Technology** 54(September 1981):358-363. (3 refs.)

This paper summarizes a study that uses the RETRAN system analysis code to simulate an overfeeding transient that occurred at the Three Mile Island Unit 2 plant on December 2, 1978 during a plant startup.

*Transients; Computer programs; RETRAN

0690
"Reductions in TMI-2 Work Force are Announced." **Public Utilities Fortnightly** 112(December 8, 1983):53. (News)

*Cleanup; Costs; Nuclear workers

0691
"Remote Survey of Reactor Building Basement Begins." **Nuclear News** 28(January

1985):39. (News)

*Remote vehicles; Reactor containment building; Cleanup

0692
"Remote Vehicles Used in TMI-2 Cleanup." **Public Utilities Fortnightly** 116(December 12, 1985):62, 64, 66. (News)

*Remote vehicles; Cleanup

0693
Rider, Robert L. "Reactor Defuelling Gets Under Way." **Nuclear Engineering International** 31(April 1986):27-29.

This update focuses on planning for the defueling operation. Removal of the damaged core is under way, and the plenum assembly has been lifted. Described are the condition of the core, preparation for defueling, and steps for core removal.

*Defueling; Reactor core analysis; Cleanup

0694
Rider, Robert L. "The Three Mile Island Situation." **Chemtech** 14(December 1984):737-739.

The initial cleanup of TMI has started with "inspection, assessment, and lots of precaution." The article stresses safety measures and precautions that have been taken to minimize radiation dosage. Worker efficiency is emphasized.

*Cleanup; Nuclear workers; Safety

0695
Rider, Robert L. "TMI Containment Entry Program." In: **Three Mile Island Cleanup: Experiences, Waste Disposal, and Environmental Impact**, edited by Lester J. King and James H. Opelka, 6-8. New York: American Institute of Chemical Engineers, 1982. (AIChE Symposium Series, no. 213, v. 78, 1982)

Rider describes a program for a controlled series of entries into the Three Mile Island Unit 2 containment building. The program is intended to extract a maximum amount of data with minimum radiation dose to workers. Briefly described are the entry planning, radiological controls, protective clothing, tasks performed, and observations.

*Containment entry; Cleanup

0696
Schwartz, Joe. "Harrisburg: Counting the Cost." **Nature** 278(April 12, 1979):589. (News)

*Cleanup; Metropolitan Edison; Management; Stock market

0697
Schwartz, Joe. "Three Mile Island: Reactor Entry Bid Aborted." **Nature** 285(May 29, 1980):278-279. (News)

*Cleanup; Containment entry

0698
Schwarz, W.F. "Three Mile Island Provides a Unique Research Opportunity." **Nuclear Engineering International** 27(July 1982):35-38.

The Instrumentation and Electrical (I&E) Programme is part of a large research and development effort conducted by four organizations (GEND agreement). This article reports on tests of the I&E Program, whose purpose is to identify instruments and electrical components that failed during or since the accident, to identify the causes of failure, to assess the physical conditions of those that did not fail, and to inform the nuclear industry of the results.

*Research and development; GEND; Instrument function; Electrical equipment; Instrument failure

0699
"SDS Completes First Task; Wall Spraying to Begin." **Nuclear News** 25(April 1982):29. (News)
*Submerged Demineralizer System; Decontamination

0700
"SDS Processing Passes 100,000-Gallon Mark." **Nuclear News** 24(December 1981):70-71. (News)
*Submerged Demineralizer System; Decontamination

0701
"Second Camera Probes Find Unit 2 Fuel Gap." **Nuclear News** 25(September 1982):42-43. (News)
*Videocamera use; Reactor core; Cleanup

0702
"Senate has Cleanup Bill." **Nuclear News** 25(May 1982):88-89. (News)
*Cleanup; Financing; Legislation

0703
"Seven Year Delay: Three Mile Island." **Economist** 282(January 9, 1982):26. (News)
*Cleanup; Cleanup; Financing

0704
Shanaman, Susan M. "The Financial Impact of Three Mile Island." **Public Utilities Fortnightly** 107(March 26, 1981):44-46.

In a statement before the President's Nuclear Safety Oversight Committee, Shanaman, Chairman of the Pennsylvania Public Utility Commission, presents TMI's cost problem in four parts: replacement power cost, cost of lost unit, toxic cleanup cost, and reconstruction cost. Shanaman asks for federal financial aid to be provided as a grant.

*Cleanup; Costs; General Public Utilities; Financing

0705
Siemens, D.H.; Knowlton, D.E.; and Shupe, M.W. "Three Mile Island Zeolite Vitrification Demonstration Program." In: **Three Mile Island Cleanup: Experiences, Waste Disposal, and Environmental Impact**, edited by Lester J. King and James H. Opelka, 41-44. New York: American Institute of Chemical Engineers, 1982. (AIChE Symposium Series, no. 213, v. 78, 1982)

In cooperation with the Department of Energy, the Pacific Northwest Laboratory is taking "a portion of the zeolites from the SDS process and demonstrat[ing], on a production scale, that these zeolites can be vitrified using the in-can melting process." This paper briefly overviews the TMI Zeolite Vitrification Program.

*Zeolites; Vitrification

0706
"16 Percent of Unit 2 Core Canned, Removed." **Nuclear News** 29(June 1986):39-40.
*Reactor core; Waste shipment; Fuel debris; Reactor core analysis

0707
"Sloppy Clean-up." **New Scientist** 98(April 14, 1983):62. (News)
*Cleanup; General Public Utilities; Management

0708
"Some Cleanup Waste Seen as Misclassified." **Nuclear News** 28(November 1985):119. (News)
*Waste shipment; Waste classification

0709
"Some Fuel Melting Occurred in Accident." **Nuclear News** 28(April 1985):44. (News)
*Videocamera use; Fuel melting; Reactor core analysis; Cleanup

0710
"Some TMI-2 Rods Insert, Others Fail to Move." **Nuclear News** 25(August 1982):41-42. (News)
*Cleanup; Reactor core; Fuel rods

0711
Stambler, Irwin. "Researchers Get First Look into Disabled TMI Reactor." **Industrial Research & Development** 24(October 1982):58, 60. (News)
*Reactor core; Videocamera use; Cleanup

0712
"Stier Finds No Coverup on Crane Charges." **Nuclear News** 29(April 1986):36-37. (News)
*Cleanup; Safety; Polar crane

0713
"Studies Show 35 Percent of Unit 2 Core Melted." **Nuclear News** 30(June 1987):38. (News)
*Reactor core analysis; Fuel melting; Cleanup

0714
"Submerged Demineralization System to be Used to Process TMI Wastewater." **Industrial Water Engineering** 18(July/August 1981):12-13. (News)
*Submerged Demineralizer System; Cleanup; Decontamination

0715
"System Installed to Treat `Hot' Water." **Nuclear News** 24(May 1981):34-35. (News)
*Submerged Demineralizer System; Decontamination

0716
Tanabe, F.; Matsumoto, K.; Yoshida, K.; and Shimooke, T. "Post-Facta Analysis of the TMI Accident (II): Analysis of Fuel Rod Behavior and Core Damage Estimation by Use of TOODEE2-J." **Nuclear Engineering and Design** 69(1, 1982):37-42. (10 refs.)

The TOODEE2-J computer program was used to analyze fuel rod behavior and core damage of TMI-2 during the first three hours of the accident. Results suggest that almost all except the peripheral low-powered rods burst, that the zircaloy cladding formed liquid Zr-U-O, and that 43.5% of the zircaloy was converted to zirconium-dioxide in the first three hours, with "major damage to the fuel rods."

*Computer programs; Reactor core damage; Zircaloy; Fuel rods

0717
Tanabe, F.; Yoshida, K.; Matsumoto, K.; and Shimooke, T. "Post-Facta Analysis of the TMI Accident (I): Analysis of Thermal Hydraulic Behavior by Use of RELAP4/MOD6/U4/J2." **Nuclear Engineering and Design** 69(1, 1982):3-35. (16 refs.)

The RELAP4/MOD6/U4/J2 computer program was used to analyze the thermal-hydraulic behavior of the TMI-2 reactor in the first three hours of the accident. Results show "that the tops of active fuel rods would have begun to be exposed at 6490 s into the accident and the core mixture level remained at a height of 3.4 to 4 ft. from the bottom of the core for about 40 min."

*Computer programs; Reactor core damage; Fuel rods; Reactor core uncovery

0718
Tarpinian, James E., and Hollenbeck, Peter. "A Summary of the Radiological and Physical Conditions of the Three Mile Island Unit-2 Reactor Containment and Their Impact on Recovery." In: **Energy, Environment, and the Economy**, edited by Shyamal K. Majumdar, 219-228. Easton, PA: Pennsylvania Academy of Science, 1981.

Tarpinian (Bechtel Northern Corp.) and Hollenbeck (GPU) discuss the initial stage of the recovery effort at TMI. The physical conditions of the reactor building are described along with the radiological conditions. Radiological protection of personnel entering the reactor building is also described.

*Cleanup; Reactor building

0719
Taylor, Gregg M. "Remote Handling Systems Help TMI-2 Cleanup Efforts." **Nuclear News** 27(December 1984):52-56.

Cleanup activities at TMI are described. The focus is on the remote handling systems, which will be used in areas where radioactivity does not permit human entry. Vehicles described include: the Remote Reconnaissance Vehicle, nicknamed Rover; a prototype named Moose; early remote vehicles used at TMI; and second-generation vehicles.

*Remote vehicles; Cleanup

0720
"Third Small-break Test adds TMI Behavior." **Nuclear News** 23(August 1980):43-44. (News)

*Loss of coolant accident

0721
"3-D Computer Plant Model, Radiation Map to be Used." **Nuclear News** 28(June 1985):98, 101-102. (News)

*Computer models; Reactor building; Radioactivity; Cleanup

0722
"Three Mile Island Girds for a `Hot' Water Clean Up." **Chemical Week** 128(April 8, 1981):49. (News)

*Decontamination; Cleanup; Costs

0723
"Three Mile Island Opens its Doors." **New Scientist** 87(July 31, 1980):352. (News)

*Containment entry

0724
"Three Mile Island Situation Analyzed." **Public Utilities Fortnightly** 108(September 24, 1981):41. (News)

*Cleanup; Costs; General Public Utilities; Financing; General Accounting Office

0725
"TMI Cleanup Cost Estimates Raised." **Public Utilities Fortnightly** 106(December 18, 1980):47. (News)

*Cleanup; Costs

0726
"TMI Cleanup Costs Mount." **Engineering News-Record** 205(November 20, 1980):18. (News)

*Cleanup; Costs

0727
"TMI Clean-up Moves to Phase II." **Nuclear Engineering International** 27(November

1982):9. (News)
*Cleanup; Decontamination

0728
"TMI Cleanup Pushed Back." **Engineering News-Record** 210(February 17, 1983):11. (News)
*Cleanup; Financing

0729
"TMI Core Damage Check Reveals No Major Melt." **Engineering News-Record** 209(July 29, 1982):19. (News)
*Cleanup; Reactor core damage; Fuel melting

0730
"TMI Core Damage Explained." **Machine Design** 57(June 20, 1985):8. (News)
*Reactor core analysis

0731
"TMI Cost-Share Plan Pushed by Utility Group." **Engineering News-Record** 207(September 17, 1981):40. (News)
*Cleanup; Financing; Edison Electric Institute

0732
"TMI Needs Your Help." **Nuclear Engineering International** 28(January 1983):8. (News)
*Cleanup; Financing; Tours

0733
"TMI Recommissioning Goes to Bechtel Power." **Engineering News-Record** 204(February 21, 1980):14. (News)
*Cleanup; Bechtel Corp.

0734
"TMI Tests Turn Up Good News." **Engineering News-Record** 209(July 8, 1982):18. (News)
*Control rods; Reactor core

0735
"TMI Vents its Krypton." **New Scientist** 86(June 19, 1980):297. (News)
*Venting; Krypton

0736
"TMI 2 Allegations Evaluated ... While Core Surveys Proceed." **Nuclear Engineering International** 28(November 1983):8-10. (News)
*GPU Nuclear; Management; Cleanup; Reactor head lift; Reactor core analysis

0737
"TMI-2 Cleanup Closes in on Defueling Stage." **Engineering News-Record** 215(July 18, 1985):30-32. (News)
*Cleanup; Defueling

0738
"TMI-2 Cleanup Costs, Schedule Now in Doubt." **Engineering News-Record** 210(May 19, 1983):24. (News)
*Cleanup; Financing

0739
"TMI-2 Fuel Hotter than Expected ... and Clean-up Could be Delayed." **Nuclear**

**Engineering International** 30(January 1985):12. (News)
*Reactor core analysis; Research and development; Remote vehicles; Cleanup

0740
"TMI-2 Head is Lifted." **Nuclear Engineering International** 29(September 1984):2-3. (News)
*Reactor head lift; Defueling; Cleanup; Financing; Restart

0741
Torrey, Lee. "Clean-up Activity at Three Mile Island." **New Scientist** 83(August 23, 1979):571-572. (News)
*Cleanup; EPICOR; Law suits; Lancaster, PA; Susquehanna River

0742
Torrey, Lee. "Hazards of Cleaning up Three Mile Island." **New Scientist** 87(September 11, 1980):766. (News)
*Cleanup; Safety

0743
Torrey, Lee. "Is Three Mile Island Leaking Radioactive Water?" **New Scientist** 86(April 17, 1980):133. (News)
*Water contamination, radioactive; Containment leak

0744
Tuite, Peter. "Generation, Classification, Treatment and Disposal of Solid Waste Forms Resulting from Cleanup of TMI-2." In: **Three Mile Island Cleanup: Experiences, Waste Disposal, and Environmental Impact**, edited by Lester J. King and James H. Opelka, 22-26. New York: American Institute of Chemical Engineers, 1982. (AIChE Symposium Series, no. 213, v. 78, 1982)

Primary and secondary radioactive wastes from the TMI accident and from decontamination activities following the accident are discussed. Included are information on: disposal regulations and criteria, accident sludge, ion exchange media, and evaporator concentrates. The EPICOR II and SDS ion exchange systems are discussed.
*Waste disposal; Decontamination; EPICOR; Submerged Demineralizer System

0745
"TV Camera Tapes TMI Unit 1 Reactor's Innards: Unit 2 Next." **Electrical World** 196(May 1982):115. (News)
*Videocamera use; Reactor core; Cleanup

0746
"TV Shows TMI-2 Core Damage." **Nuclear Engineering International** 30(April 1985):5. (News)
*Videocamera use; Reactor core analysis

0747
"Unit-2 Core Drilling Shows Debris Can be Cut." **Nuclear News** 29(September 1986):36-37. (News)
*Fuel debris; Reactor core analysis

0748
"Unit 2 Core Sampling to Start This Month." **Nuclear News** 26(September 1983):36-37. (News)
*Reactor core analysis

0749
"Unit 2 Head Lifted." **Nuclear News** 27(September 1984):39. (News)
*Reactor head lift; Reactor head

0750
"Unit 2 Water Clarity Improved by Coagulant." **Nuclear News** 30(March 1987):30. (News)
*Defueling; Water filtration

0751
"USE Would Rather Leave TMI-2 Waste in Trenches." **Nuclear News** 28(December 1985):81. (News)
*Waste shipment; Waste classification; U.S. Ecology, Inc.

0752
"`Utilities' Cleanup Fund Ready to Pay Out -- EEI." **Nuclear News** 28(January 1985):30-31. (News)
*Cleanup; Financing; Edison Electric Institute

0753
"Utility Health Hinges Partly on TMI Cleanup." **Electrical World** 195(October 1981):28-29. (News)
*Cleanup; Financing

0754
Vaughan, James W., Jr. "Developments from TMI." **IEEE Spectrum** 21(December 1984):16. (Letter)
*Cleanup

0755
"Venting Done in 2 Weeks; Entry Next." **Nuclear News** 23(August 1980):36, 38. (News)
*Venting; Krypton; Containment entry

0756
"Venting of Krypton-85 Begun at Unit 2." **Nuclear News** 23(April 1980):33-34. (News)
*Venting; Krypton

0757
"Venting Plan Approved; Could Finish by August." **Nuclear News** 23(July 1980):34-35. (News)
*Venting; Krypton

0758
"What Happened at Three Mile Island?" **Analytical Chemistry** 56(January 1984):35A. (News)
*Reactor core analysis; Cleanup

0759
Winkleblack, R.K. "Remote Systems for TMI-2 Surveillance and Characterization." **EPRI Journal** 10(March 1985):56.
The remote reconnaissance vehicle (RRV-1) has been used by GPU for two missions into the basement of TMI-2. Television pictures and videotapes were taken of the basement, and radioactivity was measured. Two other remotely controlled robots are under development.
*Cleanup; Remote vehicles; Television; Videocamera use

**GPU MANAGEMENT**

0760
"Criminal Misconduct." **New Scientist** 100(November 10, 1983):401. (News)
*Metropolitan Edison; Management

0761
David, Peter. "Comfort for Three Mile Island." **Nature** 306(December 15, 1983):631. (News)
*General Public Utilities; Management

0762
David, Peter. "Three Mile Island; Accusations Abound as Inquiries Proliferate." **Nature** 305(October 6, 1983):461. (News)
*General Public Utilities; Management; Cleanup; Nuclear Regulatory Commission

0763
David, Peter. "Three Mile Island; Utility Indicted for Criminal Misconduct." **Nature** 306(November 17, 1983):213. (News)
*Metropolitan Edison; Management; Cleanup; Financing

0764
Evanoff, Mark. "Bechtel Workers Reveal Flaws in TMI Cleanup." **Not Man Apart** 13(May 1983):19+.
Unavailable
*Cleanup; Whistleblowing; General Public Utilities; Management

0765
"Fraud Charge at Three Mile Island." **New Scientist** 98(June 2, 1983):613. (News)
*General Public Utilities; Management

0766
"GPU Faults Self on Tests; Would Accept Rehearing." **Nuclear News** 24(October 1981):30-31. (News)
*Operators; Cheating; Cleanup; Submerged Demineralizer System

0767
"Met Ed Cited, but Not Fined for Info Inadequacies." **Nuclear News** 24(March 1981):37. (News)
*Metropolitan Edison; Management; Nuclear Regulatory Commission

0768
"Met Ed Hit with Record NRC Fine, Pa. Show-Cause." **Nuclear News** 22(December 1979):28, 33-34. (News)
*Metropolitan Edison; Nuclear Regulatory Commission; Fines

0769
Marshall, Eliot. "Investigation Confirms TMI Cleanup Problems." **Science** 221(September 30, 1983):1357. (News)
*Parks, Richard D.; Whistleblowing; Cleanup; Safety

0770
"NRC Orders Hearing on Leak-Rate Falsification." **Nuclear News** 29(February 1986):32-33. (News)
*Metropolitan Edison; Management; Leak rate; Nuclear Regulatory Commission; Hearings

0771
"NRC Slams TMI Utility, Proposes Heaviest Fine." **Engineering News-Record** 203(November 1, 1979):12-13. (News)
*Metropolitan Edison; Management; Nuclear Regulatory Commission

0772
"The Nuclear Regulatory Commission's Investigation of Three Mile Island: Recent Findings." **Public Utilities Fortnightly** 114(September 13, 1984):39-41. (News)
*General Public Utilities; Management

0773
"Personal Disputes at Three Mile Island." **Nature** 305(October 6, 1983):461. (News)
*General Public Utilities; Management; Parks, Richard D.; Safety violations; Whistleblowing

0774
"Professional Misconduct; The Latest Inquiry at Three Mile Island Raises Urgent Questions about Professionals' Protection." **Nature** 305(October 6, 1983):460. (News)
*General Public Utilities; Management; Ethics

0775
"Reports: No Cancer Change, Same Leak-Test Findings." **Nuclear News** 28(October 1985):31-32. (News)
*Leak rate; Restart; Court cases; Health effects - cancer

0776
"Three Investigations Find No GPU Conspiracy." **Nuclear News** 27(June 1984):44-46.
*General Public Utilities; Management; Cheating

0777
"Two RO's Admit Cheating on Exams, are Terminated." **Nuclear News** 24(September 1981):75-76. (News)
*Operators; Cheating; Rate hearings

# 10

# Sociopolitical Response and Commentary

0778
"The Accident at Three Mile Island." **Public Utilities Fortnightly** 103(April 26, 1979):40-41.
*Public opinion; Politics

0779
Avery, George A. "The Costs of Nuclear Accidents and Abandonments in Rate Making." **Public Utilities Fortnightly** 104(November 8, 1979):17-23. (8 refs.)
Avery focuses on the effects of a nuclear accident on the utility's cost of service and plant investment, including rate base effects of extended shutdowns, the expenses of the accident itself, and the cost of replacement power. Regulatory responses to Three Mile Island are discussed.
*Rate making; Reactor accidents; Costs; Regulation

0780
Aviel, David. "Congressional Attitudes toward Nuclear Energy in the Wake of the Three Mile Island Accident." **New England Journal of Business & Economics** 9(Fall 1982):74-89.
Unavailable.
*Congress; Attitude to nuclear power

0781
Aviel, David. "Congressional Views on Nuclear Energy." **Bulletin on the Atomic Scientists** 39(February 1983):48. (Editorial)
Members of Congress and state governors were surveyed as to their attitude to nuclear power. Responses were analyzed in relation to the influence of TMI on their attitude.
*Public opinion; Politicians

0782
Barlow, Alan. "Three Mile Reactor Accident Clouds Future of Industry." **Congressional Quarterly Weekly Report** 37(April 7, 1979):627-630.
TMI is viewed as more of a problem from the political point of view than from a technical viewpoint. The accident has public policy implications for the nuclear industry and will affect future legislation.
*Public policy; Nuclear industry; Legislation; Politics

0783
Baum, Andrew; Fleming, Raymond; and Davidson, Laura M. "Natural Disaster and Technological Catastrophe." **Environment and Behavior** 15(May 1983):333-354. (27 refs.)
The authors make a distinction between natural disasters and technological catastrophes. Three Mile Island is used as an example of the latter. As evidenced by TMI and other events, data "suggest that technological catastrophes are more likely

to have long-term effects," and to affect people outside the immediate area.
*Disasters, natural; Technological catastrophes

0784
Bedell, Douglas H. "On `Meltdown of Democracy?'" **Bulletin of the Atomic Scientists** 39(August/September 1983):57. (Letter)
*Local residents; Protest groups; Restart

0785
Budnitz, Robert J. "The Response of the Nuclear Regulatory Commission to the Accident at Three Mile Island." **Annals of the New York Academy of Sciences** 365(1981):203-209.
Budnitz discusses some of the flaws that were revealed by TMI. Technical flaws discussed were safety design, human factors, emphasis on hardware, siting policy, and instrumentation and controls inadequacies. Institutional flaws were complacency, concern with regulation rather than safety, poor feedback of operating experience, lack of self-policing, disincentives to safety improvements, and inadequate emergency response.
*Nuclear Regulatory Commission; Safety; Human factors

0786
Carter, Luther J. "Political Fallout from Three Mile Island." **Science** 204(April 13, 1979):154-155.
The political ramifications of the Three Mile Island incident may be far-reaching. The accident has "given a new credibility to antinuclear groups," and members of Congress are speaking out for "more effective licensing and oversight of reactor operations by the Nuclear Regulatory Commission." Slower growth of the nuclear power industry is expected. States such as California may go further than Washington in regulatory and safety requirements for nuclear plants.
*Politics; Regulation

0787
Cutter, Susan, and Barnes, Kent. "Evacuation Behavior and Three Mile Island." **Disasters** 6(2, 1982):116-124. (35 refs.)
Residents near TMI were surveyed in April 1979 via a mailed questionnaire regarding their response to the accident. Both the decisions to evacuate or to remain were influenced by ambiguous information. Other influencing factors were "proximity to the facility, stage in life cycle and the actions of friends and neighbors." Behavior of TMI residents "serves as a case study highlighting some of the contextual, spatial, and social factors influencing voluntary evacuations from technologically hazardous situations."
*Evacuation behavior

0788
Cutter, Susan L. "Risk Cognition and the Public: The Case of Three Mile Island." **Environmental Management** 8(1, 1984):15-20. (22 refs.)
Residents within 50 miles of TMI were surveyed via questionnaire in 1979, 1980, and 1982 to assess attitudes toward nuclear power and to assess coping response. The results indicated that most thought that another accident would occur in their lifetime, that nuclear power plants should continue to operate, and that both government and utilities were experiencing an erosion of public trust.
*Public opinion; Risk perception; Attitude to nuclear power; Coping

0789
Doub, William O., and Shoemaker, James R. "Who Pays the Costs of Industrial Accidents Like Three Mile Island?" **Public Utilities Fortnightly** 115(May 16, 1985):13-16.
The TMI accident is used as a case study to show how losses from an industrial accident are allocated in our society. The public was compensated using the Price-

Anderson Act, but GPU shareholders and ratepayers had to absorb losses. Costs are usually allocated via the courts.

*Society; Legal aspects; Reactor accidents; Costs; Cost allocation

0790
Elmore, Richard F. "The Safe Operation of Nuclear Plants: Implementing Federal Policy in the Aftermath of Three Mile Island." **International Journal of Mass Emergencies and Disasters** 2(March 1984):61-87. (34 refs.)

The TMI accident has caused an intense examination of nuclear plant safety. Problems in implementing regulatory policy for the safe operation of nuclear power plants "are characterized by a low tolerance for terror, high organizational and technical complexity, and highly visible, costly outcomes." Rather than designing policies dealing with specific errors, the key is to design incentives for utilities to solve their own problems and systems. Current policy stresses technical and regulatory solutions, where financial incentives that would induce utilities to engage in safe operating practices would produce better results.

*Regulatory policy; Public policy; Safety

0791
Etzkowitz, Henry. "Corporate Induced Disaster: Three Mile Island and the Delegitimation of Nuclear Power." **Humanity and Society** 8(August 1984):228-252. (43 refs.)

After differentiating four types of disasters, the author uses Three Mile Island as a case study to analyze the type of disaster "where human causation and foreknowledge of the result can be identified." Delegitimation can occur following a corporate induced disaster. Suppression of information, "mystification," and "flow of influence" are among the topics discussed in relation to TMI.

*Delegitimation; Information acquisition; Disasters; Nuclear power

0792
Ferrar, Terry A. "Three Mile Island -- The Regulatory Challenge of 1979." **Public Utilities Fortnightly** 104(July 19, 1979):15-18.

The accident at TMI has identified a regulatory challenge that will affect the utility industry, suggests a "risk-benefit structure to the regulatory question, and identifies the stress placed on traditional institutions by catastrophic accidents." The author concludes that beneficiaries of large projects must assume the risks.

*Regulation; Risk; Nuclear industry; Management

0793
Flynn, Cynthia Bullock. "Local Public Opinion." **Annals of the New York Academy of Sciences** 365(1981):146-158. (16 refs.)

This paper reviews results of seven surveys on the public response to TMI. Emergency period behavior varied significantly from people who seemed not to be affected to those who evacuated. Also summarized from the studies are stress and psychological effects, post-accident effects, continuing stress, and movement from the area.

*Local residents; Public opinion; Evacuation; Health effects - psychological; Stress; Stress, chronic

0794
Freudenburg, William R., and Baxter, Rodney K. "Host Community Attitudes Toward Nuclear Power Plants: A Reassessment." **Social Science Quarterly** 65(December 1984):1129-1136. (20 refs.)

Pre- and post-TMI surveys show a change from at most 33 percent opposition to at least 50 percent opposition to nuclear power facilities. Available data suggest that the TMI accident has created "a tendency for host communities to oppose the construction of new nuclear power facilities."

*Public opinion; Attitude to nuclear power

0795
Gilinsky, Victor. "The Impact of Three Mile Island." **Bulletin of the Atomic Scientists** 36(January 1980):18-20.

Historical aspects of America's nuclear policy are discussed. Three Mile Island caused the nuclear industry and the general public to reevaluate the risk of new and dangerous technologies.

*Energy policy; Public policy; Nuclear industry; Risk

0796
Goldhaber, Marilyn K.; Houts, Peter S.; and DiSabella, Renee. "Moving After the Crisis; A Prospective Study of Three Mile Island Population Mobility." **Environment and Behavior** 15(January 1983):93-120. (34 refs.)

The entire population living within five miles of TMI was traced one year later in a study on residential mobility and population composition. While around 15% of the movers said TMI was the major reason for moving, the rate at which people moved was "the same the year after the accident as before." Controlling for attributes associated with mobility, "attitudes about TMI were virtually the same among movers and nonmovers." New people in the area had a more positive attitude about TMI.

*Population mobility; Local residents; Attitude to TMI

0797
Graham, John. "The Political Fallout from TMI-2." **Nuclear News** 22(May 1979):37-40, 43. (refs.)

Graham describes the reaction to TMI in Washington, from the White House to Capitol Hill, including Jimmy Carter's appointment of a Presidential Commission, Congressional hearings on the accident, some of the proposed legislation, and energy committee discussions

*Government; Politics; Carter administration; Legislation

0798
Green, Harold P. "Lessons of Three Mile Island for the Institutional Management and Regulation of Nuclear Power." **Annals of the New York Academy of Sciences** 365(1981):311-315.

Green observes that lessons learned from TMI include: people are not perfect, the public must have confidence in regulated technologies, and the nuclear industry should not be overprotected. In the long run, TMI provides an opportunity to revive the nuclear power industry.

*Regulation; Institutional response; Lessons learned

0799
Greenberg, Dan. "Accidents Will Happen." **New Scientist** 82(May 10, 1979):489-490. (Editorial)

*Public opinion

0800
Grygiel, Fred S., and Zarillo, Anthony J. "Three Mile Island: The New Jersey Regulatory Response." **Public Utilities Fortnightly** 106(December 18, 1980):23-28. (15 refs.)

The article focuses on "the New Jersey Board of Public Utilities' actions related to Jersey Central Power and Light Company," a 25 percent owner of TMI. The board dealt with five major problems: substantial increases in JCP&L's replacement power requirements; a decrease in JCP&L's need for short-term credit; the unavailability of energy from TMI; potential fault of JCP&L in the accident; and economic viability of JCP&L.

*New Jersey Board of Public Utilities; Jersey Central Power & Light Co.; Regulation

0801
Gwynne, P. "US Reactor Melts Nuclear Credibility." **New Scientist** 82(April 5, 1979):3. (News)
*Public opinion; Public relations; Attitude to nuclear power

0802
Houts, Peter S.; Lindell, Michael K.; Hu, Teh Wei; Cleary, Paul D.; Tokuhata, George; and Flynn, Cynthia B. "The Protective Action Decision Model Applied to Evacuation During the Three Mile Island Crisis." **International Journal of Mass Emergencies and Disasters** 2(March 1984):27-39. (14 refs.)

The protective action decision model was used to predict evacuation behavior during the Three Mile Island accident. Data are based on the NRC telephone survey conducted in July 1979 by Flynn. Results show that all independent variables (distance from TMI, presence of children under six years old, and a pregnant woman in the household) were significantly related to evacuation. Barriers to evacuation included inability to leave a job, lack of transportation, no place to go, and physical inability to travel. In natural disasters, conflicting information is usually given as a reason for staying; with TMI, conflicting information was considered a reason to evacuate, perhaps due to an inability for the average citizen to assess actual danger.
*Evacuation behavior; Health effects - psychological; Telephone surveys

0803
Johnson, J.H., Jr., and Zeigler, D.J. "Modelling Evacuation Behavior During the Three Mile Island Reactor Crisis." **Socio-Economic Planning Sciences** 20(3, 1986):165-171. (32 refs.)

Using the NRC sponsored telephone survey, an "empirical test of a causual model of nuclear reactor emergency evacuation decision-making" was developed. The conclusion is that evacuation behavior in nuclear emergencies is "predictable on the basis of locational, life-cycle, social status, attitudinal, and risk perception variables."
*Evacuation behavior; Telephone surveys

0804
Johnsrud, Judith. "The Reaction of the Regulatory Agency as Viewed from the Perspective of the Public Interest Organization." **Annals of the New York Academy of Sciences** 365(1981):257-262.

Johnsrud, one of two public interest intervenors in the operating license procedure for TMI, discusses the role of public interest intervenors, and their relationship with the Nuclear Regulatory Commission, particularly following the TMI accident.
*Public interest intervenors; Activism; Nuclear Regulatory Commission

0805
Kasperson, J.X.; Kasperson, R.E.; Hohenemser, C.; and Kates, R.W. "Institutional Responses to Three Mile Island." **Bulletin of the Atomic Scientists** 35(December 1979):20-24.

Kasperson et al. discuss energy policy and institutional responses to TMI, the President's Special Commission, Congressional reaction, NRC actions, and effects on the nuclear industry. Questions asked include "Is nuclear power compatible with democratic institutions?" and "Is a societal consensus on nuclear power possible?" Possible routes out of the current impasse are presented.
*Institutional response; Energy policy; Public policy; Nuclear industry

0806
Klein, Jeff. "Three Mile Island and the Kemeny Commission: Regulating the Regulators." **Stanford Lawyer** 15(Winter-Spring 1979-1980):11-14.

Klein discusses the Kemeny Commission and its recommendations, response of the Carter administration and effect on the NRC of these recommendations, involvement of the Kemeny Commission's legal staff in the investigation, and debate concerning a moratorium on nuclear plant licensing.

*President's Commission; Recommendations; Carter administration; Regulation

0807
Knox, Richard. "Three Mile Island: Worse to Come?" **Nuclear Engineering International** 24(May 1979):15. (Editorial)

The most important implications of the accident are the "effects on political and public opinion." The effect on energy policy in the U.S., Sweden and Britain are briefly discussed. The accident "will result in the plugging of yet another hole in the safety system."

*Politics; International reaction; Nuclear industry; Public opinion; Safety

0808
Krause, Carolyn. "Socioeconomic Impacts of Nuclear Power Plants." **Oak Ridge National Laboratory Review** 12(Fall 1979):36-45.

Unavailable.

*Socioeconomic effects

0809
Leventhal, Paul. "The Reaction of Congress to the Accident at Three Mile Island." **Annals of the New York Academy of Sciences** 365(1981):210-215.

Leventhal, an advisor to the Senate Subcommittee on Nuclear Regulation, characterizes congressional reaction to TMI as "one of caution, hesitation and continued sensitivity to code words, particularly moratorium and antinuclear." Congress has reacted, but not yet responded to the accident.

*Congressional reaction

0810
Lybrand, Dave. "TMI Plus 2." **IEEE Spectrum** 18(9, 1981):14. (Letter)

*Public opinion

0811
MacLeod, Gordon K. "The Politics of Public Health: Three Mile Island." **Forum on Medicine** 3(1981):152-159.

Unavailable.

*Politics; Public health

0812
Manning, D.Thompson. "Post-TMI Perceived Risk from Nuclear Power in Three Communities." **Nuclear Safety** 23(July-August 1982):379-384. (16 refs.)

As part of a larger telephone survey regarding the psychological and socioeconomic impacts of TMI, residents of two Massachusetts communities, one near a nuclear plant and one 50 miles from a nuclear plant, were surveyed regarding their attitude to nuclear power. In comparing these responses with responses to a similar survey of TMI residents, it was determined that "attitudes toward nuclear power depended on the respondent's residence," with TMI residents being most negative, and Massachusetts residents near the nuclear plant most positive; Massachusetts residents not near a nuclear plant were similar to the TMI group.

*Public opinion; Risk perception; Telephone surveys; Attitude to nuclear power

0813
Mazur, Allan. "Three Mile Island and the Scientific Community." **Annals of the New York Academy of Sciences** 365(1981):216-221. (4 refs.)

Mazur surveyed a group of prominent scientists, both pro- and anti-nuclear, and found that none of them had "realigned themselves on the nuclear issue because of TMI."

*Public opinion; Scientists

0814
Melber, Barbara D. "The Impact of TMI Upon the Public Acceptance of Nuclear Power." **Progress in Nuclear Energy** 10(3, 1982):387-398. (4 refs.)

This paper discusses long-term trends in public attitudes toward nuclear power, focusing on the extent to which the accident at Three Mile Island appears to have affected public acceptance of nuclear energy. Based on data from national surveys from the early 1970s through 1981, TMI appears to have increased the opposition to nuclear energy and has influenced the decline of construction of new nuclear power plants.

*Attitude to nuclear power; Public opinion; Nuclear industry

0815
Mitchell, Robert Cameron. "Polling on Nuclear Power: A Critique of the Polls After Three Mile Island." In: **Polling on the Issues**, edited by Albert H. Cantril, 66-68. Washington, D.C.: Seven Locks Press, 1980. (18 refs.)

Mitchell critiques the national and state/regional public polls which asked more than 150 questions regarding nuclear power immediately following the TMI accident. He feels that "the real problem with the post-Three Mile Island polling on the nuclear issue is that both the polling and the media analysis of the poll results are far more superficial than they need to be."

*Public opinion; Polls; Nuclear power

0816
Mitchell, Robert Cameron. "Public Opinion and Nuclear Power Before and After Three Mile Island." **Resources** 64(January-April 1980):5-8.

Results from several polls on nuclear power taken before and after the TMI accident indicate that while TMI had a temporary negative effect on public opinion, the public is willing to accept the risk offered by this technology. Because of the accident, public awareness of nuclear power is much higher than before.

*Public opinion; Nuclear power; Polls

0817
Morgello, Clem. "Three Mile Island Hangover: Treat the Basic Problem, Not the Symptoms." **Dun's Review** 116(1980):96. (Editorial)

Morgello suggests that the major difficulty in planned venting of radioactive gas may be credibility, as opposed to psychological stress, of the local residents.

*Credibility; Public relations; Venting; Stress

0818
Nelkin, Dorothy. "Some Social and Political Dimensions of Nuclear Power: Examples from Three Mile Island." **American Political Science Review** 75(March 1981):132-142. (44 refs.)

Three Mile Island became a focus of the criticism of nuclear power that had been building for over a decade. Repercussions of the accident revealed social risks and political difficulties of managing this technology. TMI is used here to examine nuclear technology, including costs, physical effects and indirect risks. Implications for social, political and administrative institutions are explored.

*Social sciences; Politics; Institutional response; Nuclear power

0819
"New Jersey Utility Found at Fault in TMI 2 Accident." **Public Utilities Fortnightly** 113(May 24, 1984):58. (News)

*Regulation; New Jersey Board of Public Utilities; Jersey Central Power and Light Co.

0820
"Nuclear Accident Casts a Long Shadow." **Nature** 283(January 17, 1980):235. (News)

*Public opinion

0821
"Nuclear Accident to Throw Out All America's Energy Sums." **Economist** 271(April 7, 1979):47-48. (News)
*Nuclear power; Public policy

0822
"One-Week Protest Vigil Follows TMI Observance." **Nuclear News** 23(May 1980):32. (News)
*Protest; TMI - anniversary (1)

0823
Payne, Jon. "TMI: Toward Politization." **Nuclear News** 22(June 1979):39. (Editorial)
*Politics

0824
Pelham, Ann. "After Three Mile Island: Congress More Skeptical of Nuclear Power's Future." **Congressional Quarterly Weekly Report** 37(November 24, 1979):2663-2665, 2668. (refs.)
The effect of TMI on congressional attitudes to nuclear power is discussed. A variety of legislation is pending, including a six-month moratorium on issuance of new construction permits. The nuclear industry faces an uncertain future, as public opinion has been shaken.
*Congressional reaction; Legislation; Nuclear industry; Public opinion; Politics

0825
"Post-TMI Amendments to CP Regs Published." **Nuclear News** 25(March 1982):50, 53. (News)
*Licensing; Regulation

0826
Purcell, Arthur H. "Resources Policy - the Legacy of Three Mile Island." **Resources Policy** 7(March 1981):2-3.
Purcell comments on TMI's legacy for our energy future and materials policy. He points to the concept of "appropriate technology" as an outgrowth of the accident.
*Resources policy; Energy policy; Technology

0827
"Response: The Mobilization of Industry and Government." **EPRI Journal** 5(June 1980):14-23.
Industry responded with the formation of the Nuclear Safety Analysis Center and the Institute of Nuclear Power Operations. Government also responded rapidly, with the appointment of the Kemeny Commission, investigation by the NRC, and Congressional hearings. A chronicle of events for 1979-1980 outlines industry and government activities. This is the second of four articles.
*Industry response; Government; Nuclear industry; Accident investigations

0828
Rozen, L. "Pa. Business Says Nuclear Scare was Overplayed." **Advertizing Age** 50(April 9, 1979):88. (News)
*Local business; Local residents

0829
Schulman, Mark A. "The Impact of Three Mile Island." **Public Opinion** 2(June/July 1979):7-9.
Following TMI, public opinion on nuclear power has fallen dramatically, with about 47 percent in favor and 45 percent opposed according to an ABC News/Harris survey.
*Public opinion; Nuclear power; Polls

0830
Sills, David L. "A Comment on Dorothy Nelkin's `Some Social and Political Dimensions of Nuclear Power: Examples from Three Mile Island.'" **American Political Science Review** 75(March 1981):143-145. (9 refs.)
Sills' commentary on Nelkin's article, *American Political Science Review* 75(March 1981):132-142, refutes the "assumption that energy is a technical subject to be studied by natural scientists."
*Natural scientists; Social sciences

0831
Sills, David L., moderator. "Institutional Reactions: Panel Discussion." **Annals of the New York Academy of Sciences** 365(1981):263-274.
Participants in this panel discussion on institutional reactions were Robert J. Budnitz, Judith Johnsrud, Paul Leventhal, Allan Mazur, Robert A. Szalay, Emilio E. Varanini III, and C.P. Wolf.
*Institutional response

0832
Sills, David L., moderator. "Public Reactions: Panel Discussion." **Annals of the New York Academy of Sciences** 365(1981):196-202.
Participants in the discussion were Rupert F. Chisholm, Barbara S. Dohrenwend, Cynthia B. Flynn, Robert C. Mitchell, Dorothy Nelkin, and Anne D. Trunk.
*Public opinion

0833
Soderstrom, E. Jonathan; Sorensen, John H.; Copenhaver, Emily D.; and Carnes, Sam A. "Risk Perception in an Interest Group Context: An Examination of the TMI Restart Issue." **Risk Analysis** 4(3, 1984):231-244.
Unavailable.
*Risk; Public interest groups; Restart

0834
Stallings, Robert A. "Evacuation Behavior at Three Mile Island." **International Journal of Mass Emergencies and Disasters** 2(March 1984):11-26. (29 refs.)
Evacuation behavior related to the TMI accident is described and compared with the pattern usually found in natural disasters. It is concluded that "the voluntary evacuation at Three Mile Island did not differ significantly from those taking place in natural disasters." Thus, no special plans or procedures seemed needed "over and above those in place for other kinds of disaster evacuations." The major identifying feature of the accident seemed to stem from inconsistent information supplied from official sources, resulting in voluntary evacuation. Evacuees from TMI conformed to most of the social and psychological characteristics of evacuees from other disasters.
*Evacuation behavior

0835
Starr, Phillip, and Pearman, William A. "Three Mile Island Implications for Government Policy." **Free Inquiry in Creative Sociology** 10(May 1982):35-37, 42.
Unavailable.
*Public policy; Government

0836
Sylves, Richard T. "Carter Nuclear Licensing Reform Versus Three Mile Island." **Publius: The Journal of Federalism** 10(Winter 1980):69-79. (40 refs.)
Public opinion in response to TMI has affected the Carter administration nuclear licensing reform effort. In addition to discussing lead times of plant construction and the Nuclear Siting and Licensing Act, reactions and responses to TMI by Congress, the President, environmental and public interest groups, the nuclear industry and

state government are all overviewed.

*Licensing; Public opinion; Nuclear industry; Carter administration; Politics

0837
Temples, James R. "The Nuclear Regulatory Commission and the Politics of Regulatory Reform: Since Three Mile Island." **Public Administration Review** 42(July/August 1982):355-362. (47 refs.)

The changes in the regulatory policies and organizational structure of the Nuclear Regulatory Commission as a result of Three Mile Island are examined. The reforms made thus far "may have little impact on the quality of the NRC's decision making" or on safety within the nuclear industry.

*Nuclear Regulatory Commission; Regulatory policy; Politics

0838
Temples, James R. "The Politics of Nuclear Power: A Subgovernment in Transition." **Political Sciences Quarterly** 95(Summer 1980):239-260. (59 refs.)

Temples discusses the political effects of Three Mile Island as part of a general analysis of public policy toward the nuclear power industry.

*Public policy; Politics; Nuclear industry

0839
Trunk, Anne D., and Trunk, Edward V. "Three Mile Island: A Resident's Perspective." **Annals of the New York Academy of Sciences** 365(1981):175-185.

Following historical information regarding Metropolitan Edison, the Middletown community, and its power needs, the Trunks review the main events of each day of the accident, from March 28 to April 1, 1979, along with public reaction. Brief information on events following the initial week is given.

*Local residents; Public opinion; Press coverage; Middletown, PA

0840
Varanini, Emilio E., III. "Reassessing the Nuclear Option: Three Mile Island from a State Perspective." **Annals of the New York Academy of Sciences** 365(1981):230-243. (11 refs.)

Implications of TMI for state government include an expansion of "jurisdiction in all areas of energy planning, especially those areas that directly affect the health and safety of citizens." Varanini discusses six issues that summarize state interests resulting from TMI: need for power, supply planning, economics, alternatives, land use policy and plant siting, and emergency planning. Also discussed are state inspection and audits, participation in federal proceedings, and state authority.

*State government; Nuclear industry

0841
"Voters Reject Reopening Undamaged TMI Plant." **Chemical & Engineering News** 60(May 24, 1982):8. (News)

*Referendum; Election; Restart

0842
Walsh, Edward J. "Local Community vs. National Industry: The TMI and Santa Barbara Protests Compared." **International Journal of Mass Emergencies and Disasters** 2(March 1984):147-163. (21 refs.)

Walsh focuses on the relatively successful efforts of TMI residents to block the restart of TMI Unit 1, contrasting the effort with the unsuccessful attempt of activists to prevent resumption of oil drilling at Santa Barbara in 1969. The success of the TMI protest is attributed to the importance of the national antinuclear movement.

*Activism; Local residents; Restart; Protest

0843
Walsh, Edward J. "Resource Mobilization and Citizen Protest in Communities Around Three Mile Island." **Social Problems** 29(October 1981):1-21. (45 refs.)

"The rapid growth and development of grass-roots protest organizations" following TMI included two pre-existing groups, Environmental Coalition on Nuclear Power (ECNP) and Three Mile Island Alert (TMIA), in addition to Central Pennsylvanians United, Susquehanna Valley Alliance (SVA), People Against Nuclear Energy (PANE), Newberry Township anti-TMI group, and Anti-Nuclear Group Representing York (ANGRY). The first year mobilization of the groups is summarized, and the "importance of grievances in precipitating and sustaining protest" are documented. Both conflict and cooperation among the groups is discussed. Data can be used to assess and refine theories on social movements.

*Local residents; Protest groups; Activism; Social mobilization

0844
Walsh, Edward J. "Response." **Bulletin of the Atomic Scientists** 39(August/September 1983):57-58. (Letter)

*Local residents; Protest groups; Restart

0845
Walsh, Edward J. "The Role of Target Vulnerabilities in High-Technology Protest Movements: The Nuclear Establishment at Three Mile Island." **Sociological Forum** 1(Spring 1986):199-218. (26 refs.)

Human and technological weaknesses of the "target" are discussed in "this analysis of citizens' six-year endeavor to prevent the restart of Three Mile Island's Unit 1 reactor after the Unit 2 accident." Accidents such as TMI will precipitate "a new genre of social movements pitting local residents against high technology industries."

*Local residents; Protest groups; Target vulnerabilities; Social movement organizations

0846
Walsh, Edward J. "Three Mile Island: Making the Decisions." **America** 141(September 8, 1979):89-91.

The stakes are high in the public policy question of who should be involved in the decision on the restart of TMI-1. Three individuals who are activists in keeping TMI closed and who live in the five-mile area of TMI are used as examples of how local residents have been affected by the accident.

*Activism; Public policy; Restart; Local residents

0847
Walsh, Edward J. "Three Mile Island: Meltdown of Democracy?" **Bulletin of the Atomic Scientists** 39(March 1983):57-60.

Walsh chronicles the efforts of local citizen protest groups to monitor the cleanup of Unit 2 and prevent the restart of Unit 1. The strong local opposition to the restart of TMI Unit 1 "raises fundamental questions about democratic decision-making in the age of high technology."

*Local residents; Activism; Democratic theory; Restart

0848
Walsh, Edward J. "Three Mile Island: The Battle of Unit 1." **Bulletin of the Atomic Scientists** 41(May 1985):30-31.

Walsh reviews the six-year history of the fight against the restart of TMI Unit 1. The focus is on citizen protest and on "charges against the utility and the NRC."

*Local residents; Protest groups; Restart

0849
Walsh, Edward J., and Warland, Rex H. "Social Movement Involvement in the Wake of a Nuclear Accident: Activists and Free Riders in the TMI Area." **American Sociological Review** 48(December 1983):764-780. (43 refs.)

This paper focuses on the community mobilization processes following the TMI accident. The data provide qualified support for Olson's model of social movement, involving activists and free riders. Activists in the TMI area are compared with free riders on a variety of dimensions, including solidarity, ideology and grievance.

*Social movement organizations; Activism; Free riders

0850
Wenger, Dennis, guest editor. "The Legacy of Three Mile Island: Research and Policy Implications." **International Journal of Mass Emergencies and Disasters** 2(March 1984):1-211.

This special journal issue, published five years after the accident, focuses on research implications, evacuation behavior, psychology, public policy issues, media coverage, and overall societal impact of the accident. Wenger's introduction (pp.5-9) introduces nine articles and a bibliography, which are each identified elsewhere in this bibliography (see: Stallings, Robert; Houts, Peter S. et al.; Shelton, R.E.; Elmore, Richard F.; Mileti, Dennis S. et al.; Nimmo, Dan; Walsh, Edward J.; Hultaker, Orjan; Sylves, Richard T.; and Pope, Terri and Wenger, Dennis).

*Public policy; Research and development; TMI - anniversary (5)

0851
Winfield-Laird, Idee; Hastings, Matthew; and Cawley, Margaret E. "Changes in Uranium Plant Community Leaders' Attitudes Toward Nuclear Power: Before and After TMI." **Energy** 7(May 1982):449- 455. (2 refs.)

Public and private sector officials in nuclear power plant host communities were surveyed to assess their attitudes toward nuclear power following the TMI accident. Results, compared against findings from a similar study prior to TMI, indicate that "community leaders' attitudes remained highly favorable toward the continued use of nuclear power." They still viewed the "plant presence as having a positive impact on their communities."

*Attitude to nuclear power; Public opinion; Public officials; Nuclear industry

0852
Wolf, C.P. "The Accident at Three Mile Island: Social Science Perspectives." **Items** 33(December 1979):56-61.

TMI can be viewed primarily as a social systems failure. Many consultants worked on this project to develop a social sciences perspective of TMI. The perspectives are divided into four groups: organizational behavior, regulation, public participation, and conflict and consensus.

*Social sciences

0853
Wolf, C.P. "The Accident at Three Mile Island: Social Science Perspectives." **Social Impact Assessment** 47/48(November/December 1979):3-8.

Reprinted from *Items* 33(December 1979)56-61.

*Social sciences

0854
Wolf, C.P. "Public Interest Groups." **Annals of the New York Academy of Sciences** 365(1981):244-256. (24 refs.)

TMI aroused public interest in nuclear energy. The activity of public interest groups following TMI is discussed. The accident has underscored "the importance of taking into account those people that oppose nuclear power in making energy decisions."

*Public interest groups; Public opinion

0855
Zebroski, E.L. "The Dynamics of Institutional Changes Following the TMI Accident." **Progress in Nuclear Energy** 10(3, 1982):249-257. (3 refs.)

The author describes institutional changes that have occurred in "structure, objective and means" in the following: utility industry organizations and associations, individual utilities, domestic reactor suppliers, materials supply industry, the Nuclear Regulatory Commission, and other government agencies and legislative bodies.

*Institutional response; Nuclear industry

0856

Zeigler, Donald J.; Brunn, Stanley D.; and Johnson, James H., Jr. "Evacuation from a Nuclear Technological Disaster." **Geographical Review** 71(January 1981):1-16. (25 refs.)

Using a survey questionnaire to a random sample of residents both within and outside a fifteen-mile radius, the authors studied evacuation behavior and decision making as a result of TMI. Among the variables discussed were: geography, zone of evacuation, reasons for evacuating, and destinations of evacuees. A conceptual model of stress-induced evacuation is applied to the TMI population. Data from other evacuation surveys are referenced.

*Evacuation behavior; Stress; Disasters, technological

0857

Zeigler, Donald J., and Johnson, James H., Jr. "Evacuation Behavior in Response to Nuclear Power Plant Accidents." **Professional Geographer** 36(2, 1984):207-215. (41 refs.)

In a telephone survey of residents near the Shoreham, NY reactor, it was found that "the intended behavior of Long Island residents in response to a general emergency at Shoreham is likely to parallel the actual behavior of TMI area residents during the 1979 mishap." Evacuations to nuclear emergencies differ from responses to other disasters. "Spontaneous evacuation and its geographic manifestation, the evacuation shadow phenomenon, seem to place nuclear power plant accidents in a class by themselves." This survey focuses on the Shoreham plant, with TMI as the comparative site.

*Evacuation behavior; Shoreham nuclear plant; Telephone surveys

# 11

# Restart

0858
"Another Step Forward for TMI1." **Nuclear Engineering International** 28(July 1983):6. (News)
*Restart; Atomic Safety and Licensing Appeal Board; Remote vehicles

0859
"ASLAB Clears Unit 1 Tech Issues; Restart Still Iffy." **Nuclear News** 26(July 1983):32. (News)
*Restart; Licensing

0860
"Awaiting Decision on Unit 1 Restart." **Nuclear News** 28(August 1985):45-46. (News)
*Restart; Court cases

0861
"Economic Veto Upheld, but Stress Worries Removed." **Nuclear Engineering International** 28(June 1983):4. (News)
*Restart; Court cases; Health effects - psychological

0862
"GPU Fights for Unit 1 Restart." **Nuclear Engineering International** 29(January 1984):13. (News)
*Restart; Nuclear Regulatory Commission; GPU Nuclear; Management

0863
"High Courts Clear Unit 1 Restart." **Nuclear News** 28(November 1985):51. (News)
*Restart; Court decisions

0864
Krigman, Alan. "TMI Unit 1 and Startup: Should Nice Guys Win Ballgames?" **InTech (Instrumentation Technology)** 32(July 1985):35-36. (News)
*Nuclear Regulatory Commission; Restart

0865
"Lengthy Hearings to Delay Unit 1's Restart." **Nuclear News** 22(August 1979):26-27. (News)
*Restart; Cleanup; Health effects; Research and development

0866
Leshnoff, S.D., and Erdogan, F. "Mechanical Integrity Analysis of TMI-1 OTSG Unplugged Tubes." **Nuclear Engineering and Design** 79(1, 1984):31-37. (5 refs.)
This paper evaluates the "stability of tube cracks under steady-state and anticipated transient conditions" of Once Through Steam Generator unplugged tubes

of TMI-1.
*Once through steam generator tubes; Tube cracks

0867
"Met Ed Seeks Rate Hike to Cover Unit 1 Delay." **Nuclear News** 22(November 1979):38-39. (News)
*Metropolitan Edison; Financing; Restart

0868
"Moving Closer to Unit 1 Restart Vote?" **Nuclear News** 27(September 1984):39-40. (News)
*Restart; Hearings; Licensing

0869
"No Surprise: NRC Delays Unit 1 Restart Vote -- Again." **Nuclear News** 27(October 1984):46-47. (News)
*Restart; Hearings; Licensing

0870
"NRC Hopes to Decide Unit 1 Restart in December." **Nuclear News** 25(November 1982):70. (News)
*Restart decision; Nuclear Regulatory Commission

0871
"NRC Sets Decision Date on TMI 1 Start-Up." **Nuclear Engineering International** 29(June 1984):13. (News)
*Restart decision; Nuclear Regulatory Commission; GPU Nuclear; Management

0872
"NRC Votes to Approve Unit 1 Restart." **Nuclear News** 28(July 1985):25-26, 28. (News)
*Restart decision; Nuclear Regulatory Commission; Restart; Hearings; Politics

0873
"100 Percent on TMI-1 After Final Trip Test." **Nuclear News** 29(February 1986):33-34. (News)
*Restart

0874
"Pa. Seeks Ouster of Restart ASLB Chair." **Nuclear News** 28(February 1985):40. (News)
*Restart; Atomic Safety and Licensing Board; Petitions; Smith, Ivan W.

0875
Payne, Jon. "TMI-1: Ending a 6 1/2-year Wait." **Nuclear News** 28(November 1985):47. (Editorial)
Payne comments on the restart of TMI-1 and its importance to nuclear energy.
*Restart; Nuclear industry

0876
"Plan for TMI Restart Follows Owner Shakeup." **Engineering News-Record** 211(December 15, 1983):15-16. (News)
*Restart; General Public Utilities; Management

0877
"Prehearing Conference on TMI-1 Restart Set." **Nuclear News** 23(March 1980):49-50. (News)
*Restart

0878
"Restart Approved, With Conditions." **Nuclear Engineering International** 27(February 1982):5-6. (News)
*Restart decision; Health effects - psychological; Court cases

0879
"Smith Retains Chair on Restart ASLB." **Nuclear News** 28(May 1985):42. (News)
*Restart; Smith, Ivan W.; Atomic Safety and Licensing Board; Petitions

0880
"Smooth Restart of TMI 1 Shows Value of Preventive Maintenance." **Nuclear Engineering International** 30(December 1985):24-25. (News)
*Restart

0881
"Three Mile Island Restart Appears Uncertain Now." **Public Utilities Fortnightly** 114(November 22, 1984):49-50. (News)
*Restart

0882
"TMI-1 Intervention Deadline Extended; Creitz Resigns." **Nuclear News** 22(October 1979):23. (News)
*Restart; Metropolitan Edison; Management; Creitz, Walter M.

0883
"TMI-1 is Back; PJM is Happy." **Electrical World** 199(November 1985):28. (News)
*Restart

0884
"TMI-1 Restart Approved." **Engineering News-Record** 214(June 6, 1985):13-14. (News)
*Restart; Nuclear Regulatory Commission

0885
"TMI 1 Restart Decision by June?" **Nuclear Engineering International** 29(March 1984):10-11. (News)
*Restart decision; Nuclear Regulatory Commission; Cleanup; Financing; Reactor core analysis; Radiation dosage

0886
"Tube Cracks May Add Year to Unit 1 Outage." **Nuclear News** 25(March 1982):47. (News)
*Restart; Steam tubes

0887
"$25-Million Program Set for Unit 1 Tube Repair." **Nuclear News** 25(May 1982):35. (News)
*Restart; Financing; Steam tubes

0888
"Unit 1 Dangles on Technicalities." **Nuclear News** 27(July 1984):36-38. (News)
*Restart; Hearings; Court cases

0889
"Unit 1 Record Reopened." **Nuclear News** 26(October 1983):34-35. (News)
*Restart; Hearings; Nuclear Regulatory Commission

0890
"Unit 1 Restart Delayed to 1985, NRC Predicts." **Nuclear News** 26(November 1983):59-60. (News)
*Restart decision; Nuclear Regulatory Commission; Hearings; General Public Utilities; Management

0891
"Unit 1 Startup Proceeds to On-line Operation." **Nuclear News** 28(November 1985):51-52. (News)
*Restart

0892
"Unit 1 to 75 Percent Despite s.g. Deposits." **Nuclear News** 29(January 1986):38. (News)
*Restart

0893
"Unit 1 Tube Repair Techniques Endorsed." **Nuclear News** 27(December 1984):34. (News)
*Restart; Steam tubes

0894
"Unit Restart Proceeding Slowed." **Nuclear News** 28(June 1985):57-58. (News)
*Restart; Nuclear Regulatory Commission; GPU Nuclear; Management

0895
"Utility Pushes TMI Restart." **Engineering News-Record** 211(December 8, 1983):16. (News)
*Restart; General Public Utilities; Management

# 12

# Legal Ramifications

0896
Annas, George J. "Nuclear Power: Psychology and Statistics." **American Journal of Public Health** 73(November 1983):1327-1328. (2 refs.)
Court decisions involving Three Mile Island (*Metropolitan Edison Co. v. People Against Nuclear Energy*, and *U.S. Nuclear Regulatory Commission v. People Against Nuclear Energy*) and Maine Yankee (*Maine Yankee Atomic Power Co. v. Maine Nuclear Referendum Committee*) are discussed in terms of the use of statistics in court cases.
*Statistics; Metropolitan Edison Co. v. People Against Nuclear Energy; U.S. Nuclear Regulatory Commission v. People Against Nuclear Energy; Court cases

0897
Ashman, Allan. "Environment ... `Psychological Health'." **American Bar Association Journal** 68(September 1982):1159-1160. (News)
*People Against Nuclear Energy v. U.S. Nuclear Regulatory Commission; Health effects - psychological; National Environmental Policy Act; Restart; Court cases

0898
Bauser, Deborah B. "Through the Looking Glass and Back Again: The *PANE* Case, a Rebuttal." **Harvard Environmental Law Review** 9(1985):211-235. (127 refs.)
Bauser examines "the facts and law at issue in *PANE*, with particular reference to the misconceptions of the case presented in the article by Jordan in *Harvard Environmental Law Review* 8(Winter 1984):55-87. In *Metropolitan Edison Co. v. People Against Nuclear Energy*, "the Supreme Court soundly rejected the proposition that a federal agency is required to assess the psychological health impacts of a federal project." In a five-part analysis of the TMI-related case, the author shows that it is unrealistic to consider psychological impacts under NEPA.
*Metropolitan Edison Co. v. People Against Nuclear Energy; National Environmental Policy Act; People Against Nuclear Energy; Health effects - psychological; Court cases

0899
Blake, E. Michael. "Three Mile Island; GPU Claims NRC Allowed Accident, Files Suit." **Nuclear News** 24(January 1981):50-51. (News)
*General Public Utilities; Nuclear Regulatory Commission; Law suits

0900
Bommarito, Michael J. "*Metropolitan Edison Co. v. People Against Nuclear Energy* and the Cognizability of Psychological Stress Under NEPA: A Stress Test for the United States Supreme Court." **Detroit College of Law Review** 1983(Fall 1983):943-964. (121 refs.)
"This Note discusses the Supreme Court's finding that psychological health effects might not be cognizable under NEPA." Sections include: the background for the case; the relevance of an environmental impact statement; and the cognizability of psychological stress under NEPA, with prior case law, the Court of Appeals' earlier

decision, an analysis of the Supreme Court's decision, and problems in dealing with psychological health under NEPA.

*Metropolitan Edison Co. v. People Against Nuclear Energy; National Environmental Policy Act; Health effects - psychological; Court cases

0901
Brill, Steven. "Three Mile Island: How Davis Polk's Zone Defense Bested Kaye, Scholer." **American Lawyer** 5(April 1983):1, 6, 8-12.

The case of General Public Utilities, defended by Kaye, Scholer, versus Babcock & Wilcox, defended by Davis Polk, is discussed in an extensive news item. In addition to the specifics of the case and its settlement, the attorneys involved and their specific assignments are given.

*General Public Utilities v. Babcock & Wilcox; Davis Polk; Kaye, Scholer; Court cases

0902
Caccavaro, Louis J., Jr. "*People Against Nuclear Energy v. United States Nuclear Regulatory Commission*: Redefining Environmental Policy in the Stressful Aftermath of a Nuclear Accident." **New England Law Review** 18(April 1983):449-486. (315 refs.)

This Comment examines issues related to Three Mile Island as presented in *People Against Nuclear Energy v. United States Nuclear Regulatory Commission*. Included are a review of NEPA, a chronology of the TMI accident, the NRC's role at TMI, a description of the stress following TMI, an overview of the case (*PANE v. NRC*), a consideration of stress under NEPA, and an analysis of the implications of the case.

*People Against Nuclear Energy v. U.S. Nuclear Regulatory Commission; National Environmental Policy Act; Court cases; Stress; Accident sequence

0903
Chrastil, Rockford R. "Damages from Three Mile Island: Balancing Equities." **Trial** 17(February 1981):14-18, 58-59. (58 refs.)

This paper examines the "question of who should bear the loss that would result from the accident at TMI if the reactor were to be permanently retired and decommissioned." It discusses factors considered by state and federal utility commissions, and analyzes reasons why the federal government should support some or all of the losses.

*Costs; Financing; Equity; Rate making

0904
"Class Action Suit Filed in Pa. Against 3 Mile Island Facility." **National Underwriter (Prop Ed)** 83(July 20, 1979):57. (News)

*Class action suits

0905
"Court Blocks Restart." **Nuclear News** 25(February 1982):37. (News)

*Restart; Court rulings; Health effects - psychological

0906
"Court Tries Combining Three Mile Island Suits." **National Underwriter (Prop Ed)** 83(July 27, 1979):2. (News)

*Class action suits

0907
Dickson, David. "Mental Stress Given Environmental Status; Court Delays Start at Three Mile Island." **Nature** 295(January 21, 1982):179. (News)

*Restart; National Environmental Policy Act; People Against Nuclear Energy; Court cases; Stress

0908
Dougherty, J.B. "Weighing Human Impacts Under NEPA: NRC to Study Psychological Fallout of Three Mile Island." **Environmental Law Reporter** 12(July 1982):10065-10069. (40 refs.)

The case of *People Against Nuclear Energy v. Nuclear Regulatory Commission*, centering on evaluating the psychological impact of the TMI accident under the National Environmental Policy Act, is discussed.

*People Against Nuclear Energy v. U.S. Nuclear Regulatory Commission; National Environmental Policy Act; Health effects - psychological; Court cases

0909
Elliott, Jon F. "The Kemeny Report on the Accident at Three Mile Island." **Ecology Law Quarterly** 8(Summer 1980):810-817. (51 refs.)

The findings and recommendations of the Kemeny Report are summarized from a legal perspective, along with an analysis of the "Report's major conceptual and analytical shortcomings."

*President's Commission; Recommendations; Legal aspects

0910
Emmrich, Stuart. "GPU Sues Supplier for Nuclear Accident." **Business Insurance** 14(March 31, 1980):1, 25. (News)

*General Public Utilities v. Babcock & Wilcox; Law suits

0911
"Federal Grand Jury Indicts Met Ed." **Nuclear News** 26(December 1983):31-32. (News)

*Metropolitan Edison Co.; Court cases; GPU Nuclear; Management; Nuclear Regulatory Commission; Restart decision

0912a
"File 3 Mile Island Class-Action Suit." **National Underwriter (Prop Ed)** 83(April 27, 1979):62. (News)

*Class action suits

0712b
"Godzilla Meets Goliath: Babcock & Wilcox Square Off Against TMI Reactor Owner in $Billion Damage Suit." **Not Man Apart** 12(December 1982):14+.

Unavailable.

*Law suits; Babcock & Wilcox v. General Public Utilities

0913
Goodwin, Karen. "Searching for a Bright Line Test: *People Against Nuclear Energy v. Nuclear Regulatory Commission*." **Connecticut Law Review** 15(Winter 1983):303-325. (138 refs.)

"This note examines the *People Against Nuclear Energy* decision in light of the National Environmental Policy Act (NEPA) and the Atomic Energy Act." It "focuses on statutory construction, legislative history, previous NEPA and AEA decisions, and tort law trends." A factual background of the TMI accident is given, along with a procedural history and an examination of the majority and dissenting opinions.

*People Against Nuclear Energy v. U.S. Nuclear Regulatory Commission; National Environmental Policy Act; Court cases

0914
"GPU Asks Supreme Court to Review `Stress' Judgment." **Nuclear News** 25(August 1982):42. (News)

*General Public Utilities; Court cases; People Against Nuclear Energy; Health effects - psychological

0915
"GPU, B&W Settle Suits; Cleanup Stretched Out." **Nuclear News** 26(March 1983):43-44. (News)
*General Public Utilities v. Babcock & Wilcox; Court cases; Cleanup

0916
"GPU Not Obligated to Pay for State, Local Response." **Nuclear News** 25(October 1982):38. (News)
*General Public Utilities; Court cases; Government

0917
"GPU Settles Lawsuit." **Nuclear News** 26(May 1983):36. (News)
*General Public Utilities; Law suits

0918
"GPU Suit Against B&W Goes to Trial in N.Y." **Nuclear News** 25(December 1982):40-41. (News)
*General Public Utilities v. Babcock & Wilcox; Court cases

0919
"GPU Takes Suit Against NRC to Federal Court." **Nuclear News** 25(January 1982):49. (News)
*General Public Utilities v. Nuclear Regulatory Commission; Court cases

0920
Groy, Jeffrey B., and Johnston, Frederick P., Jr. "NEPA and the Development of an `Impact Cognization' Standard: *Metropolitan Edison Co. v. People Against Nuclear Energy*." *Washington University Journal of Urban and Contemporary Law* 26(1984):157-175. (96 refs.)

In *Metropolitan Edison Co. v. People Against Nuclear Energy*, the Supreme Court, reversing a lower court decision, ruled that psychological damage did not have to be considered under NEPA in the restart of unit 1. Earlier court decisions under NEPA are presented and the rationale for this decision is discussed.

*Metropolitan Edison Co. v. People Against Nuclear Energy; National Environmental Policy Act; Court cases; Health effects - psychological

0921
Gutierrez, Erenio, Jr. "Denial of Hearing at Three Mile Island Held to be Error." **Natural Resources Journal** 21(October 1981):911-918. (45 refs.)

In *Sholly v. NRC* the essential issues are the "interpretation of section 189(a) of the Atomic Energy Act and its relation to the NRC decision-making process and to subsequent judicial review." "The court held that the NRC was required to hold a hearing at the plaintiffs' request."

*Sholly v. Nuclear Regulatory Commission; Court cases

0922
Gwynne, Peter. "Three Mile Island's Courtroom Fall-Out is Halted." **New Scientist** 97(February 3, 1983):290. (News)
*General Public Utilities; Court cases

0923
Hanson, Paula Z. "District Court Review of NRC's Three Mile Island Decisions Upheld." **Natural Resources Journal** 21(October 1981):919-924. (19 refs.)

Hanson discusses the Third Circuit Court of Appeals ruling that "the district court has subject matter jurisdiction over actions by citizen group plaintiffs against the Nuclear Regulatory Commission arising out of the nuclear accident at Three Mile Island, even when the plaintiffs have not exhausted all remedies with the Nuclear Regulatory Commission." The cases in reference were *Susquehanna Valley Alliance v.*

*Three Mile Island Nuclear Reactor* and *General Public Utilities Corp. v. Susquehanna Valley Alliance*.

*Susquehanna Valley Alliance v. Three Mile Island Nuclear Reactor; General Public Utilities v. Susquehanna Valley Alliance; Court cases

0924
Hickey, Richard J.; Bowers, Evelyn J.; and Allen, I. Elaine. "TMI Psychological Impact." **Nuclear News** 25(May 1982):28, 31. (Letter)

*Restart; Court rulings; Health effects - psychological

0925
"High Court Hears About `Psychological Stress'." **Nuclear News** 26(April 1983):36, 38. (News)

*Restart; Court rulings; Health effects - psychological; National Environmental Policy Act

0926
Hoff, Ellen Harris. "Psychological Health Damage as an Environmental Effect: *Metropolitan Edison Co. v. People Against Nuclear Energy*." **Arizona Law Review** 26(1984):497-505. (49 refs.)

Hoff discusses *Metropolitan Edison Co. v. People Against Nuclear Energy*, in which the court concluded that psychological health damage was not an environmental effect under NEPA.

*Metropolitan Edison Co. v. People Against Nuclear Energy; National Environmental Policy Act; Court cases; Health effects - psychological; Stress

0927
"Injury Claims Ousted from TMI Action." **National Underwriter (Prop Ed)** 84(July 25, 1980):4, 6. (News)

*Liability insurance; Class action suits

0928
Jordan, William S., III. "Psychological Harm After *PANE*: NEPA's Requirement to Consider Psychological Damage." **Harvard Environmental Law Review** 8(Winter 1984):55-87. (221 refs.)

Jordan examines "the scope of federal responsibility to consider the probable psychological health impacts of a variety of federal projects under NEPA following the *PANE* decision." Part I overviews NEPA's coverage of psychological health concerns in general, and reviews the Three Mile Island litigation in particular; and Part II analyzes the scope of NEPA consideration of psychological effects in different contexts, including "the example of the Nuclear Regulatory Commission's responsibility to consider psychological health effects before licensing new nuclear reactors."

*Metropolitan Edison Co. v. People Against Nuclear Energy; People Against Nuclear Energy; National Environmental Policy Act; Health effects - psychological; Court cases

0929
Kellman, Barry. "Anxiety Over the TMI Accident: An Essay on NEPA's Limits of Inquiry." **George Washington Law Review** 51(January 1983):219-253. (160 refs.)

The accident at TMI is discussed in terms of the psychological stress it caused, and in relation to two court cases involving NEPA's limits of inquiry. In *People Against Nuclear Energy v. U.S. Nuclear Regulatory Commission*, the D.C. District Court held that under NEPA, psychological health must be evaluated as an environmental impact; the Supreme Court reversed that decision in *Metropolitan Edison Co. v. People Against Nuclear Energy*.

*National Environmental Policy Act; Health effects - psychological; Stress; People Against Nuclear Energy v. U.S. Nuclear Regulatory Commission; Metropolitan Edison Co. v. People Against Nuclear Energy; Court cases

0930
Kellman, Barry. "Anxiety Over the TMI Accident: An Essay on NEPA's Limits of Inquiry." **Land Use and Environmental Law Review** 15(1984):655-689. (160 refs.)

This is a reprint of an article from the *George Washington Law Review* 51(January 1983):219-253, dealing with the D.C. District Court of Appeals' decision regarding consideration of psychological effects in the decision on the restart of TMI-1, according to the National Environmental Policy Act.

*National Environmental Policy Act; Health effects - psychological; Stress; People Against Nuclear Energy v. U.S. Nuclear Regulatory Commission; Metropolitan Edison Co. v. People Against Nuclear Energy; Court cases

0931
Knowles, Linda. "$225,000 TMI Settlement Awaits Approval." **Business Insurance** 18(August 20, 1984):78. (News)

*Liability insurance; Class action suits

0932
"Lawsuit Against TMI Supplier Begins." **Public Utilities Fortnightly** 110(November 11, 1982):55. (News)

*General Public Utilities v. Babcock & Wilcox; Court cases

0933
"Legal Actions Explosion Clouds Three Mile Island." **National Underwriter (Prop Ed)** 85(April 10, 1981):42-43. (News)

*Court cases; Class action suits

0934
Lyons, Lois J. "U.S. Nuclear Pools Reward Good Record; Three-Mile Island Accident Claims Hit 14,000." **National Underwriter (Prop Ed)** 86(July 30, 1982):1, 25. (News)

*Liability insurance; Class action suits

0935
McLeod, Douglas. "Courts Deciding Who Will Pay for Three Mile Island." **Business Insurance** 17(January 17, 1983):1, 14-15.

*Cleanup; Financing; Court cases

0936
Mannenbach, Stephen F. "The Decision to Choose Nuclear Power Before and After Three Mile Island." **Environmental Law** 11(Winter 1981):421-434. (57 refs.)

The perception of the safety of nuclear power is discussed in terms of the "immense economic and social costs associated with the accident at TMI." The regulatory and statutory incentives which promoted the development of nuclear energy before TMI are analyzed in this Comment, along with the risks and uncertainties caused by TMI which "should cause utilities to reconsider nonnuclear alternatives."

*Regulation; Nuclear power; Safety; Economic effects; Legal aspects

0937
Marshall, Eliot. "Carving Up TMI's Class Action Fund." **Science** 222(October 14, 1983):142-144.

A $25 million trust fund was awarded to local citizens of TMI in a class action suit. The law firm that sued the utility, David Berger, Attorneys at Law, has been "criticized by local groups for its management of a $5-million research fund," which was part of the settlement. The firm has been slow to develop plans to use the funds.

*Class action suits; David Berger, Attorneys at Law; Local residents; Citizen groups

0938
Marshall, Eliot. "Fear as a Form of Pollution." **Science** 215(January 29, 1982):481. (News)
*People Against Nuclear Energy v. U.S. Nuclear Regulatory Commission; People Against Nuclear Energy; Court cases; Health effects - psychological; Restart

0939
Marshall, Eliot. "NRC Must Weigh Psychic Costs." **Science** 216(June 11, 1982):1203-1204.
The ruling by the U.S. Court of Appeals, that psychological stress must be considered in the restart of TMI-1, is reviewed. The NRC had originally turned down the request by a citizens' group, People Against Nuclear Energy (PANE), but PANE took its case to the courts to get an injunction against restart.
*People Against Nuclear Energy v. U.S. Nuclear Regulatory Commission; Restart; People Against Nuclear Energy; Stress; Court cases

0940
Mulholland, John. "Emergency Payments Top $1 Million." **National Underwriter (Prop Ed)** 83(April 20, 1979):1, 52. (News)
*Liability insurance; Class action suits

0941
Nelson, Mary Elizabeth. "Rejection of Risk Under NEPA: Stress and People Against Nuclear Energy." **American University Law Review** 33(Winter 1984):535-558. (181 refs.)
Covered in this legal note are: the procedural history of PANE, including earlier legal decisions related to TMI; the operation of, and judicial decisions related to, NEPA; and the Supreme Court opinion on *Metropolitan Edison Co. v. People Against Nuclear Energy*.
*Metropolitan Edison Co. v. People Against Nuclear Energy; People Against Nuclear Energy; Court cases; National Environmental Policy Act; Stress

0942
"NRC Rejects Damage Claim; GPU to Take it to Court." **Nuclear News** 24(July 1981):35-36. (News)
*General Public Utilities v. Nuclear Regulatory Commission; Liability; Court cases

0943
"OK Settlement for Three Mile Island." **National Underwriter (Prop Ed)** 85(September 25, 1981):4, 57. (News)
*Class action suits

0944
Osbakken, Mary. "TMI Public Health Fund." **Science** 222(November 25, 1983):872, 874. (Letter)
*TMI Public Health Fund; Class action suits

0945
Overdorff, John S. "The `Extraordinary Nuclear Occurrence' Provision of the Price-Anderson Act: Its Evolution and Application at Three Mile Island." **Trial** 16(November 1980):96-99, 110-111. (72 refs.)
The application of the Price-Anderson Act to the Three Mile Island accident is discussed.
*Price-Anderson Act; Extraordinary nuclear occurrence

0946
"Payout at Three Mile Island." **New Scientist** 105(February 28, 1985):5. (News)
*Compensation; Liability insurance

0947
"*People Against Nuclear Energy (PANE) v. Nuclear Regulatory Commission* -- The Cognizability of Psychological Impacts Under NEPA." **Maryland Law Review** 42(Summer 1983):619-638. (110 refs.)
The D.C. Circuit Court of Appeals held that the NRC must consider the psychological health of residents in the TMI area in its environmental assessment to restart TMI-1. This Note discusses the legal definition of psychological health under NEPA, with extensive references to previous court cases.
*People Against Nuclear Energy v. U.S. Nuclear Regulatory Commission; Court cases; National Environmental Policy Act; Health effects - psychological

0948
"Pro-GPUN Rulings Left in Effect by Courts." **Nuclear News** 29(May 1986):30-31. (News)
*GPU Nuclear; Court cases; Restart

0949
"Psychological Effects of Three Mile Island Must Be Evaluated Under Environmental Policy Act." **Mental Disability Law Reporter** 6(July-August 1982):234-235.
A D.C. Circuit Court of Appeals ruled that psychological stress from the restart of TMI must be evaluated as an environmental condition based on the National Environmental Policy Act (NEPA). Arguments for and against this ruling are discussed.
*National Environmental Policy Act; Court cases; Health effects - psychological

0950
"Psychological Risks are not an Environmental Impact." **Mental Disability Law Reporter** 7(May-June 1983):239-240.
In *Metropolitan Edison Company v. People Against Nuclear Energy*, the Supreme Court reversed its ruling on evaluating psychological damage as an environmental impact of the National Environmental Policy Act.
*Metropolitan Edison v. People Against Nuclear Energy; National Environmental Policy Act; Health effects - psychological; Court cases

0951
"Psychological Stress Litigation Continues." **Nuclear News** 25(October 1982):39. (News)
*U.S. Nuclear Regulatory Commission v. People Against Nuclear Energy; National Environmental Policy Act; Health effects - psychological; Court cases

0952
Rush, Susan B. "D.C. Circuit Broadly Extends NEPA Coverage for Nuclear Power Plant Accidents." **Natural Resources Journal** 23(April 1983):459-466. (27 refs.)
This case note discusses the legal implications of the decision on *People Against Nuclear Energy v. Nuclear Regulatory Commission*, which requires psychological health to be assessed in the restart of TMI-1. However, while this was in press, the Supreme Court reversed its decision in *Metropolitan Edison Co. v. People Against Nuclear Energy*.
*People Against Nuclear Energy v. U.S. Nuclear Regulatory Commission; Health effects - psychological; Court cases; National Environmental Policy Act

0953
Staenberg, Marc R. "Financial and Legal Implications of the Three Mile Island Accident." **Nuclear Law Bulletin** no. 24(December 1979):65-71. (5 refs.)
This paper focuses on the financial theories and mechanisms that deal with liability of nuclear accidents. The Price-Anderson Act is overviewed, the situation at TMI is examined in relation to direct costs incurred, and other actions and costs

arising out of the accident are presented, including regulatory actions, litigation, and congressional activity.
*Liability; Price-Anderson Act; Court cases; Financial impact; Regulation; Congress

0954
Sun, Marjorie. "NRC Puts Limit on Mental Stress Tests." **Science** 217(July 30, 1982):431. (News)
*Health effects - psychological; Court cases; Restart; Stress

0955a
"Supreme Court Allows TMI Cleanup Trial." **Engineering News-Record** 206(January 22, 1981):46, 48. (News)
*Cleanup; Court cases

0955b
"Three-Mile Island Settlement Gets a Preliminary Approval." **National Underwriter (Prop Ed)** 88(September 28, 1984):36. (News)
*Law suits; Local government; Emergency preparedness

0956
"TMI Cost to Area Residents Pegged at $27.1 Million." **National Underwriter (Prop Ed)** 85(September 11, 1981):56. (News)
*Local residents; Law suits; Economic effects; Costs

0957
"TMI Damage Claims Dismissed." **Public Utilities Fortnightly** 110(November 25, 1982):56. (News)
*General Public Utilities; Law suits; Pennsylvania; Costs

0958
"TMI Faces Another Class Action Suit." **National Underwriter (Prop Ed)** 85(March 13, 1981):47. (News)
*Class action suits

0959
"TMI Ruling Challenged by Utility." **Public Utilities Fortnightly** 110(August 5, 1982):56-57. (News)
*Restart; People Against Nuclear Energy; Court cases; Health effects - psychological

0960
"TMI Settlement Sustains Concern for Industry." **Industrial Research & Development** 25(March 1983):53, 58. (News)
*General Public Utilities v. Babcock & Wilcox; Court cases

0961
"TMI Suit Calls B&W Inept." **Engineering News-Record** 204(April 3, 1980):21. (News)
*General Public Utilities v. Babcock & Wilcox; Court cases

0962
"TMI Unit One Startup Delayed by Court Order." **Engineering News-Record** 208(January 14, 1982):11-12. (News)
*Restart; Court cases; Health effects - psychological

0963
"Top Court Rules For, Against Nuclear Industry." **Chemical & Engineering News** 61(May 2, 1983):35. (News)

*People Against Nuclear Energy v. U.S. Nuclear Regulatory Commission; Court cases

0964
"$25 Million Claims in TMI Suit Given Tentative Approval." **National Underwriter (Prop Ed)** 85(February 27, 1981):2, 66. (News)
*Liability Insurance; Class action suits

0965
"Utility Files Suit Against NRC Over TMI." **Public Utilities Fortnightly** 109(January 7, 1982):32. (News)
*General Public Utilities v. Nuclear Regulatory Commission; Law suits

0966
"Venting a Violation." **Nuclear News** 24(January 1981):51. (News)
*People Against Nuclear Energy; Court cases; Venting

0967
Vogler, Robert T. "The Supreme Court as Guardian of the Environment: The Metropolitan Edison Decision in Perspective." **Duquesne Law Review** 22(Winter 1984):479-490. (72 refs.)
The Supreme Court Decision on *Metropolitan Edison Co. v. People Against Nuclear Energy* is discussed in relation to the National Environmental Policy Act. The decision "imposes a standard which limits NEPA consideration to those effects having a reasonably close relationship to the physical environment."
*Metropolitan Edison v. People Against Nuclear Energy; National Environmental Policy Act; People Against Nuclear Energy; Court cases

0968
Von Hippel, Frank. "TMI Public Health Fund." **Science** 222(November 25, 1983):874. (Letter)
*TMI Public Health Fund

0969
Walsh, Edward, and Cable, Sherry. "Litigation and Citizen Protest After the Three Mile Island Accident." **Research in Political Sociology** 2(1986):293-316. (24 refs.)
This study overviews more than six years of litigation and citizen protest related to Three Mile Island and includes a schematic summary of social movement organization (SMO) litigation processes. SMO litigation is discussed in regard to dumping of radioactive water, rate base decisions by PUC, krypton venting, and psychological stress. Secondary effects of SMO litigation are discussed, along with the unit 1 restart case. "A model of critical variables influencing outcomes is presented."
*Legal aspects; Protest groups; Social movement organizations

0970
Widoff, Mark P. "The Accident at Three Mile Island." **Western New England Law Review** 4(Fall 1981):223-238. (116 refs.)
Legal aspects of TMI and the law suits from the accident are discussed in the following categories: financial ramifications on the nuclear power industry; civil liability, including the class action suit, the relationship of state and federal law and individual lawsuits by Class III plaintiffs; federal health and safety regulation, including judicial and administrative developments; and state utility regulation.
*Legal aspects; Liability; Class action suits; Regulation

0971
Woodwell, G.M. "TMI Public Health Fund." **Science** 222(November 25, 1983):872. (Letter)
*TMI Public Health Fund; Class action suits

0972
Ziolko, Lynn Thomas. "Environmental: Three Mile Island in Your Neighborhood?" **Arizona State Law Journal** (Fall 1982):1009-1029. (186 refs.)

Ziolko discusses *People Against Nuclear Energy v. U.S. Nuclear Regulatory Commission*, including the background, legal precedents, factors involved in the decision, and an analysis of the issues. She concludes that the NRC should consider psychological stress in all of its proceedings.

*People Against Nuclear Energy v. U.S. Nuclear Regulatory Commission; Health effects - psychological; National Environmental Policy Act; Stress; Court cases

13

# Federal Documents: President's Commission on the Accident at Three Mile Island

0973
U.S. President's Commission on the Accident at Three Mile Island. **The Need for Change; The Legacy of TMI. The Report of the President's Commission on the Accident at Three Mile Island.** Washington, DC: GPO, October 1979. 201 p. (125 notes) glossary, illus.

The charge to the President's Commission was to conduct a comprehensive study of Three Mile Island including technical assessment of events, analysis of the role of the managing utility, assessment of emergency preparedness/emergency response, evaluation of NRC's licensing, inspection, operations and enforcement, assessment of how the public's right to know was served, and to make appropriate recommendations. In fulfilling this charge, the Commission examined the sequence of events (causes and severity), analyzed radiation releases, examined the role of the utility, examined emergency plans, examined coverage of the news media, and investigated the NRC.

The final document, better known as the Kemeny Report, issued findings on health effects, public health, emergency response, training of operations personnel at TMI. In addition, the Commission wrote recommendations in each area they examined. This report also includes supplemental reports on views of the Commissioners and an account of the TMI accident.

*Accident investigations; Technical assessment; Emergency planning; Emergency preparedness; Licensing; Communications; Right to know; Radioactive releases; Reactor accident analysis; Media; News coverage; Health effects; Operators; Recommendations; Public opinion

0974
U.S. President's Commission on the Accident at Three Mile Island. **Public's Right to Information Task Force. Report of the Public's Right to Information Task Force.** D.N. Rubin and others. Washington, DC: GPO, October 1979. 262 p. bibliography.

The examination of information flow following the accident that is discussed in this report lead to the conclusion that "the public's right to know was not served because conditions under which all parties operated were such that the public's right to know could not be served." The Task Force believes that the public's right to know was ill-served by all institutions involved whether because of lack of planning, failure to convey clear, candid information or just general ignorance.

*Communications; Information flow; Media; News coverage; Right to know

0975
U.S. President's Commission on the Accident at Three Mile Island. **Report of the Office of Chief Counsel on the Nuclear Regulatory Commmission.** S.M. Gorinson and others. Washington, DC: GPO, October 1979. 139 p. (712 notes)

This report looks at the Nuclear Regulatory Commission in an attempt to evaluate the Commmission's activities of licensing, inspection, operation and enforcement as they apply to Three Mile Island. A short history of the NRC precedes the actual body of the report which deals with plant licensing, inspection and enforcement, operator licensing and training and the emergency response.

*Regulation; Licensing; Nuclear Regulatory Commission

0976
U.S. President's Commission on the Accident at Three Mile Island. **Reports of the Office of Chief Counsel on Emergency Preparedness, Emergency Response.** S.M. Gorinson and K.P. Kane and others. Washington, DC: GPO, October 1979. 229 p. (806 notes)

The two reports that comprise this volume examine the state of the emergency preparedness (federal, state, local and utility) in place at the time of the accident at Three Mile Island and the emergency response subsequent to the incident. The Response section is a narrative account of the actions of "individuals and organizations that responded to the accident at Three Mile Island."

*Emergency preparedness

0977
U.S. President's Commission on the Accident at Three Mile Island. Emergency Preparedness and Response Task Force. **Report of the Emergency Preparedness Response Task Force.** R.R. Dynes and others. Washington, DC: GPO, October 1979. 168 p. (147 refs.)

This report examines the emergency plans in effect for local, state, and federal levels at the time of the Three Mile Island nuclear accident and analyzes the responses of each.

*Emergency planning; Emergency preparedness

0978
U.S. President's Commission on the Accident at Three Mile Island. Office of Chief Counsel. **The Role of the Managing Utility and Its Suppliers.** S.M. Gorinson and others. Washington, DC: GPO, October 1979. 231 p. footnoted.

This document represents a review of approximately 100,000 documents and nearly 70 depositions pertinent to the management and function of Metropolitan Edison and its suppliers before the accident at Three Mile Island. Facility design, management within the utility, emergency procedures, emergency management and TMI-2 recovery program are a few of the many topics covered.

*Management; Metropolitan Edison; Accident investigations

0979
U.S. President's Commission on the Accident at Three Mile Island. Public Health and Safety Task Force. **Reports of the Public Health and Safety Task Force.** Washington, DC: GPO, October 1979. 423 p. illus. bibliography.

This report consists of five studies on the following topics: (1) public health and safety summary, (2) health physics and dosimetry, (3) radiation health effects, (4) behavioral effects, and (5) public health and epidemiology. Each is referenced or footnoted.

*Health effects; Health effects - psychological; Safety; Radiation effects

0980
U.S. President's Commission on the Accident at Three Mile Island. Technical Assessment Task Force. **Reports of the Technical Assessment Task Force.** Washington, DC: GPO, October 1979. 4 volumes. illus. bibliography.

This four-volume report of the President's Commission Technical Assessment Task Force discusses various technical aspects of the TMI accident. Volume 1 includes a descriptive sequence of events. Volume 2 reviews chemistry, thermal hydraulics, core damage, the WASH-1400 reactor safety study, and alternative sequences. Volume 3 looks at selection, training, quality and licensing of TMI reactor operating personnel, technical assessment of operating abnormal and emergency procedures, and control room design and performance. Volume 4, the largest of the reports, covers a number of topics from quality assurance to the TMI cleanup program.

*Technical assessment; Thermal hydraulics; Reactor core damage; Reactor accident analysis; Operators; Licensing; Emergency preparedness; Control room design; Cleanup; Decontamination

# 14

# Federal Documents: Nuclear Regulatory Commission

0981
U.S. Nuclear Regulatory Commission. **N.R.C. Action Plan Developed as a Result of the TMI-2 Accident.** TMI Action Plan Steering Group. Washington, DC: GPO, May 1980. 2 volumes. (NUREG-0660)

This action plan was developed to provide a comprehensive and integrated plan for the actions now judged necessary by the NRC to correct or improve the regulation and operation of nuclear facilities based on the experience from the accident at TMI-2 and studies and investigations following the accident. The plan covers all aspects of plant operations including operational safety, siting and design, emergency preparedness and radiation effects, and practices and procedures. Volume 2 consists of a short introduction and eight detailed comparative tables of recommendations.

*Action plan; Safety; Emergency preparedness; Radiation effects; Operators; Nuclear reactor design; Recommendations

0982
U.S. Nuclear Regulatory Commission. **Revision 1 of "NRC Action Developed as a Result of the TMI-2 Accident" (NUREG-0660).** Washington, DC: GPO, August 1980. various numbering. (NUREG-0660-rev.1)

NUREG-0660 in two volumes is a loose-leaf reference work. The replacement pages contained in revision 1 are primarily for Chapter V which deals with NRC policy, organization, and management.

*Action plan

0983
U.S. Nuclear Regulatory Commission. **Three Mile Island Number 2 Dockets (Pre- and Post-Accident, 1964 - July 1979).** Springfield, VA: NTIS, 1979. 12 issues. subscription. (PB80-106958)

Information relative to TMI-2 contained in NRC Dockets from 1964 to July 1979 including licensing and operating materials and engineering drawing is reported to be available by subscription.

*Subscription

0984
U.S. Nuclear Regulatory Commission. **Three Mile Island Number 2, Part 2, Dockets (Post-Accident, July 1, 1979 to October 31, 1979).** Springfield, VA: NTIS, 1979. 1722 microfiche sheets. (PB80-168461)

This document is comprised of NRC material applicable to TMI-2 operation from July 1, 1979 through October 31, 1979.

*Accident investigations

0985
U.S. Nuclear Regulatory Commission. **TMI-2 and Its Impact on the Regulatory Process.** S. Israel. Springfield, VA: NTIS, 1979. 14 p. CSNI Specialist Meeting on

Regulatory Review in the Licensing Process. Madrid, Spain, November 7, 1979. (INIS-mf-5700; CONF-7911101-5)

The NRC investigated the TMI accident with regard to reactor design and operations and recommended improvements in systems and equipment. These changes should be reflected in safer operations.

*Nuclear reactor design; Accident investigations; Regulation

0986
U.S. Nuclear Regulatory Commission. Advisory Committee on Reactor Safeguards. **Comments on the NRC Safety Research Program Budget.** Springfield, VA: NTIS, July 1979. 47 p. (NUREG-0603)

Budgetary considerations of the NRC as a result of the TMI accident and attendant necessary research are presented.

*Safety; Research and development; Financing

0987
U.S. Nuclear Regulatory Commission. Argonne National Laboratory. Illinois. **TMI-2 Accident: Postulated Heat Transfer Mechanisms and Available Data Base (Technical Report).** P. Viskanta and A.K. Mohanty. Springfield, VA: NTIS, April 1981. 119 p. (NUREG/CR-2121; ANL-81-26)

Literature relative to the formulation of a data base to better understand heat transfer and fluid friction activities in loss of coolant accidents like TMI-2 was examined.

*Loss of coolant accident; Literature; Data bases

0988
U.S. Nuclear Regulatory Commission. Battelle Columbus Laboratories. Columbus, Ohio. **An Analysis of the Three Mile Island Accident and Alternative Sequences.** R.O. Wooten; R.S. Denning; and P. Cybulskis. Springfield, VA: NTIS, January 1980. 113 p. bibliography. (NUREG-CR-1219; CONTRACT NRC-04-76-293-08)

Using the MARCH computer code and the accident sequence, core damage was evaluated. Hypothetical core melt data were also studied to determine containment capabilities.

*Computer simulation; Reactor core damage; Decision making, computer assisted; Accident sequence

0989
U.S. Nuclear Regulatory Commission. Brookhaven National Laboratories. Upton, New York. **Analysis of the TMI-2 Source Range Detector Response.** J.F. Carew; D.J. Diamond; and J.M. Eridon. Springfield, VA: NTIS, 1980. 15 p. American Nuclear Sociaty Thermal Reactor Safety Meeting. Knoxville, Tennessee, April 8, 1980. (BNL-NUREG-26762; CONF-800403-8; CONTRACT EV-76-C-02-0016)

The TMI-2 source range detectors showed a wide variation in reading post accident. This report attempts to analyze the responses in an effort to explain the variations.

*Source range detectors; Instrument function

0990
U.S. Nuclear Regulatory Commission. Division of Site Safety and Environmental Analysis. **Non-Radiological Consequences of the Aquatic Biota and Fisheries of the Susquehanna River from the 1979 Accident at the Three Mile Island Nuclear Station. Regulatory Report. March - July 1979.** C.R. Hickey, Jr. and R.B. Samworth. Springfield, VA: NTIS, November 1979. 117 p. illus. (NUREG-0596)

Non-radiologic effects, i.e. discharge of industrial wastes, from TMI-2 into the Susquehanna River during the period March 28 to July 1979 was evaluated for effects on fish and aquatic plants. No adverse effects were detected.

*Industrial waste; Environmental effects; Fish; Susquehanna River

0991
U.S. Nuclear Regulatory Commission. Division of Systems Safety. **Evaluation of Long-Term Post-Accident Core Cooling of Three Mile Island Unit 2. Docket 50-320. Staff Report.** Springfield, VA: NTIS, May 1979. 216 p. (NUREG-0557)

Current data were evaluated to determine probable core damage, and the effect of natural convection cooling considered. NRC staff recommendations are included.

*Reactor core damage; Reactor core temperature; Recommendations

0992
U.S. Nuclear Regulatory Commission. Division of Technical Information and Document Control. **TMI-Related Requirements for New Operating Licenses.** Springfield, VA: NTIS, June 1980. 38 p. (NUREG-0694)

Examination of the requirements for qualifications for a new operating license for a nuclear power facility in light of the TMI accident are reviewed. "This report summarizes the several parts of the list of TMI-related requirements approved by the Commission for new operating licenses."

*Licensing; Regulations

0993
U.S. Nuclear Regulatory Commission. E, G, and G Idaho. Idaho Falls, Idaho. **EPICOR II Resin Degradation Results from First Resin Samples of PF-8 and PF-20. Technical Report.** J.W. McConnell and R.D. Sanders. Springfield, VA: NTIS, July 1985. 55 p. (EGG-2376; NUREG/CR-4150/GAR)

Decontamination of the water on the floor of the Auxiliary and Fuel Handling Buildings at TMI-2 was accomplished by EPICOR II demineralization system. Analyses of the synthetic ion exchange resins in prefilters PF-8 and PF-20 are reported and summarized to determine if degradation of resins occurred with high internal radiation exposure.

*EPICOR; Ion exchange; Filters; Filtration; Reactor auxiliary building; Water pollutants, radioactive; Decontamination

0994
U.S. Nuclear Regulatory Commission. Essex Corporation. Alexandria, Virginia. Special Inquiry Group. **Human Factors Evaluation of Control Room Design and Operator Performance at Three Mile Island-2. Volume 1. Final Report.** T.B. Malone; M. Kirkpatrick; K. Mallory; D. Eike; and J.H. Johnson. Springfield, VA: NTIS, January 1980. 154 p. (NUREG/CR-1270-v-1; CONTRACT NRC-04-79-209)

The first volume of a three-volume report on the degree to which human factors engineering contributed to the TMI accident contains the major portion of the report. It examines the reactor operators performance and how it may have been affected by the control room design, training and emergency response.

*Human factors; Operators; Control room design; Emergency preparedness

0995
U.S. Nuclear Regulatory Commission. Essex Corporation. Alexandria, Virginia. Special Inquiry Group. **Human Factors Evaluation of Control Room Design and Operator Performance at Three Mile Island 2. Volume II. Final Report.** T.B. Malone; M. Kirkpatrick; K. Mallory; D. Eike; and J.H. Johnson. Springfield, VA: NTIS, January 1980. 486 p. (NUREG/CR-1270-v-2; CONTRACT NRC-04-79-209)

The second volume of a three-volume report on the degree to which human factors engineering contributed to the TMI accident contains appendices to the report. They include information on the accident progression, safety factors, operator training, etc.

*Human factors; Safety; Operators; Control room design

0996
U.S. Nuclear Regulatory Commission. Essex Corporation. Alexandria, Virginia. Special Inquiry Group. **Human Factors Evaluation of Control Room Design and Operator Performance at Three Mile Island-2. Volume III. Final Report.** T.B.

Malone; M. Kirkpatrick; K. Mallory; D. Eike; and J.H. Johnson. Springfield, VA: NTIS, January 1980. 166 p. (NUREG/CR-1270-v-3; CONTRACT NRC-04-79-209)

The third volume of a three-volume report on the degree to which human factors engineering contributed to the TMI accident contains appendices to the report. They include control room design, industry standards, human factors, engineering, etc.

*Human factors; Control room design

0997
U.S. Nuclear Regulatory Commission. Human Design Group. Olney, Maryland. **Psychological Stress for Alternatives of Decontamination at the TMI-2 Reactor Building Atmosphere. Technical Report.** A. Baum; R. Gatchel; S. Streufert; C.S. Baum; and R. Fleming. Springfield, VA: NTIS, August 1980. 34 p. (NUREG-CR-1584)

This report reviews the literature on stress and discusses the level of psychological stress that may be generated by the TMI-2 cleanup/decontamination processes. Long-term vs. short-term cleanup activities are weighed against the stress they may cause.

*Health effects - psychological; Stress; Cleanup; Decontamination

0998
U.S. Nuclear Regulatory Commission. Idaho National Engineering Laboratory. Idaho Falls, Idaho. **Water Reactor Safety Programs Sponsored by the Nuclear Regulatory Comission's Division of Reactor Safety Research, April - June 1979. Quarterly Technical Progress Report April - June 1979.** L.J. Ybarrondo; D.J. Calflin; and E.L. Wells. Springfield, VA: NTIS, September 1979. 55 p. (NUREG/CR-0871; TREE-1300; CONTRACT DE-AC07-761D01570)

This report summarizes activities of E, G, and G's various computer simulations of the TMI-2 accident.

*Computer simulation; Safety

0999
U.S. Nuclear Regulatory Commission. Office of Administration. Division of Technical Information and Document Control. **Title List of Publicly Available Documents Pertaining to Three Mile Island Unit 1, Docket 50-289, Cumulated to November 16, 1979.** Springfield, VA: NTIS, December 1979. 241 p. (NUREG-0631)

This report lists the publicly available documents pertaining to TMI filed with the NRC cumulated to November 16, 1979. Topics include safety, security, medical and environmental reports.

*Bibliographies

1000
U.S. Nuclear Regulatory Commission. Office of Administration. Division of Technical Information and Document Control. **Title List of Publicly Available Documents, Three Mile Island Unit 2, Docket 50-320. Cumulated to May 21, 1979. Interim Report.** Springfield, VA: NTIS, May 1979. 163 p. (NUREG-0568)

This title list includes both pre- and post-accident documents relating to the TMI-Unit 2 that are publicly available cumulated to May 21, 1979.

*Bibliographies

1001
U.S. Nuclear Regulatory Commission. Office of Administration. Division of Technical Information and Document Control. **Title List of Publicly Available Documents, Three Mile Island Unit 2, Docket 50-320 Cumulated to June 30, 1979.** Springfield, VA: NTIS, July 1979. 307 p. (NUREG-0568-rev-1; supercedes NUREG-0568)

This title list includes both pre- and post-accident documents relating to the TMI-Unit 2 that are publicly available cumulated to June 30, 1979.

*Bibliographies

1002
U.S. Nuclear Regulatory Commission. Office of Administration. Division of Technical Information and Document Control. **Title List of Publicly Available Documents, Three Mile Island Unit 2, Docket 50-320, July 1, 1979 to October 31, 1979. Revision 1, Supplement 1.** Springfield, VA: NTIS, November 1979. 106 p. (NUREG-0568-rev-1-s-1; supplement to NUREG-0568-rev-1)

This report lists the publicly available documents pertaining to TMI filed with the NRC from July 1, 1979 to October 31, 1979. Documents cover such various topics as safety factors, medical aspects, etc.

*Bibliographies

1003
U.S. Nuclear Regulatory Commission. Office of the Executive Director for Operations. **Report to the Nuclear Regulatory Commission from the Staff Panel on the Commissioner's Determination of an Extraordinary Nuclear Occurrence (ENO). Staff Report.** L.V. Gossick. Springfield, VA: NTIS, January 1980. 255 p. (NUREG-0637)

The NRC commissioned a panel on August 17, 1979, to evaluate the TMI accident to determine whether it constituted an extraordinary nuclear occurrence (ENO). The findings presented in this report do not indicate that the accident was an ENO.

*Extraordinary nuclear occurrence; Research and development

1004
U.S. Nuclear Regulatory Commission. Office of Inspection and Enforcement. **Investigation into the March 28, 1979 Three Mile Island Accident by the Office of Inspection and Enforcement (Investigative Report No. 50-320/79-10).** Springfield, VA: NTIS, July 1979. 788 p. illus. (NUREG-0600)

A report from the Office of Inspection and Enforcement of the NRC on the accident at TMI-2 addresses the actions taken by the utility in operating the reactor and attempts to control the radioactive releases as a result of the emergency.

*Radioactive releases; Accident investigations

1005
U.S. Nuclear Regulatory Commission. Office of Inspection and Enforcement. **Report of the Special Review Group. Office of Inspection and Enforcement on Lessons Learned from Three Mile Island. Regulatory Report.** Springfield, VA: NTIS, December 1979. 212 p. (NUREG-0616)

The Special Review Group, Office of Inspection and Enforcement, made 219 recommendations for change as a result of the lesssons learned from the accident at TMI-2.

*Reactor accidents; Reactor accident analysis; Accident investigations; Recommendations; Lessons learned

1006
U.S. Nuclear Regulatory Commission. Office of Inspection and Enforcement. Division of Emergency Preparedness. **Emergency Preparedness Evaluation for TMI-1: Metropolitan Edison Company, et al. Three Mile Island Nuclear Station, Unit 1, Docket 50-289.** Washington, DC: GPO, December 1980. 31 p. (NUREG-0746)

This report reviewed the TMI Unit 1 Emergency Plan and found it to be adequate and acceptable pending the implementation of five additional planning standards which are detailed in the document.

*Emergency planning; Emergency preparedness

1007
U.S. Nuclear Regulatory Commission. Office of Inspection and Enforcement. Division of Emergency Preparedness. Bolt, Beranek, and Newman, Inc. Pacific Northwest Laboratory. **Evaluation of the Prompt Alerting Systems at Four Nuclear**

**Power Stations.** D.A. Towers; G.S. Anderson; D.N. Keast (Bolt, Beranek, and Newman, Inc.); J.L. Kenoyer; and A.E. Desrosiers (Pacific Northwest Laboratory). Springfield, VA: NTIS, September 1982. 259 p. illus. (NUREG/CR-2655; NRC FIN B2311; PNL-4226)

This report evaluates the prompt notification siren systems at the following four nuclear facilities: Trojan, Three Mile Island, Indian Point and Zion. The objective of the evaluation of these was to predict siren-system effectiveness under specific conditions in the ten-mile emergency planning zone surrounding the nuclear power plants.

*Emergency planning; Emergency preparedness; Nuclear power plants; Siren system

1008
U.S. Nuclear Regulatory Commission. Office of Inspection and Enforcement. Division of Program Development and Appraisal. **Investigation into Information Flow During the Accident at Three Mile Island.** Springfield, VA: NTIS, January 1981. 447 p. (NUREG-0760)

The investigation by the NRC Office of Inspection and Enforcement regarding information exchange following the TMI-2 accident reveals that problems did exist in communications. It was felt that these were largely caused by the lack of a system.

*Communications; Information flow; Reactor accidents; Accident investigations

1009
U.S. Nuclear Regulatory Commission. Office of Management and Program Analysis. **Report to Congress on Abnormal Occurrences. January - March 1979. Volume 2, Number 1. Quarterly Report.** Springfield, VA: NTIS, July 1979. 44 p. (NUREG-0090-v-2-1)

The NRC report to Congress for the period January-March 1979 includes four abnormal occurrences at nuclear power plants throughout the United States. One of the four was the accident at Three Mile Island.

*Abnormal occurrences; Nuclear power plants

1010
U.S. Nuclear Regulatory Commission. Office of Nuclear Material Safety and Safeguards. Division of Waste Management. Brookhaven National Laboratory. Upton, New York. **Evaluation of the Three Mile Island Unit 2 Reactor Building Decontamination Process. Technical Report.** D. Dougherty and J.W. Adams. Springfield, VA: NTIS, August 1983. 66 p. (NUREG/CR-3381; BNL-NUREG-51689; CONTRACT NRC FIN A3162)

The disposal of low grade radioactive waste materials resulting from TMI-2 cleanup is discussed especially with regard to the stability of the radionuclides and leaching agents.

*Waste disposal; Cleanup; Decontamination; Radionuclides

1011
U.S. Nuclear Regulatory Commission. Office of Nuclear Reactor Regulation. **Answers to Frequently Asked Questions about Cleanup Activities at Three Mile Island Unit 2.** Springfield, VA: NTIS, March 1984. 58 p. (NUREG-0732-rev-1)

This document updates the original "Answers..." prepared in 1981 and provides additional information on cleanup at TMI-2.

*Cleanup

1012
U.S. Nuclear Regulatory Commission. Office of Nuclear Reactor Regulation. **Answers to Questions about Updated Estimates of Occupational Radiation Doses at Three Mile Island Unit 2.** Springfield, VA: NTIS, December 1983. 27 p. (NUREG-1060)

Questions about occupational radiation dose estimates in the TMI-2 cleanup are addressed.

*Radiation dosage; Radiation exposure, occupational; Cleanup

1013
U.S. Nuclear Regulatory Commission. Office of Nuclear Reactor Regulation. **Decontamination of the Three Mile Island Unit 2 Reactor Building Atmosphere. Final Environmental Assessment. Final NRC Staff Report.** Springfield, VA: NTIS, May 1980. 102 p. (NUREG-0662-v.1)

The final environmental assessment includes input from government and state agencies, as well as private citizens, concerning decontamination of the TMI-2 Unit. The NRC staff recommendation is included.

*Reactor building; Decontamination; Environmental assessment; Recommendations

1014
U.S. Nuclear Regulatory Commmission. Office of Nuclear Reactor Regulation. **Decontamination of the Three Mile Island Unit 2 Reactor Building Atmosphere (Final Environmental Assessment). Volume 2. Final Staff Report.** Springfield, VA: NTIS, June 1980. 539 p. (NUREG-0660-v-2)

Concerning the TMI-2 cleanup activities, including krypton purging, this report is an accumulation of comments and letters on proposed decontamination plans or alternative suggestions. Comments are from a broad cross-section of the business, scientific and private communities.

*Cleanup; Krypton; Venting; Decontamination; Public opinion; Environmental assessment

1015
U.S. Nuclear Regulatory Commission. Office of Nuclear Reactor Regulation. **GPU (General Public Utilities) Corporation versus B and W (Babcock and Wilcox) Company Lawsuit Review and Its Effect on TMI-1 (Three Mile Island-1) Nuclear Station Unit 1, Docket 50-289, Volume 1.** Springfield, VA: NTIS, September 1983. 112 p. (NUREG-1020-v.1)

A review of the *General Public Utilities versus Babcock and Wilcox* lawsuit is presented with special reference to whether any of the information would preclude restart of TMI-1.

*General Public Utilities v. Babcock & Wilcox; Reactor accidents; Court cases

1016
U.S. Nuclear Regulatory Commission. Office of Nuclear Reactor Regulation. **GPU (General Public Utilities) Corporation Versus B and W (Babcock and Wilcox) Company Lawsuit Review and Its Effect on TMI-1 (Three Mile Island - 1) Nuclear Station Unit 1, Docket 50-289, Volume 2. Appendices.** Springfield, VA: NTIS, September 1983. 837 p. (NUREG-1020-v.2-App)

The details of the *GPU vs. B&W* lawsuit are delineated as they impact on the restart of TMI-1.

*General Public Utilities v. Babcock & Wilcox; Restart; Court cases

1017
U.S. Nuclear Regulatory Commission. Office of Nuclear Reactor Regulation. **NRC Plan for Cleanup Operations at Three Mile Island Unit 2.** R. Lo and B.J. Snyder. Springfield, VA: NTIS, July 1980. 24 p. (NUREG-0698)

The NRC plan reported in this document outlines the TMI-2 cleanup to assure that all the Commission's regulatory responsibilities and objectives will be met. Such areas as overseeing the activities, planning, reviews and coordination are reported.

*Cleanup

1018
U.S. Nuclear Regulatory Commission. Office of Nuclear Reactor Regulation. **NRC Plan for Cleanup Operations at Three Mile Island Unit 2. Revision 1.** R. Lo and B.J. Snyder. Washington, DC: GPO, February 1982. 33 p. 2 appendices. references. (NUREG-0698-rev.1)

This revision reflects policy developments in the area of the NRC's review and approval process in regard to cleanup operations. Also included are the updated cleanup schedule and the progress that has taken place. The NRC's role in the ongoing and future cleanup activities is discussed.

*Cleanup

1019
U.S. Nuclear Regulatory Commmission. Office of Nuclear Reactor Regulation. **NRC Views and Analysis of the Recommendations of the President's Commission on the Accident at Three Mile Island. Technical Report.** Springfield, VA: NTIS, November 1979. 53 p. (NUREG-0632)

This report reviews the NRC's opinions on the recommendations made by the President's Commission on the TMI accident and includes some thoughts of the NRC Commissioners.

*Reactor accidents; President's Commission; Recommendations

1020
U.S. Nuclear Regulatory Commission. Office of Nuclear Reactor Regulation. **Operation of the Submerged Demineralizer System at Three Mile Island Nuclear Station. Unit No. 2, Docket 50-320. Metropolitan Edison Company, Jersey Central Power and Light Company, Pennsylvania Electric Company.** Springfield, VA: NTIS, June 1981. 83 p. bibliography. (NUREG-0796)

As a result of the March 28, 1979 accident at TMI-2, thousands of gallons of contaminated water spilled into the reactor and auxiliary buildings. This safety evaluation report looks at the Submerged Demineralizer System as a means to decontaminate the water by filtering and ion exchange.

*Submerged Demineralizer System; Filtration; Ion exchange; Contaminated water; Reactor building; Reactor auxiliary building

1021
U.S. Nuclear Regulatory Commission. Office of Nuclear Reactor Regulation. **Proceedings of the Workshop on Psychological Stress Associated with the Proposed Restart of the Three Mile Island, Unit 1: Held at McLean, Virginia, February 4-5, 1982.** P. Walker; W.E. Fraize; J.J. Gordon; and R.C.Johnson. Proceedings prepared by the Energy and Resources Division of the MITRE Corporation. Washington, DC: GPO, April 1982. 153 p. bibliography. (NUREG/CP-0026; MTR-82W26)

This document reports the results of a two-day conference, convened by the MITRE Corporation at the request of the NRC, to consider the psychological stress that might be associated with the restart of TMI Unit 1. Eleven experts remarks are included and a seven-page bibliography completes the report.

*Health effects - psychological; Stress; Restart

1022
U.S. Nuclear Regulatory Commission. Office of Nuclear Reactor Regulation. **Programmatic Environmental Impact Statement (Final) Related to Decontamination and Disposal of Radioactive Wastes Resulting from March 28, 1979, Accident at Three Mile Island Nuclear Station Unit 2. Docket 50-320. Environmental Impact Statement (Final). Metropolitan Edison Company, Jersey Central Power and Light Company, Pennsylvania Electric Company.** Springfield, VA: NTIS, March 1981. 590 p. (NUREG-0683-v-1)

Volume 1 of a two-volume report contains material on TMI-2 options from no action to complete cleanup. A review of necessary steps to accomplish cleanup is reported.

*Cleanup; Decontamination; Environmental impact; Radioactive waste; Waste disposal

1023
U.S. Nuclear Regulatory Commission. Office of Nuclear Reactor Regulation. **Programmatic Environmental Impact Statement (Final) Related to Decontamination and Disposal of Radioactive Wastes Resulting from March 28, 1979 Accident at Three Mile Island Nuclear Station Unit 2. Docket 50-320. Appendices A-Z. Environmental Impact Statement (Final). Metropolitan Edison Company, Jersey Central Power and Light Company, Pennsylvania Electric Company.** Springfield, VA: NTIS, March 1981. 665 p. (NUREG-0683-v-2-app)

Volume 2 of a two-volume NRC report on TMI-2 decontamination and cleanup contains the appendices. Options from no action to complete cleanup are considered.

*Cleanup; Decontamination; Environmental impact; Radioactive waste; Waste disposal

1024
U.S. Nuclear Regulatory Commission. Office of Nuclear Reactor Regulation. **Programmatic Environmental Impact Statement Related to Decontamination and Disposal of Radioactive Wastes Resulting from March 28, 1979 Accident, Three Mile Island Nuclear Station, Unit 2. Docket 50-320. Final Supplement Dealing with Occupational Radiation Dose. Final Report.** GPU Nuclear, Inc. Springfield, VA: NTIS, October 1984. 177 p. (NUREG-0683-sup-n-1/GAR)

This supplement was issued when it was anticipated that occupational radiation doses as a result of TMI-2 cleanup activities might exceed the original estimates. The original exposure estimates were projected to be around 2000 to 8000 person-rem (occupational) but were revised upward to 13,000 to 46,000 person-rem.

*Cleanup; Decontamination; Radiation dosage; Radiation exposure, occupational; Radioactive waste; Waste disposal

1025
U.S. Nuclear Regulatory Commission. Office of Nuclear Reactor Regulation. **Safety Evaluation and Environmental Assessment. Docket 50-320, Three Mile Island Nuclear Station, Unit No. 2. Technical Report.** Metropolitan Edison Company, Jersey Central Power and Light Company, Pennsylvania Electric Company. Springfield, VA: NTIS, February 1980. 23 p. (NUREG-0647)

A directive from the NRC, this document details the requirements that TMI-2 be maintained with respect to the Technical Specifications and that the operating license include these technical specifications to ensure safety.

*Licensing; Regulations; Safety; Environmental assessment

1026
U.S. Nuclear Regulatory Commission. Office of Nuclear Reactor Regulation. **Safety Evaluation Report Related to Steam Generator Tube Repair and Return to Operations, Three Mile Island Nuclear Station, Unit No. 1, Docket 50-289.** GPU Nuclear Corporation. H. Silver. Springfield, VA: NTIS, November 1983. 197 p. (NUREG-1019)

Steam generator tube repair processes have been deemed acceptable and thus, do not constitute cause to deny TMI-1 restart.

*Once through steam generator tubes; Restart

1027
U.S. Nuclear Regulatory Commission. Office of Nuclear Reactor Regulation. **Staff Report on the Generic Assessment of Feedwater Transients in Pressurized Water Reactors Designed by Babcock and Wilcox Company. Technical Report.** Springfield, VA: NTIS, May 1979. 401 p. (NUREG-05060; PB-294 874/0GA)

An analysis of the feedwater transient at the TMI-2 power plant which resulted in damage to the reactor core is discussed and the general conclusions are that the reactor design, equipment failure, and operator error contributed. The NRC task group assessments are presented for each area.

*Transients; Reactor core damage; Nuclear reactor design Equipment failure; Human error

1028

U.S. Nuclear Regulatory Commission. Office of Nuclear Reactor Regulation. **Three Mile Island, Unit 2, Radiation Protection Program: Report of the Special Panel.** C.B. Meinhold; T.D. Murphy; D.R. Neely; R.L. Kathren; and B.L. Rich. Springfield, VA: NTIS, December 1979. 50 p. (NUREG-0640)

The radiation protection plan at TMI was reviewed and found to have deficiencies in management and technical aspects. However, management attitudes have demonstrated positive changes and, as cleanup activities progress, the reviewing panel believes the program will be sufficiently strong to support these activities.

*Radiation protection; Cleanup; Management

1029

U.S. Nuclear Regulatory Commission. Office of Nuclear Reactor Regulation. **TMI-1 Restart: An Evaluation of the RHR (Rohrer, Hibler and Repogle, Inc.) BETA (Basic Energy Technology Associates, Inc.) and Draft INPO Reports as They Affect Restart Issues at Three Mile Island Nuclear Station Unit 1, Docket 50-289.** Springfield, VA: NTIS, October 1983. 188 p. (NUREG-0680-sup-4)

An evaluation of the management's integrity as it relates to TMI-1 restart is presented with reference to accusations of data falsification concerning leak rates at TMI-2.

*Restart; Regulations; Management

1030

U.S. Nuclear Regulatory Commission. Office of Nuclear Reactor Regulation. **TMI-1 Restart: An Evaluation of the Licensee's Management Integrity as It Affects the Restart of Three Mile Island Nuclear Station Unit 1, Docket 50-289.** Springfield, VA: NTIS, July 1984. 180 p. (NUREG-0680-sup-5)

Supplement five to the Safety Evaluation Report on TMI-1 Restart indicates that present management at Three Mile Island will be able to safely generate power without undue risk.

*Restart; Regulations; Management

1031

U.S. Nuclear Regulatory Commission. Office of Nuclear Reactor Regulation. **TMI-2 (Three Mile Island) Lessons Learned Task Force Final Report.** Springfield, VA: NTIS, October 1979. 50 p. (NUREG-0585)

This final report of lessons learned from TMI-2 contains suggested areas of reactor design and construction improvement and changes in operations management.

*Nuclear reactor design; Control room design; Operators; Lessons learned

1032

U.S. Nuclear Regulatory Commission. Office of Nuclear Reactor Regulation. **TMI-2 Lessons Learned Task Force: Status Report and Short-Term Recommendations. Technical Report.** Springfield, VA: NTIS, July 1979. 103 p. (NUREG-0578)

The TMI-2 Lessons Learned Task Force, in evaluating the accident, recommended the institution of a number of safety precautions that are applicable to all operating nuclear power plants. Areas of general safety, system design, plant operations and licensing are addressed.

*Accident investigations; Safety; Nuclear reactor design; Operators; Licensing; Lessons learned

1033
U.S. Nuclear Regulatory Commission. Office of Nuclear Reactor Regulation. Division of Engineering. **Impact of the 1979 Accident at Three Mile Island Nuclear Station on Recreational Fishing in the Susquehanna River.** C.R. Hickey, Jr. Washington, DC: GPO, March 1981. 43 p. (27 refs.) (NUREG-0754)

An examination of fish harvests in the Susquehanna River near TMI after the accident indicates that record low levels were reached for the five months following the accident. This was not related to any physical change in the environment or fish but rather seemed attributable to anglers' perception of reduced environmental quality. The effect was temporary and by September 1979 fish harvest had attained normal levels.

*Environmental effects; Fish; Susquehanna River

1034
U.S. Nuclear Regulatory Commission. Office of Nuclear Reactor Regulation. Division of Engineering. Utility Finance Branch. **Potential Impact of Licensee Default on Cleanup of TMI-2.** J.O. Roberts and J. Slatzman. Washington, DC: GPO, November 1980. 145 p. illus. (NUREG-0689)

The financial situation of Metropolitan Edison relative to the cleanup operations necessary at TMI-2 are evaluated. The possiblity of Metropolitan Edison's default is discussed and alternatives are considered especially federal responsibilities. Ten appendices complete this document.

*Cleanup; Metropolitan Edison; Financing; Economic effects

1035
U.S. Nuclear Regulatory Commission. Office of Nuclear Reactor Regulation. Division of Human Factors Safety. **Control Room Design Review Report of TMI-1. Metropolitan Edison Company, et al. Three Mile Island Nuclear Station Unit 1. Docket 50-289.** Springfield, VA: NTIS, December 1980. 41 p. (NUREG-0752)

Modifications made in the TMI-1 control room as a result of observations at TMI-2 have helped to ensure the safe operation of TMI-1.

*Nuclear reactor design; Control room design; Safety

1036
U.S. Nuclear Regulatory Commission. Office of Nuclear Reactor Regulation. Division of Human Factors Safety. **Control Room Design Review Report for TMI-1. Metropolitan Edison Company, et al. Three Mile Island Nuclear Station, Unit No. 1, Docket 50-289.** Washington, DC: GPO, April 1981. 15 p. (9 refs.) (NUREG-0752-sup-n-1)

After correction of deficiencies herein noted, it is reported that TMI Unit 1 could be safely restarted and operated.

*Control room design; Restart

1037
U.S. Nuclear Regulatory Commission. Office of Nuclear Reactor Regulation. Division of Licensing. **Clarification of TMI Action Plan Requirements: Requirements for Emergency Response Capability.** Washington, DC: GPO, January 1983. 28 p. 14 reference documents. (NUREG-0737-sup-n-1)

This document is a letter from the Director of Licensing, Nuclear Reactor Regulation, detailing post-TMI requirements for emergency response capability which have been approved. The report supplies additional clarification of emergency response/emergency operation procedures.

*Action plan; Emergency planning; Emergency preparedness

1038
U.S. Nuclear Regulatory Commission. Office of Nuclear Reactor Regulation. Division of Licensing. **Clarification of TMI Action Plan Requirements. Technical Report.** Springfield, VA: NTIS, November 1980. 290 p. (NUREG-0737)

This NRC document is a letter to operators of nuclear power plants which details post-TMI changes and conditions for licensure.
*Action plan; Licensing; Regulations

1039
U.S. Nuclear Regulatory Commission. Office of Nuclear Reactor Regulation. Division of Licensing. **Comparison of Implementation of Selected TMI Action Plan Requirements on Operating Plants Designed by Babcock and Wilcox.** J.O. Thoma; R. Hernan; N.Prasad Kadambi; G. Vissing; S. Miner; and J. VanVliet. Washington, DC: GPO, May 1984. 186 p. illus. (NUREG-1066)
"This report provides the results of a study conducted by the U.S. Nuclear Regulatory Commission staff to compare the degree to which eight Babcock and Wilcox designed licensed nuclear power plants have complied with the requirements in NUREG-0737, `Clarification of TMI Action Plan Requirements.'"
*Action plan; Nuclear reactor design; Regulations

1040
U.S. Nuclear Regulatory Commission. Office of Nuclear Reactor Regulation. Division of Licensing. **TMI-1 Restart: Evaluation of Licensee's Compliance with Short and Long-Term Items of Section II of the Nuclear Regulatory Commission Order Dated August 9, 1979, Metropolitan Edison Company, et al. Three Mile Island Nuclear Station, Unit 1, Docket 50-289.** Washington, DC: GPO, November 1980. 202 p. 4 appendices. (NUREG-0680-sup-1)
This supplement to NUREG-0680 (TMI restart evaluation) addresses the management-related requirements and indicates that management capabilities and resources are substantially improved and efforts continue. The conclusion is that the licensee is not yet in full compliance with order item six.
*Restart: Regulations

1041
U.S. Nuclear Regulatory Commission. Office of Nuclear Reactor Regulation. Division of Licensing. **TMI-1 Restart: Evaluation of Licensee's Compliance with Short and Long Term Items of Section II of the Nuclear Regulatory Commission Order Dated August 9, 1979, Metropolitan Edison Company, et al. Three Mile Island Nuclear Station, Unit 1. Docket 50-289.** Springfield, VA: NTIS, March 1981. 28 p. (NUREG-0680-sup-2)
This supplement to an earlier NRC report on TMI-1 restart indicates that all changes but one have been implemented as ordered on the August 9, 1979 order.
*Restart; Regulations

1042
U.S. Nuclear Regulatory Commission. Office of Nuclear Reactor Regulation. Division of Licensing. **TMI-1 Restart: Evaluation of Licensee's Compliance with Short and Long Term Items of Section II of the Nuclear Regulatory Commission Order Dated August 9, 1979, Metropolitan Edison Company, et al. Three Mile Island Nuclear Station, Unit 1, Docket 50-289.** Washington, DC: GPO, April 1981. 126 p. (NUREG-0680-sup-3)
Supplement number three evaluates the additional information submitted by the Metropolitan Edison Company since the issuance of NUREG-0680, supplements 1 and 2 for compliance. It was judged that all short-term requirements were met and all but one long-term requirement.
*Restart; Regulations

1043
U.S. Nuclear Regulatory Commission. Office of Nuclear Reactor Regulation. Division of Project Management. **TMI-1 Restart: Evaluation of Licensee's Compliance with Short and Long Term Items of Section II of the Nuclear Regulatory Commission Order Dated August 9, 1979, Metropolitan Edison Company, et al.**

**Three Mile Island Nuclear Station, Unit 1, Docket 50-289.** Springfield, VA: NTIS, June 1980. 253 p. illus. references. (NUREG-0680)

Several items of the NRC order of August 9, 1979, regarding TMI-1 restart remain open and are the subject of additional review. Also noted are the possibility of additional requirements that may be forthcoming as well as necessity of supplying supportive data for completion of the review.

*Restart; Regulations

1044
U.S. Nuclear Regulatory Commission. Office of Nuclear Reactor Regulation. Division of Systems Safety and Division of Operating Reactors. **Staff Report on the Generic Assessment of Feedwater Transients in Pressurized Water Reactors Designed by Babcock and Wilcox Company.** Springfield, VA: NTIS, May 1979. 350 p. (NUREG-0560)

This document reports the assessment and evaluation of feedwater transients (which caused core damage) at TMI-2 in an effort to determine the necessary basis for continued safe operation of other pressurized water reactors designed by the Babcock and Wilcox Company. The TMI-2 problems were related to design error, equipment malfunction and operator error.

*Transients; Reactor core damage; Nuclear reactor design; Pressurized water reactors; Equipment failure; Human error

1045
U.S. Nuclear Regulatory Commission. Office of Nuclear Reactor Regulation. Office of Inspection and Enforcement. **The Public Whole Body Counting Program Following the Three Mile Island Accident. Technical Report. April - September 1979.** R.L. Gotchy and R.J. Bores. Springfield, VA: NTIS, December 1980. 61 p. (NUREG-0636)

In an effort to determine whether radioactive releases from TMI-2 had accumulated in the population surrounding the plant, 753 people were included in the public whole body counting program. No radioactivity attributable to the accident was discovered.

*Radioactive releases; Radiation dosage; Population exposure; Whole body count

1046
U.S. Nuclear Regulatory Commission. Office of Nuclear Reactor Regulation. Three Mile Island Program Office. **Answers to Frequently Asked Questions about Cleanup Activities at Three Mile Island Unit 2. Public Information Report.** B.J. Snyder. Springfield, VA: NTIS, September 1980. 40 p. (NUREG-0732)

This report deals with responses to questions about TMI-2 cleanup activities.

*Cleanup

1047
U.S. Nuclear Regulatory Commission. Office of Nuclear Reactor Regulation. Three Mile Island Program Office. **Answers to Questions about Removing Krypton from the Three Mile Island Unit 2 Reactor Building. Public Information Report.** Springfield, VA: NTIS, May 1980. 24 p. forms. (NUREG-0673)

This report summarizes information relative to krypton venting at TMI-2 and possible alternatives.

*Krypton; Venting; Reactor building

1048
U.S. Nuclear Regulatory Commission. Office of Nuclear Reactor Regulation. Three Mile Island Program Office. **Environmental Assessment of Radiological Effluents from Data Gathering and Maintenance Operation on Three Mile Island Unit 2; Interim Criteria Approved by the Commission on April 7, 1980.** Springfield, VA: NTIS, May 1980. 15 p. forms. (NUREG-0681)

This report discusses the data gathering and maintenance operations at TMI-2 as they apply to the cleanup of the damaged reactor. Cleanup operations that are "environmentally acceptable" will be considered.

*Cleanup; Environmental pollutants; Radioactive releases

1049
U.S. Nuclear Regulatory Commission. Office of Nuclear Reactor Regulation. Three Mile Island Program Office. **NRC Plan for Cleanup Operations at Three Mile Island Unit 2. Revision 2.** M.T. Masnik and B.J. Snyder. Washington, DC: GPO, March 1984. 32 p. (NUREG-0698-rev.2)

This revision provides a discussion of the developments that have occurred that will have an impact on the cleanup operations at TMI. "Because of major uncertainties in the funding of the cleanup, portions of this plan including the estimated schedule are likely to require further changes as availability of funding and other factors affect the pace of cleanup."

*Cleanup; Financing

1050
U.S. Nuclear Regulatory Commission. Office of Nuclear Reactor Regulation and Environmental Protection Agency. Environmental Monitoring Systems Laboratory. **Investigation of Reported Plant and Animal Health Effects in the Three Mile Island Area.** G.E. Gears; G. LaRoche; J. Cable; B. Jaroslow; and D. Smith. Washington, DC: GPO, October, 1980. 40 p. 1 map. (20 refs.) (NUREG-0738; EPA 600/4-80-049)

A study of reported problems with plants and animals in the area of TMI both during operations and after the accident was conducted. None of the problems appeared to be TMI-related and no general pattern was detected.

*Environmental effects; Wildlife; Plants

1051
U.S. Nuclear Regulatory Commission. Office of Nuclear Regulatory Research. **Characterization of TMI-Type Wastes and Solid Products. Quarterly Progress Report April - September 1981.** A.J. Weiss (principle investigator). Springfield, VA: NTIS, December 1981. 47 p. (NUREG/CR-2516; BNL-NUREG-51499-v-1-nos 1,2,3; CONTRACT AC02-76CH00016)

This quarterly progress report examines the type of radiation waste products which are expected as a result of the TMI cleanup operation. Of particular interest are the ion exchange media wastes and their modification due to heavy radiation doses.

*Radioactive waste; Cleanup; Ion exchange

1052
U.S. Nuclear Regulatory Commission. Office of Nuclear Regulatory Research. Division of Health, Siting, and Waste Management. Mountain West Research; Social Impact Research. **Socioeconomic Impacts of Nuclear Generating Stations: Three Mile Island Case Study. Docket Nos. 50-289 and 50-320. Volume 12.** C.B. Flynn. Springfield, VA: NTIS, July 1982. 239 p. (NUREG/CR-2749-v-12)

The purpose of this series of reports was to assess the socioeconomic effects of nuclear power stations on the residents of the surrounding areas. Each volume is the study of an individual power plant. All reports contain ten chapters which discuss the social and economic impact; the TMI report, volume 12, also includes chapters on the effects of the accident (evacuation, economic, health, etc.) long term effects, and identification of issues which will surface in the future.

*Socioeconomic effects; Nuclear power plants

1053
U.S. Nuclear Regulatory Commission. Office of Nuclear Regulatory Research. Division of Reactor Safety Research. Sandia Laboratories. Albuquerque, New Mexico. **Generation of Hydrogen During the First Three Hours of the Three Mile**

**Island Accident.** R.K. Cole, Jr. Springfield, VA: NTIS, October 1979. 40 p. illus. (NUREG/CR-0913; SAND-79-1357)

This report provides an estimate of the amount of hydrogen generated from the accident at TMI.

*Hydrogen; Radiolytic gas

1054
U.S. Nuclear Regulatory Commission. Office of Nuclear Regulatory Research. Division of Safeguards, Fuel Cycle and Environmental Research. Mountain West Research, Inc. Tempe, Arizona. **The Social and Economic Effects of the Accident at Three Mile Island: Findings to Date.** C.B. Flynn and J.A. Chalmers. Springfield, VA: NTIS, January 1980. 114 p. illus. maps. (prepared in cooperation with Social Impact Research, Inc.) (NUREG/CR-1215; CONTRACT NRC-04-78-192)

The social and economic impact of the TMI accident was reviewed for the first six months. Influence was felt in the areas of local business and farming, emergency preparedness, temporary evacuation, and stress.

*Socioeconomic effects; Local business; Emergency preparedness; Evacuation

1055
U.S. Nuclear Regulatory Commission. Office of Nuclear Regulatory Research. Division of Safeguards, Fuel Cycle and Environmental Research. Mountain West Research, Inc. Tempe, Arizona. **Three Mile Island Telephone Survey: Preliminary Report on Procedures and Findings.** C.B. Flynn. Springfield, VA: NTIS, October 1979. 90 p. (NUREG/CR-1093; CONTRACT NRC-04-78-192)

A telephone survey was conducted during July and August 1979 of approximately 1500 households in the TMI area to determine the attitudes toward nuclear power and effects of the accident.

*Public opinion; Attitude to nuclear power; Telephone surveys

1056
U.S. Nuclear Regulatory Commission. Office of Nuclear Regulatory Research. Los Alamos Scientific Laboratory. Los Alamos, New Mexico. **Preliminary Calculations Related to the Accident at Three Mile Island.** J.R. Ireland; P.K. Mast; T.R. Wehner; P.B. Bleiweis; and W.L. Kirchner. Springfield, VA: NTIS, April 1980. 135 p. illus. (NUREG/CR-1353; LA-8273-MS; CONTRACT W-7405-ENG-36)

Using the Transient Reactor Analysis Code (TRAC-P1A) an understanding of the thermal-hydraulic and thermal-mechanical responses were obtained for the first three hours post accident. Additionally, core damage estimates were made.

*Computer simulation; Reactor core damage; Thermal hydraulics

1057
U.S. Nuclear Regulatory Commission. Office of Nuclear Regulatory Research. UNC Nuclear Industries, Inc. Richland, Washington. **Evaluation of Nuclear Facility Decommissioning Projects: Summary Status Report - Three Mile Island Unit 2 Reactor Coolant System and Systems Decontamination. Technical Report. April 5, 1979 - December 19, 1984.** D.H. Doerge; R.L Miller; and K.S. Scotti. Springfield, VA: NTIS, May 1986. 43 p. (NUREG/CR-4315-v-1/GAR)

A summary of decontamination and restoration of TMI-2 coolant system is provided. The data were collected and computerized to allow ready access should a loss of coolant accident occur in another nuclear facility.

*Loss of coolant accident; Decontamination; Computer programs

1058
U.S. Nuclear Regulatory Commission. Office of Nuclear Regulatory Research. UNC Nuclear Industries, Inc. Richland, Washington. **Evaluation of Nuclear Facility Decommissioning Projects: Summary Status Report - Three Mile Island Unit 2 Reactor Building Decontamination. Technical Report. June 1, 1979 - March 29, 1985.** D.H. Doerge; R.L. Miller; and K.S. Scotti. Springfield, VA: NTIS, May 1986. 164 p. (NUREG/CR-4315-v-2/GAR)

A summary of the decontamination activities of the TMI-2 reactor building is provided. The data were collected and computerized to allow ready access should a like accident occur at another nuclear facility.

*Decontamination; Reactor building; Computer programs

1059
U.S. Nuclear Regulatory Commission. Office of Nuclear Regulatory Research. UNC Nuclear Industries, Inc. Richland, Washington. **Evaluation of Nuclear Facility Decommissioning Projects: Summary Status Report - Three Mile Island Unit 2 Reactor Defueling and Disassembly. Technical Report. April 23, 1979 - April 16, 1985.** D.H. Doerge; R.L. Miller; and K.S. Scotti. Springfield, VA: NTIS, May 1986. 83 p. (NUREG/CR-4315-v-3/GAR)

A summary of the defueling and disassembly activities at TMI-2 is provided. The data were collected and computerized to allow ready access should recovery activities be necessary at another nuclear facility.

*Defueling; Computer programs

1060
U.S. Nuclear Regulatory Commmission. Office of Nuclear Regulatory Research. UNC Nuclear Industries, Inc. Richland, Washington. **Evaluation of Nuclear Facility Decommissioning Projects: Summary Status Report - Three Mile Island Unit 2 Radioactive Waste and Laundry Shipments. Technical Report. April 5, 1979 - December 19, 1984.** D.H. Doerge; R.L. Miller; and K.S. Scotti. Springfield, VA: NTIS, May 1986. 40 p. (NUREG/CR-4315-v-9/GAR)

A summary of radioactive waste shipment to disposal sites and protective clothing shipment to decontamination laundry facilities is provided. The data were collected and computerized to allow easy access.

*Radioactive waste; Waste shipment; Laundry, contaminated; Computer programs

1061
U.S. Nuclear Regulatory Commission. Pennsylvania State University. University Park Institute for Research on Land and Water Resources. **Effects of the Accident at Three Mile Island on Residential Property Values and Sales. Final Report. April 1980 - January 1981.** H.B. Gamble and R.H. Downing. Springfield, VA: NTIS, April 1981. 137 p. (NUREG/CR-2063)

The study, conducted by the Pennsylvania State University, examined the effect of the TMI-2 accident on residential property values in the twenty-five mile radius of the nuclear facility and determined that no long term effect resulted. The real estate market seemed to recover in about eight weeks.

*Residential property values; Reactor accidents

1062
U.S. Nuclear Regulatory Commission. Three Mile Island Special Inquiry Group. Rogovin, Stern and Huge. Washington, D.C. **Three Mile Island: A Report to the Commissioners and to the Public. Volume 1.** M. Rogovin; G.T. Frampton, Jr.; E.K. Cornell; R.C. DeYoung; and R. Budnitz. Springfield, VA: NTIS, January 1980. 202 p. (NUREG/CR-1250-v-1)

This report is the NRC's special inquiry into the accident at TMI. It is a technical assessment designed to fulfill the Commission's regulatory responsibilities.

*Nuclear power plants; Regulation; Accident investigations

1063
U.S. Nuclear Regulatory Commission. Three Mile Island Special Inquiry Group. **Three Mile Island Report to the Commissioners and to the Public. Volume 2, Part 1. Final Version.** M. Rogovin and G.T. Frampton. Springfield, VA: NTIS, April 5, 1979. 319 p. (NUREG/CR-1250-v-2-p-1)

Part 1 of three parts of volume 2 of the NRC's Special Inquiry group report to the Commissioners, this section examines the pre-accident licensing and regulation. It includes pre-accident history of TMI-2.

*Nuclear power plants; Licensing; Regulation; Accident investigations

1064
U.S. Nuclear Regulatory Commission. Three Mile Island Special Inquiry Group. **Three Mile Island Report to the Commissioners and to the Public. Volume 2, Part 2. Final Version.** M. Rogovin and G.T. Frampton. Springfield, VA: NTIS, April 5, 1979. 615 p. (NUREG/CR-1250-v-2-p-2)

Volume 2, part 2 of a three-part report discusses the technical aspect of the accident at Three Mile Island.

*Reactor accident analysis; Accident investigations

1065
U.S. Nuclear Regulatory Commission. Three Mile Island Special Inquiry Group. **Three Mile Island Report to the Commissioners and to the Public. Volume 2, Part 3. Final Version.** M. Rogovin and G.T. Frampton. Springfield, VA: NTIS, April 5, 1979. 477 p. (NUREG/CR-1250-v-2-p-3)

Volume 2, part 3, of a three-part report discusses the response to the accident by the utilities and the government agencies.

*Accident investigations; Reactor accidents; Industry response

1066
U.S. Nuclear Regulatory Commission. UNC Nuclear Industries, Inc. Richland, Washington. **Evaluation of Nuclear Facility Decommissioning Projects: Summary Report - Three Mile Island Unit 2 Polar Crane Recovery. Technical Report. May 14, 1981 - December 19, 1983.** D.H. Doerge and R.L. Miller. Springfield, VA: NTIS, August 1984. 63 p. (NUREG/CR-3884)

This report reviews information relative to the restoration to operation of the TMI-2 polar crane including manpower expended and radiation exposure experienced.

*Radiation exposure, occupational; Radiation dosage; Polar crane

# 15

# Federal Documents: United States Department of Energy

1067
U.S. Department of Energy. **Institutional Agreements and Interaction for Research and Disposition of Nuclear Waste from Three Mile Island.** J.K. Reilly; T.W. McIntosh; D.J. McGoff; P.J. Grant; and L.H. Barrett. Springfield, VA: NTIS, 1985. 2 p. Waste Management '85. Tucson, Arizona, March 24, 1985. (CONF-850314-27; EGG-TIO-M-00585; DE85008929/GAR; CONTRACT AC07-761D01570)

Agreements for the distribution of TMI-2 nuclear wastes for the purpose of research and development were made among various co-operating institutions.

*Radioactive waste; Waste handling; Research and development

1068
U.S. Department of Energy. **Radiation Effects on Resins and Zeolites at Three Mile Island Unit II.** J.K. Reilly; P.J. Grant; G.J. Quinn; and K.J. Hofstetter. Springfield, VA: NTIS, 1984. 11 p. International Symposium on Effects of Radiation on Materials. Williamsburg, Virginia, June 18, 1984. (CONF-840604-21; EGG-TIO-M-00484; DE84015478; CONTRACT AC07-761D01570)

The effects of radiation on zeolite and resin used in TMI-2 cleanup are discussed especially with reference to the generation of hydrogen and other gases. It appears that gas production is directly related to the amount of radioactivity and that the use of sorbents in the water decontamination systems lends a high degree of stability even under high radiation conditions.

*Hydrogen; Radiolytic gas; Cleanup; Zeolites; Resins; Contaminated water; Decontamination

1069
U.S. Department of Energy. **Summary of the Three Mile Island Accident: From Zero Hours to Lessons for the Future.** L.F.S. de Oliveira and B.A.C. de Oliveira. Springfield, VA: NTIS, 1982. 45 p. (INIS-mf-7388; DE83780248)

The TMI-2 accident was studied and the features of TMI-2 and Angra-1 (in Brazil) are compared.

*Accident investigations; Reactor accident analysis

1070
U.S. Department of Energy. Argonne National Laboratory. **Neutronic Analysis of the Three Mile Island Unit 2 Ex-Core Detector Response.** D.J. Malloy and Y.I. Chang. Springfield, Va: NTIS, October 1981. 37 p. (ANL-81-75; DE82004040; CONTRACT W-31-109-ENG-38)

Explanation of the anomalous behavior of ex-core neutron detectors at TMI-2 is speculated relative to combinations of reactor coolant voids and influx of fission products into the coolant.

*Fission product release; Coolant; Instrument function; Cooling system

1071
U.S. Department of Energy. Argonne National Laboratory. **Prediction of TMI-2 Core Temperatures from the Fission-Product Release History. Final Report.** J. Rest and C.E. Johnson. Springfield, VA: NTIS, November 1980. 23 p. (EPRI-NSAC-12; DE82004492; CONTRACT W-31-109-ENG-38)

By examining the fission gases and fission products post TMI-2 accident and correlating them to the GRASS-SST computer code, the approximate core temperatures were calculated.

*Computer simulation; Fission product release; Radiolytic gas; Reactor core temperature

1072
U.S. Department of Energy. Atomic Energy Control Board. Ottawa, Ontario. **Three Mile Island -- A Review of the Accident and Its Implications for CANDU Safety.** R.J. Pannell and F.R. Campbell. Springfield, VA: NTIS, March 1980. 81 p. (INFO-0003; DE85780518/GAR)

A collection of data on the TMI-2 accident and a survey of licensees of Canadian nuclear power plants, this report describes the March 28th accident, opinions on cause, and adequacy of safety procedures at Canadian plants.

*Reactor accident analysis; Safety; Accident - causes; Canada

1073
U.S. Department of Energy. Atomic Industrial Forum, Inc. Washington, D.C. **Lessons Learned from Three Mile Island (TMI), USNRC Long-Term Trends, Industry Responses.** R.A. Szalay. Springfield, VA: NTIS, 1981. 9 p. (10 refs.) International Study Meeting on Modern Power Stations. Liege, Belgium, October 26, 1981. (INIS-mf-7348; CONF-8110236-1; DE83780287)

As a result of the TMI-2 accident, a number of safety procedures and instructions, some changes in design, and better operator and managerial training have been implemented.

*Operators; Safety; Lessons learned; Industry response

1074
U.S. Department of Energy. Atomics International Division. Energy Systems Group. Richland, Washington. **Analysis of the Three Mile Island (TMI-2) Hydrogen Burn.** J.O. Henrie and A.K. Postma. Springfield, VA: NTIS, July 1982. 37 p. International Topical Meeting on Nuclear Reactor Thermal Hydraulics. Santa Barbara, California, January 11, 1982. (RHO-RE-SA-8-rev. 1; CONF-830103-21-rev.1; DE82020725; CONTRACT AC06-77RL01030)

A study of the hydrogen burn during the TMI-2 accident is presented and conclusions are enumerated. This burn probably began at the lowest level and spread to all three levels of containment. Temperatures probably reached 760 degrees centigrade (1400 degrees F). Burn damage was primarily in the upper portions of the containment.

*Hydrogen burn; Reactor core damage; Reactor core temperature; Temperature variation; Thermal damage

1075
U.S. Department of Energy. Atomics International Division. Rockwell Hanford Operations. Richland, Washington. **Analysis of the Three Mile Island Unit 2 Hydrogen Burn.** Volume 4. J.O. Henrie and A.K. Postma. Springfield, VA: NTIS, March 1983. 55 p. (GEND-INF-023-v.4; DE83012186; CONTRACT AC07-761D01570)

Records of temperature, pressure and time, tabulated and correlated, provide the basis of analysis of the effect of the hydrogen burn at TMI-2.

*Hydrogen burn; Temperature variation

1076
U.S. Department of Energy. Atomics International Division. Rockwell Hanford Operations. Richland, Washington. **Proposed Methods for Defueling the TMI-2 Reactor Core.** J.O. Henrie. Springfield, VA: NTIS, May 1983. 27 p. (RHO-RE-PL-9P; DE83012923; CONTRACT AC06-77RL01030)

Core removal recommendations for TMI-2 are presented by the Debris Defueling Working Group.

*Defueling; Reactor core; Recommendations

1077
U.S. Department of Energy. Australian Atomic Energy Commission Research Establishment. Lucas Heights. **Significance of the Human Factor in the Safety of Nuclear Reactors: The French Experience and the Lessons of Three Mile Island.** C. Houze and J.M. Oury. Springfield, VA: NTIS, May 1982. 19 p. Translated from *Revue Generale Nuclaire*, September-October 1981) (AAEC-LIB/Trans-762; DE83700479)

The importance of human error in the operation of a nuclear power facility is discussed as it relates to the French nuclear safety policies in the light of lessons learned from TMI-2.

*Operators; Human error; Human factors; Safety; France; Lessons learned

1078
U.S. Department of Energy. Babcock and Wilcox Company. Lynchburg, Virginia. **Equipment for Removal of the TMI-2 Plenum Assembly.** M.W. Ales; J.D. Connell; R.V. DeMars; and D.A. Nitti. Springfield, VA: NTIS, June 1984. 45 p. (GEND-INF-051; DE84014442; CONTRACT AC07-761D01570)

Because of the TMI-2 accident, the internal core (plenum) became misshapen with some fuel assemblies adhering to the containment vessel sides. This report describes the design and preparation of special equipment to free the plenum from its vessel.

*Plenum; Reactor core damage

1079
U.S. Department of Energy. Babcock and Wilcox Company. Lynchburg, Virginia. **Equipment for Removal of TMI-2 Plenum Assembly. Revision 1.** W.H. Abbott; M.W. Ales; S.K. Brown; J.D. Connell; and R.V. DeMars. Springfield, VA: NTIS, April 1985. 72 p. (GEND-INF-051-Rev 1; DE85014607/GAR; CONTRACT AC07-761D01570)

Because of the TMI-2 accident, the internal reactor (plenum) became misshapen with some of the fuel assemblies adhering to the containment vessel sides. This report describes the design and preparation of special equipment to free the plenum for removal from its vessel and represents a revision of a previous report on this subject.

*Plenum; Reactor core damage; Fuel assemblies; Fuel damage

1080
U.S. Department of Energy. Babcock and Wilcox Company. Lynchburg, Virginia. Nuclear Power Generation Division. **Flow-Induced Vibration Analysis of Three Mile Island Unit 2 Once-Through-Steam Generator Tubes. Volume 1. Final Report.** J.R. Johnson; C.E. Harris; E.J. McGuinn; and J.C. Simonis. Springfield, VA: NTIS, June 1981. 85 p. (EPRI-NP-1876-V.1; DE81903813)

Response of the once through steam generator tubes to flow vibration was measured. Analysis indicates that "steady-state tube response increases with increasing reactor power."

*Once through steam generator tubes; Vibration; Equipment function

1081
U.S. Department of Energy. Babcock and Wilcox Company. Lynchburg, Virginia. Nuclear Power Generation Division. **In-Air Vibration Analysis of Three Mile Island**

**Unit-2 Once-Through-Steam Generator Tubes. Volume 1.** J.C. Simonis. Springfield, VA: NTIS, October 1982. 87 p. (EPRI-NP-2692-V.1; DE83900564)

Resonant frequency tests in the Once-through-steam generator tubes (OSTG) are reported for TMI-2.

*Once through steam generator tubes; Vibration; Equipment function

1082
U.S. Department of Energy. Babcock and Wilcox Company. Lynchburg, Virginia. Nuclear Power Generation Division. **In-Vessel Inspection Before Head Removal. TMI II, Phase III (Tooling and Systems Design and Verification).** G.S. Carter; R.F. Ryan; A.W. Pieleck; and H.Q. Bibb. Springfield, VA: NTIS, September 1982. 152 p. (GEND-010-Vol.3; DE82022519; CONTRACT AC07-761D01570)

This document describes the planning and development of equipment to assess core damage at TMI-2 prior to head removal. Several methods for inserting videocameras are reviewed.

*Reactor core damage; Videocamera use; Reactor core analysis

1083
U.S. Department of Energy. Babcock and Wilcox Company. Lynchburg, Virginia. Nuclear Power Generation Division. **Reactor-Core-Physics and Operating Data for Cycles 1 and 2 of Three Mile Island Unit 1 PWR Power Plant. Volume 2, Appendices A and B. Final Report.** Springfield, VA: NTIS, August 1980. 303 p. (EPRI-NP-1410-v.2, app A and B.; DE82900261)

Information on operation of the pressurized water reactor at TMI-1, Cycles 1 and 2 is detailed.

*Pressurized water reactors; Reactor core

1084
U.S. Department of Energy. Babcock and Wilcox Company. Lynchburg, Virginia. Nuclear Power Generation Division. **Static Strain Analysis of TMI-2 OSTG (Once-Through-Steam Generation) Tubes.** C.E. Harris and J.C. Brown. Springfield, VA: NTIS, December 1981. 94 p. (EPRI-NP-2146; DE82901329)

The amount of strain on the Once through steam generator (OSTG) tubes at TMI-2 -- under various conditions -- is presented and results applied to possible damage incurred in the accident.

*Once through steam generator tubes; Reactor core damage

1085
U.S. Department of Energy. Babcock and Wilcox Company. Lynchburg, Virginia. Utility Power Generation Division. **TMI-2 Gadolinia Demonstration Assembly Technical Report.** L.W. Newman; H.M. Jones; T.A. Thornton; W.A. Wittkopf; and M.E. Aldrich. Springfield, VA: NTIS, July 1983. 75 p. (DOE/ET/34212-35; BAW-1703; DE83016095; CONTRACT AC02-78ET34212)

Demonstration of four UO (two)-Gd (two)O (two) fuel rods under test conditions indicated that the assemblies operated well in the TMI-2 reactor.

*Gadolinium; Fuel rods; Fuel assemblies; Equipment function; Experimental data

1086
U.S. Department of Energy. Battelle Columbus Laboratories. Ohio. **Characteristics of EPICOR II Prefilter Liner 3.** N.L. Wynhoff and V. Pasupathi. Springfield, VA: NTIS, April 1983. 65 p. (END-027; DE83012616; CONTRACT AC07-761D01570)

This report discusses the safe handling of highly radioactive contents of the prefilter liners and the liners themselves from the EPICOR II demineralizer system. The specific liner examined for this study was prefilter liner 3 from TMI-2.

*Radioactive waste; Waste handling; EPICOR; Filters; Prefilter liners

1087
U.S. Department of Energy. Battelle Columbus Laboratories. Ohio. **Characterization of the Contamination in the TMI-2 Reactor Coolant System. Final Report.** J.C. Cunnane and S.L. Nicolosi. Springfield, VA: NTIS, November 1982. 82 p. (EPRI-NP-2722; DE83900578)

Assessment of the radioactive contamination on surfaces of systems at TMI-2 is reported in preparation for the decontamination effort.

*Cooling system; Contamination

1088
U.S. Department of Energy. Battelle Columbus Laboratories. Ohio. **Characterization of EPICOR II Prefilter Liner 16.** J.D. Yesso; V. Pasupathi; and L. Lowry. Springfield, VA: NTIS, August 1982. 62 p. (GEND-015; DE82022228; CONTRACT AC07-761D01570)

Examination and analysis of EPICOR II Prefilter Liner 16 is presented as it applies to the TMI-2 cleanup program.

*Prefilter liner; EPICOR; Cleanup; Decontamination

1089
U.S. Department of Energy. Battelle Pacific Northwest Laboratories. Richland, Washington. **Assessment of Chemical Processes for the Post-Accident Decontamination of Reactor Coolant Systems. Final Report.** L.F. Munson; C.J. Card; and J.R. Divine. Springfield, VA: NTIS, February 1983. 259 p. (EPRI-NP-2866; DE83901442)

Fourteen chemical decontamination processes were evaluated for nuclear fuel accidents. Applicability and cost were considered for TMI-2 and specific recommendations made.

*Decontamination; Cooling system; Recommendations

1090
U.S. Department of Energy. Battelle Pacific Northwest Laboratories. Richland, Washington. **Comparison of Occupational Dose Estimates for TMI-2 Recovery Operation.** G.R. Hoenes. Springfield, VA: NTIS, October 1982. 8 p. (PNL-SA-10818; DE83015088; CONTRACT AC06-76RL01830)

The estimated occupational radiation doses made by Bechtel for General Public Utilities and Argonne National Laboratories for the Nuclear Regulatory Commission differ markedly primarily due to time allotted for cleanup of the TMI-2 reactor building. This document analyzes and evaluates the two reports.

*Radiation exposure, occupational; Radiation dosage; Nuclear workers; Decontamination; Cleanup

1091
U.S. Department of Energy. Battelle Pacific Northwest Laboratories. Richland, Washington. **Feasibility of Vitrifying EPICOR II Organic Resins.** J.L. Buelt. Springfield, VA: NTIS, December 1982. 15 p. (GEND-012; DE83002733; CONTRACT AC07-761D01570)

The disposal of EPICOR II resins, which were used to remove radioactive wastes from water at the TMI-Unit 2 Reactor Auxiliary Building by the vitrification process, is presented. Developed by the Pacific Northwest Laboratory under a grant from the Department of Energy, this report claims that the process effectively deals with the resins and produces only minimal wastes.

*Vitrification; Radioactive waste; EPICOR; Water pollutants, radioactive; Decontamination

1092
U.S. Department of Energy. Battelle Pacific Northwest Laboratories. Richland, Washington. **In-Situ Zeolite Drying.** G.H. Bryan; L.L. Burger; and D.E. Knowlton. Springfield, VA: NTIS, January 1984. 10 p. (GEND-INF-048; DE84006122; CONTRACT AC07-761D01570)

A method for drying zeolite liners from the TMI-2 submerged demineralizer system is described and data concerning hydrogen generation during the process are presented. Preparation of dried zeolite for shipment within the existing regulations is noted.

*Hydrogen; Radiolytic gas; Submerged Demineralizer System; Zeolites; Waste shipment

1093
U.S. Department of Energy. Battelle Pacific Northwest Laboratories. Richland, Washington. **Summary of Radioactive Operations for Zeolite Vitrification Demonstration Program.** G.H. Bryan; R.W. Goles; C.A. Knox; and D.H. Siemens. Springfield, VA: NTIS, January 1984. 36 p. (GEND-038; DE84006260; CONTRACT AC07-761D01570)

The conversion of radioactive zeolites from the submerged demineralizer system at TMI-2 into a borosilicate glass product is described. The only significant radioactive gas generated from the process which requires heat was tritium.

*Vitrification; Zeolites; Submerged Demineralizer System; Radiolytic gas; Tritium; Waste handling

1094
U. S. Department of Energy. Battelle Pacific Northwest Laboratories. Richland, Washington. **Three Mile Island Zeolite Vitrification Demonstration Program.** D.H. Siemens; D.E. Knowlton; and M.W. Shupe. Springfield, VA: NTIS, June 1981. 12 p. American Institute of Chemical Engineers National Meeting. Detroit, Michigan, August 16, 1981. (PNL-SA-9642; CONF-810814-5; DE81028010; CONTRACT AC06-76RL01830)

Purification of contaminated water at TMI-2 will be accomplished in part by the Submerged Demineralizer System process. This system was chosen because cesium and strontium present in the water can be absorbed by the zeolite in the liners. The liners can then be immobilized by vitrification, as described in this paper.

*Vitrification; Zeolites; Filters; Submerged Demineralizer System; Contaminated water; Cesium radioisotopes; Strontium radioisotopes; Decontamination

1095
U.S. Department of Energy. Battelle Pacific Northwest Laboratories. Richland, Washington. **TMI-2 (Three Mile Island-2) Leadscrew Debris Pyrophoricity Study.** R.L. Clark; R.P. Allen; and M.W. McCoy. Springfield, VA: NTIS, April 1984. 34 p. (GEND-INF-044; DE84008590; CONTRACT AC07-761D01570)

Debris on leadscrews removed from TMI-2 was examined for combustion potential and, based on analysis, was not found to be pyrophoric.

*Leadscrews; Debris, radioactive; Combustion potential

1096
U.S. Department of Energy. Battelle Pacific Northwest Laboratories. Richland, Washington. **Use of Multi-Element Beta Dosimeters for Measuring Dose Rates in the TMI-2 Containment Building.** R.I. Scherpelz; G.W.R. Endres; and L.A. Rathbun. Springfield, VA: NTIS, July 1983. 109 p. (PNL-4714; DE83015711; CONTRACT AC06-76RL01830)

In view of the recognized inaccuracies of the thermoluminescent dosimeters, the development of aluminum shielded dosimeters is described. This report evaluates the use of the new shielded dosimeters in the TMI-2 containment building.

*Dosimeters; Instrument function

1097
U.S. Department of Energy. Battelle Pacific Northwest Laboratories. Richland, Washington. **Use of Multi-Element Beta Dosimeters for Measuring Dose Rates in the TMI-2 Containment Building.** R.I. Scherpelz; G.W.R. Endres; and L.A. Rathbun.

Springfield, VA: NTIS, September 1983. 108 p. (GEND-033; DE84000519; CONTRACT AC07-761D01570)

In view of the recognized inaccuracies of the thermoluminescent dosimeters, the development of several thickness aluminum shielded dosimeters used in four element and seven element configurations is described. The report evaluates the use of the new shielded dosimeters in the TMI-2 containment building.

*Dosimeters; Instrument function

1098
U.S. Department of Energy. Battelle Pacific Northwest Laboratories. Richland, Washington. **Vitrification Process for the Volume Reduction and Stabilization of Organic Resins.** J.L. Buelt. Springfield, VA: NTIS, October 1982. 28 p. (GEND-023; DE83002730; CONTRACT AC07-761D01570)

Pacific Northwest Laboratories has experimented with the possibility of a one-step vitrifying process for resins similar to those produced by the TMI-2 water decontamination cleanup.

*Vitrification; Experimental data; Resins; Decontamination; Cleanup

1099
U.S. Department of Energy. Battelle Pacific Northwest Laboratories. Richland, Washington. **Zeolite Vitrification Demonstration Program: Characterization of Nonradioactive Demonstration Product.** J.L. Daniel. Springfield, VA: NTIS, September 1982. 43 p. (GEND-025; DE82022513; CONTRACT AC07-761D01570)

Trial of the vitrification program, using non-radioactive products and two different formulas for glass composition, was studied for strength, impact behavior, and temperature vaporization.

*Vitrification; Zeolites; Experimental data; Waste handling

1100
U.S. Department of Energy. Battelle Pacific Northwest Laboratories. Richland, Washington. **Zeolite Vitrification Demonstration Program: Nonradioactive-Process Operations Summary.** G.H. Bryan; C.A. Knox; R.G. Goles; L.J. Ethridge; and D.H. Siemens. Springfield, VA: NTIS, September 1982. 44 p. (GEND-024; DE82021310; CONTRACT AC07-761D01570)

The conversion, immobilization and containment of the radioactive zeolite from the Submerged Demineralizer System used for decontamination of water at TMI-2 into a glass receptacle is being developed. This report describes the trial of the vitrification process using non-radioactive components.

*Vitrification; Zeolites; Submerged Demineralizer System; Waste handling

1101
U.S. Department of Energy. Bechtel National, Inc. Oak Ridge, Tennessee. **Gross Decontamination Experiment Report.** R. Mason; K. Kinney; J. Dettorre; and V. Gilbert. Springfield, VA: NTIS, July 1983. 680 p. (GEND-034; DE83017046; CONTRACT AC07-761D01570)

Low and high pressure water flushing, mechanical scrubbing devices, and strippable materials were all tested and evaluated in the gross decontamination efforts at TMI-2. This document details those findings as well as the training of personnel to accomplish the tasks.

*Decontamination; Nuclear power plants; Nuclear workers

1102
U.S. Department of Energy. Bechtel North American Power Corporation. Gaithersburg, Maryland. **TMI-2 Defueling Tools Engineering Report.** Springfield, VA: NTIS, February 1986. 180 p. (GEND-INF-073; DE86009068/GAR; CONTRACT AC07-761D01570)

The design of the TMI-2 defueling system is presented, and a summary of in-vessel tools to be used to accomplish defueling, appended.

*Defueling

1103
U.S. Department of Energy. Bechtel Northern Corporation. Gaithersburg, Maryland. **Quick Look Inspection: Report on the Insertion of a Camera into the TMI-2 Reactor Vessel Through a Leadscrew Opening.** Springfield, VA: NTIS, March 1983. 169 p. (GEND-030-Vol. 1; DE83009354; CONTRACT AC07-761D01570)

The insertion of a camera through an opening established by removing a leadscrew is described at the TMI-2 reactor. Core examination via this "quick look" method reveals that, although there is a large amount of debris and rubble, the plenum does not appear distorted.

*Videocamera use; Quick look; Photography; Photographs

1104
U.S. Department of Energy. Bechtel Northern Corporation. Gaithersburg, Maryland. **Quick Look Inspection: Results. Volume II.** Springfield, VA: NTIS, April 1983. 63 p. (GEND-030-Vol. 2; DE83012813; CONTRACT AC07-761D01570)

Three videocamera inspections of the TMI-2 reactor performed during July and August of 1982 permitted evaluation of fuel damage and other structures within the reactor. The six plus hours of videotapes were reviewed by the Videotape Review Group and their conclusions form the basis of this report.

*Videocamera use; Reactor core analysis; Quick look; Photography; Photographs

1105
U.S. Department of Energy. Brookhaven National Laboratory. Upton, New York. **Characterization of TMI-Type Wastes and Solid Products.** K.J. Swyler; C.E. Dodge; and R. Dayal. Springfield, VA: NTIS, 1983. 12 p. Annual Participants Information Meeting of the Department of Energy Low Level Waste Management Program. Denver, Colorado, August 30, 1983. (BNL-NUREG-33618; CONF-8308106-6; DE83018291; CONTRACT AC02-76CH00016)

Expected radiolytic effects of high level radionuclides, like TMI wastes, on ion exchange resins are presented and discussed.

*Radioactive waste; Radiolytic gas; Ion exchange; Resins

1106
U.S. Department of Energy. Brookhaven National Laboratory. Upton, New York. **Evaluation of SCC Test Methods for Inconel 600 in Low Temperature Aqueous Solutions.** R.C. Newman; R. Roberge; and R. Bandy. Springfield, VA: NTIS, April 1982. 24 p. American Society for Testing Materials Conference on Environment-Sensitive Fracture: Evaluation and Comparison of Test Methods. Washington, D.C., April 1, 1982. (BNL-NUREG-32798; CONF-8204128-1; DE83011358; CONTRACT AC02-76CH00016)

Leakage in the TMI-1 steam generator tubing in late 1981 was discovered to be caused by stress corrosion cracking (SCC) as a result of the contamination of the coolant by air and sodium thiosulfate.

*Coolant; Cooling system; Contamination

1107
U.S. Department of Energy. Brookhaven National Laboratory. Upton, New York. **Mechanism of TMI-1 Steam-Generator Failures.** R.C. Newman; R. Bandy; and R. Roberge. Springfield, VA: NTIS, 1983. 22 p. Environmental Degradation of Materials in Nuclear Power Systems-Water Reactors. Myrtle Beach, South Carolina, August 22, 1983. (BNL-NUREG-33385; CONF-830845-1; DE83015770; CONTRACT AC02-76CH00016)

Stress corrosion cracking was studied as it relates to steam generator failure at TMI-1. Borated sulfur compound solutions, sodium thiosulfate and sodium tetrathionate effects were reported.

*Equipment function; Steam generator

1108
U.S. Department of Energy. Brookhaven National Laboratory. Upton, New York. **Strippable Coating Used for the TMI-2 Reactor Building Decontamination.** J.W. Adams; D.R. Dougherty; and R.E. Barletta. Springfield, VA: NTIS, 1984. 5 p. Waste Management '84. Tucson, Arizona, March 11, 1984. (BNL-NUREG-34430; CONF-840307-29; DE84010516; CONTRACT AC02-76CH00016)

This report details the experiments including leaching, immersion, and biodegradation conducted on the strippable coating material used for TMI-2 reactor building decontamination.

*Reactor building; Decontamination

1109
U.S. Department of Energy. C & W Associates. Westmoreland, New Hampshire. **Crisis Contained: The Department of Energy at Three Mile Island: A History.** P.L. Cantelon and R.C. Williams. Springfield, VA: NTIS, December 1980. 243 p. (DOE/EV/10278-TI; CONTRACT AC02-79EV10278)

The report of the Department of Energy's work at the TMI-2 accident is related.

*Reactor accidents; Reactor accident analysis; Department of Energy

1110
U.S. Department of Energy. California University. Berkley, California. Lawrence Berkley Laboratory. San Francisco, Department of Radiology. **Effects of the Accident at Three Mile Island on the Mental Health and Behavior Responses of the General Population and the Nuclear Workers.** J.I. Fabrikant. Springfield, VA: NTIS, February 1982. 26 p. American Psychological Association Mid-Winter Convention. Monterey, California, February 25, 1982. (LBL-14846; CONF-820265-1; DE83000494; CONTRACT AC03-76SF00098)

The analysis of the Behavioral Effects Task Force Report to the President's Commission provides the scope of this report which cites mental health and stress as the most serious health effect of the TMI-2 accident. The groups at highest risk were nuclear workers and families with pre-school children.

*Health effects - psychological; Stress; Nuclear workers; Preschool children

1111
U.S. Department of Energy. Canadian Nuclear Association. Toronto, Ontario. **Decontamination of Nuclear Facilities.** Springfield, VA: NTIS, 1982. 633 p. International Conference on Decontamination of Nuclear Facilities. Niagra Falls, Canada, September 19, 1982. (INIS-mf-9524-v 1-2; CONF-820973-vols 1-2; DE85781029/GAR)

Thirty-seven papers on decontamination were presented at the conference; the TMI experience was cited.

*Decontamination

1112
U.S. Department of Energy. Centre d'Etude de l'Energie Nuclaire. Brussels, Belgium. **Case of Three Mile Island.** D. Hennart. Springfield, VA: NTIS, December 1984. 23 p. SFEN (French Nuclear Energy Society) Meeting on Decommissioning of Facilities: Balance and Prospects. Paris, France, December 6, 1984. (FRNC-CONF-218; CONF-841281-3; DE86750331/GAR)

This report discusses dismantling of nuclear reactors and how the work at TMI compares to other dismantling operations.

*Decontamination; Nuclear reactors; Shutdown

1113
U.S. Department of Energy. DuPont de Nemours (E.I.) and Company. Savannah River Laboratory. Aiken, South Carolina. **Effects of High Radiation Doses on Linde Ion Siv IE-95.** N.E. Bibler; R.M. Wallace; and M.A. Ebra. Springfield, VA: NTIS, 1981. 21 p. American Institute of Chemical Engineers National Meeting. Detroit,

Michigan, August 16, 1981. (DP-MS-81-18; CONF-810814-9; DE81028380; CONTRACT AC09-76SR00001)

This report examines the feasibility of using Linde Ion Siv IE-95, a zeolite, in the cleanup of TMI-2 contaminated water. Both cesium amounts capable of being removed and radiolytic gas production were investigated.

*Zeolites; Decontamination; Contaminated water; Cesium radioisotopes; Radiolytic gas

1114
U.S. Department of Energy. DuPont de Nemours (E.I.) and Company. Savannah River Laboratory. Aiken, South Carolina. **Role of Ion Exchange in the Three Mile Island Cleanup.** R.M. Wallace. Springfield, VA: NTIS, 1985. 13 p. International Meeting on Solvent Extraction and Ion Exchange in the Nuclear Fuel Cycle. Harwell, United Kingdom, September 3, 1985. (DP-MS-85-18; CONF-850947-1; DE85008259/GAR; CONTRACT AC09-76SR00001)

This paper summarizes the performance of the EPICOR II system and the Submerged Demineralizer System (SDS) ion exchange systems in the decontamination of the escape water in the TMI-2 reactor buildings.

*Submerged Demineralizer System; EPICOR; Ion exchange; Decontamination; Contaminated water; Water pollutants, radioactive

1115
U.S. Department of Energy. E,G and G Idaho, Inc. Idaho Falls, Idaho. **Acquisition of TMI-2 Core Debris Samples.** D.E. Owen; D.S. Fee; and J.M. Bower. Springfield, VA: NTIS, 1983. 6 p. American Nuclear Society Winter Meeting. San Francisco, California, October 30, 1983. (EGG-TMI-M00583; CONF-831047-111; DE84002893; CONTRACT AC07-761D01570)

The first samples of the TMI-2 core debris were obtained. Equipment used to gather the samples is discussed and plans for analysis of the samples noted.

*Debris, radioactive; Fuel debris

1116
U.S. Department of Energy. E,G and G Idaho, Inc. Idaho Falls, Idaho. **Analysis Data on Samples from the TMI-2 Reactor-Coolant System and Reactor-Coolant Bleed Tank.** R.L. Nitschke. Springfield, VA: NTIS, May 1982. 31 p. (GEND-INF-021; DE82016227; CONTRACT AC07-761D01570)

Samples of liquid obtained from the reactor coolant system and the reactor coolant bleed tanks were analyzed by three laboratories. The results of these analyses are presented.

*Coolant; Cooling system

1117
U.S. Department of Energy. E,G and G Idaho, Inc. Idaho Falls, Falls, Idaho. **Analysis of TMI-2 Samples Using the Molecular Optical Laser Examiner.** T.E. Doyle and J.L. Alvarez. Springfield, VA: NTIS, 1982. 11 p. American Nuclear Society Winter Meeting. Washington, D.C., November 14, 1982. (EGG-TIO-M-1882; CONF-821103-71; DE83005286; CONTRACT AC07-761D01570)

Using a Molecular Optical Laser Examiner (MOLE), TMI-2 samples were analyzed to learn what chemicals result from a nuclear accident, how and what chemicals escaped to the atmosphere, and what really happened with the TMI accident.

*Chemical releases; Atmospheric releases; Reactor accident analysis; Molecular optical laser examiner

1118
U.S. Department of Energy. E, G and G Idaho, Inc. Idaho Falls, Idaho. **Application of the MOLE in Post-Nuclear Accident Characterization.** S.J. Johnson and L.J. Alvarez. Springfield, VA: NTIS, 1981. 12 p. Conference on Analytical Chemistry

and Nuclear Technology, Gatlinburg, Tennessee, October 6, 1981. (EGG-M-08881; CONF-811025-23; DE82005428; CONTRACT AC07-761DO1570)

Application of the molecular optical laser examiner (MOLE) to TMI-2 for the purpose of identifying the chemical composition of contaminants in sold, liquid, and gas states is discussed.

*Chemical releases; Molecular optical laser examiner; Reactor accident analysis

1119
U.S. Department of Energy. E, G and G Idaho, Inc. Idaho Falls, Idaho. **Applications of Iron Enriched Basalt to TMI Radioactive Waste Disposal.** D.E. Owen; J.M. Welch; P.V. Kelsey; and J.E. Flynn. Springfield, VA: NTIS, 1982. 15 p. American Nuclear Society Topical Meeting on Treatment and Handling of Radioactive Wastes. Richland, Washington, April 19, 1982. (EGG-M-08282; CONF-820424-28; DE82017764; CONTRACT AC07-761D01570)

The use of iron-enriched basalt as a medium in which to seal and dispose of TMI-2 wastes is investigated.

*Radioactive waste; Waste disposal; Waste handling; Experimental data

1120
U.S. Department of Energy. E, G, and G Idaho, Inc. Idaho Falls, Idaho. **Assessment of the TMI-2 Axial Power-Shaping-Rod Dynamic Test Results.** R.W. Garner; D.E. Owen; and M.R. Martin. Springfield, VA: NTIS, March 1983. 46 p. (GEND-INF-038; DE83011489; CONTRACT AC07-761D01570)

Preliminary to reactor head removal at TMI-2, all leadscrews in control rod assemblies and axial power shaping rods (APSRs) must be removed. This report describes a test of the APSR insertion performance to determine the condition of both the rods and the core and the possibility of full insertion to facilitate leadscrew removal.

*Leadscrews; Axial power shaping rod; Equipment function; Reactor core damage

1121
U.S. Department of Energy. E, G, and Idaho, Inc. Idaho Falls, Idaho. **Characterization of Radioactive Ion Exchange Media Waste Generated at Three Mile Island.** T.C. Runion; R.E. Holzworth; R.E. Ogle; H.M. Burton; and W.W. Bixby. Springfield, VA: NTIS, October 1981. 10 p. Seminar on the Management of Radioactive Waste from Nuclear Power Plants. Karlsruhe, F.R. Germany, October 5,1981. (EGG-M-08181; CONF-811056-2; DE82007337; CONTRACT AC07-761D01570)

The EPICOR II system chosen for decontamination of the water at the TMI-2 Auxiliary and Fuel Handling Building consists of three basic stages: (1) prefilter (ion exchange) to remove primarily cesium and strontium and (2 and 3) demineralization.

*EPICOR; Ion exchange; Water pollutants, radioactive; Decontamination

1122
U.S. Department of Energy. E, G, and G Idaho, Inc. Idaho Falls, Idaho. **Characterization of the Three Mile Island Unit 2 Reactor Building Atmosphere Prior to the Reactor Building Purge.** J.K. Hartwell; J.W. Mandler; S.W. Duce; and B.G. Motes. Springfield, VA: NTIS, May 1981. 48 p. (GEND-005; CONTRACT AC07-761D01570)

The atmosphere in the TMI-2 reactor building was sampled prior to purge and analyzed by two laboratories at the Idaho National Engineering Laboratory.

*Reactor building; Air pollution, radioactive; Venting

1123
U.S. Department of Energy. E, G, and G Idaho, Inc. Idaho Falls, Idaho. **Citizen Radiation Monitoring Program for the TMI Area.** A.J. Baratta; B.G. Gricar; and W.A. Jester. Springfield, VA: NTIS, July 1981. 419 p. (GEND-008; DE81025487; CONTRACT AC07-761D01570)

A program to train citizens to monitor radiation levels is reported. Actual results of the program are presented.

*Radiation monitoring; Citizen monitoring

1124
U.S. Department of Energy. E, G, and G Idaho, Inc. Idaho Falls, Idaho. **Color Photography of the Three Mile Island Unit 2 Reactor Containment Building. Volume 1. Entries 1,2,4,5,6.** G.R. Eidam and J.T. Horan. Springfield, VA: NTIS, October 1981. 220 p. (GEND-006-Vol. 1; DE82007901; CONTRACT AC07-761D01570; CONTRACT AC07-761D7575)

This document consists of 308 photographs of the inside of the TMI-2 reactor containment building taken during entries made between July 1980 through February 1981.

*Photographs; Reactor containment building; Containment entry

1125
U.S. Department of Energy. E, G, and G Idaho, Inc. Idaho Falls, Idaho. **Computer Code Calculations of the TMI-2 Accident: Initial and Boundary Conditions.** S.R. Behling. Springfield, VA: NTIS, May 1985. 49 p. (EGG-TMI-6859; DE85012613/GAR; CONTRACT AC07-761D01570)

The TMI-2 physical plant, the accident, and conditions during the accident are discussed and calculated by computer codes.

*Computer simulation; Reactor accident analysis; Accident sequence

1126
U.S. Department of Energy. E, G, and G Idaho, Inc. Idaho Falls, Idaho. **Controlled Air Incinerator Conceptual Design Study.** Springfield, VA: NTIS, January 1982. 101 p. (GEND-021; DE82010039/GAR; CONTRACT AC07-761D01570)

The disposal of low level combustable wastes from TMI-2 by incineration through a controlled air incineration facility is proposed and discussed.

*Waste, low level; Waste disposal

1127
U.S. Department of Energy. E, G, and G Idaho, Inc. Idaho Falls, Idaho. **Data Integrity Review of Three Mile Island Unit 2. Hydrogen Burn Data. Volume 3.** J.K. Jacoby; R.A. Nelson; C.L. Nalexny; and R.H. Averill. Springfield, VA: NTIS, September 1983. 69 p. (GEND-INF-023-v. 3; DE84004115; CONTRACT AC07-761D01570)

The material obtained from several different switches, transmitters and detectors concerning the hydrogen burn at the TMI-2 reactor after the accident is examined in order to assess the extent and nature of the burn inside the building.

*Hydrogen burn; Instrument function

1128
U.S. Department of Energy. E, G, and G Idaho, Inc. Idaho Falls, Idaho. **Design and Operation of the Core Topography Data Acquisition System for TMI-2.** L.S. Beller and H.L. Brown. Springfield, VA: NTIS, May 1984. 103 p. (GEND-INF-012; DE84012366; CONTRACT AC07-761D01570)

As a part of the TMI-2 accident recovery process, ultrasonographic techniques were used to determine the configuration of the reactor core.

*Reactor core analysis; Ultrasonic technique

1129
U.S. Department of Energy. E, G, and G Idaho, Inc. Idaho Falls, Idaho. **Development of a Prototype Gas Sampler for EPICOR II Prefilter Liners.** J.M. Bower. Springfield, VA: NTIS, September 1982. 30 p. (GEND-INF-025; DE83000758; CONTRACT AC07-761D01570)

This report describes the development of a gas sampler system to be used with the EPICOR II prefilter liners at TMI-2 to safely deal with combustible gases produced by cleanup operations.

*Combustion potential; Radiolytic gas; EPICOR; Prefilter liners

1130
U.S. Department of Energy. E, G, and G Idaho, Inc. Idaho Falls, Idaho. **Development of High-Integrity Container for Intermediate-Depth Burial of Special Waste.** R.L. Chapman; A.L. Ayers, Jr.; and R.T. Haelsig. Springfield, VA: NTIS, 1982. 6 p. International Conference on Radioactive Waste Management. Winnipeg, Manitoba, Canada, September 12, 1982. (EGG-M-0618; CONF-820933-5; DE83000593; CONTRACT AC07-761D01570)

The need for development of high integrity containers to confine the radioactive EPICOR II prefilter liners for later shipment and waste disposal is discussed. Development of the containers is detailed.

*Cannister, high integrity; EPICOR; Prefilter liners; Waste shipment; Waste disposal

1131
U.S. Department of Energy. E, G, and G Idaho, Inc. Idaho Falls, Idaho. **Disposal Demonstration of a High Integrity Container (HIC) Containing an EPICOR II Prefilter from Three Mile Island.** J.W. McConnell; M.J. Tyacke; R.C. Schmitt; and H.W. Reno. Springfield, VA: NTIS, February 1985. 101 p. (GEND-045; DE85008325/GAR; CONTRACT AC07-761D01570)

This report describes the development and testing by E,G and G Idaho, Inc. of a high integrity container for the disposal of low-level (EPICOR II prefilters) waste from TMI-2.

*Cannister, high integrity; EPICOR; Prefilter liners; Waste disposal

1132
U.S. Department of Energy. E, G, and G Idaho, Inc. Idaho Falls, Idaho. **Dose Reduction and Contamination Control in the TMI-2 Reactor Building.** G.R. Eidam and D.W. Leigh. Springfield, VA: NTIS, 1982. 4 p. American Nuclear Society Winter Meeting, Washington, D.C., November 14, 1982. (EGG-TIO-M-00882; CONF-821103-64; DE83005541; CONTRACT AC07-761D01570)

The first three-week gross decontamination of TMI-2 by high pressure water was completed on March 24, 1982. Several other decontamination techniques were tested.

*Decontamination; Reactor building

1133
U.S. Department of Energy. E, G, and G Idaho, Inc. Idaho Falls, Idaho. **EPICOR and Waste Research and Disposition Program: FY-1984 Annual Report.** J.W. McConnell; A.L. Ayers; H.W. Reno; and R.C. Schmitt. Springfield, VA: NTIS, June 1985. 46 p. (EGG-2353; DE85014515/GAR; CONTRACT AC07-761D01570)

This document details the progress in 1984 toward achieving the safe removal and treatment of nuclear wastes (EPICOR II prefilters) from TMI-2 to eventual commerical disposal facilities.

*Radioactive waste; Waste shipment; Waste handling; Waste disposal; EPICOR; Prefilter liners; Research and development

1134
U.S. Department of Energy. E, G, and G Idaho, Inc. Idaho Falls, Idaho. **EPICOR II Research and Disposition Program: FY-1982 Annual Report (Idaho National Engineering Laboratory).** H.W. Reno and R.L. Dodge. Springfield, VA: NTIS, March 1983. 32 p. (EGG-2250; DE83010979; CONTRACT AC07-761D01570)

Safe disposal of the EPICOR II liners from the TMI-2 cleanup is important since the liners contain radioactive wastes. This report summarizes the progress made in 1982 by the Idaho National Engineering Laboratory toward safe storage and disposal.

*Radioactive waste; Waste handling; Waste disposal; EPICOR; Prefilter liners; Research and development

1135
U.S. Department of Energy. E, G, and G Idaho, Inc. Idaho Falls, Idaho. **EPICOR II Research and Disposition Program: FY-1983 Annual Report.** H.W. Reno; J.W. McConnell, Jr.; and R.C. Schmitt. Springfield, VA: NTIS, April 1984. 35 p. (EGG-2287; DE84011793; CONTRACT AC07-761D01570)

This document details the progress in 1983 toward removal and interim storage of nuclear wastes (EPICOR II prefilters) from TMI-2 to commercial disposal facilities. Development and testing of high integrity containers and negotiations with the State of Washington for use of these cannisters are among the items discussed.

*Cannisters, high integrity; Waste handling; Waste disposal; EPICOR; Prefilter liners

1136
U.S. Department of Energy. E, G, and G Idaho, Inc. Idaho Falls, Idaho. **EPICOR II Resin Degradation Results from First Resin Samples of PF-8 and PF-20.** J.W. McConnell and R.D. Sanders. Springfield, VA: NTIS, December 1985. 49 p. (GEND-053; DE86003799/GAR; CONTRACT AC07-761D01570)

Decontamination of the approximately 560,000 gallons of water which spilled into the Auxiliary and Fuel Handling Buildings at TMI-2 was accomplished using an EPICOR II demineralizing system. A study of the resins in the first samples from EPICOR II prefilters (PF-8 and PF-20) is reported.

*Submerged Demineralizer System; EPICOR; Contaminated water; Cleanup; Decontamination; Reactor auxilliary building

1137
U.S. Department of Energy. E, G, and G Idaho, Inc. Idaho Falls, Idaho. **Estimated Source Terms for Radionuclides and Suspended Particulates During TMI-2 Defueling Operations Report on Phase I.** P.G. Voilleque. Springfield, VA: NTIS, January 1982. 26 p. (GEND-INF-019; DE82010611; CONTRACT AC07-761D01570)

The determination of radionuclides as well as particulate matter in suspension in the TMI-2 reactor coolant affects the defueling operations and the filter systems necessary to create desirable operating conditions. This report is an accumulation of data on particulate suspensions in the reactor coolant.

*Coolant; Contamination; Defueling; Filtration

1138
U.S. Department of Energy. E, G, and G Idaho, Inc. Idaho Falls, Idaho. **Estimated Temperatures of Organic Materials in the TMI-2 Reactor During Hydrogen Burn.** H.W. Schutz and P.K. Nagata. Springfield, VA: NTIS, December 1982. 46 p. (GEND-INF-023-v.2; DE83005355; CONTRACT AC07-761D01570)

Assessment of damage to the TMI-2 Reactor Building during the hydrogen burn is speculated from photographs (depicting melting) and chemical analysis of structural materials.

*Hydrogen burn; Photographs; Reactor building; Fuel melting

1139
U.S. Department of Energy. E, G, and G Idaho, Inc. Idaho Falls, Idaho. **Evaluation of Past Nuclear Events and CORRAL-2 Analyses of Fission-Product Transport at TMI-2 and PRTR (Plutonium Recycle Test Reactor).** D.E. Kudera; K. Vinjamuri; and D.W. Croucher. Springfield, VA: NTIS, 1983. 7 p. International Meeting on Light-Water Reactor Severe Accident Evaluation. Cambridge, Massachusetts, August 28, 1983. (EGG-M-16583; CONF-830816-43; DE84000889; CONTRACT AC07-761D10570)

Using the CORRAL-2 computer code, the TMI-2 accident and past plutonium recycle test reactor accidents are reviewed and calculated to assess fission product behavior codes.

*Computer programs; Computer simulation; Fission product release; Decision making, computer assisted

1140
U.S. Department of Energy. E, G, and G Idaho, Inc. Idaho Falls, Idaho. **Evaluation of TMI-2 Pressure Switches NM-PS-1454 and NM-PS-4174.** Springfield, VA: NTIS, November 1983. 45 p. (GEND-INF-041; DE84004468; CONTRACT AC07-761D01570)

Evaluation of two pressure switches, NM-PS-1454 and NM-PS-4174, in the TMI-2 Reactor Building nitrogen manifold system is discussed relative to survival through the accident and cleanup. Except for corrosion in one switch, probably due to moisture from decontamination exercises, both were relatively intact. One switch survived and functioned while the other, damaged in handling, could not be properly evaluated.

*Pressure switches; Instrument function

1141
U.S. Department of Energy. E, G, and G Idaho, Inc. Idaho Falls, Idaho. **Examination of H8 and B8 Leadscrews from Three Mile Island Unit 2 (TMI-2).** K. Vinjamuri; D.W. Akers; and R.R. Hobbins. Springfield, VA: NTIS, September 1985. 297 p. (GEND-INF-052; DE86002327/GAR; CONTRACT AC07-761D01570)

Examination of leadscrews from the control rod drive at TMI-2 indicates significant temperature differences between the top and bottom of the plenum assembly.

*Leadscrews; Temperature variation; Plenum; Reactor core

1142
U.S. Department of Energy. E, G, and G Idaho, Inc. Idaho Falls, Idaho. **Facility Decontamination Technology Workshop. Volume II. Discussion Groups.** Springfield, VA: NTIS, January 1982. 198 p. Facility Decontamination Technology Workshop. Hershey, Pennsylvania, November 27, 1979. (GEND-002-v.2; CONF-7911104-vol. 2; DE82010114; CONTRACT AC07-761D01570)

The various possible decontamination techniques applicable to TMI-2 are reported in this document.

*Decontamination; Cleanup

1143
U.S. Department of Energy. E, G, and G Idaho, Inc. Idaho Falls, Idaho. **Field Measurements and Interpretation of TMI-2 Instrumentation: CF-1-PT3.** J.E. Jones; J.T. Smith; and M.V. Mathis. Springfield, VA: NTIS, November 1981. 42 p. (GEND-INF-017-v.1; DE82003586; CONTRACT AC07-761D01570)

An analysis of function of Core Flood Tank 1B pressure monitor CF-1-PT4 is presented. Its output was compared with unit CF-1-PT3 and found to be similar. Therefore, it is believed that unit CF-1-PT3 is working properly.

*Core flood tank monitors; Instrument function

1144
U.S. Department of Energy. E, G, and G Idaho, Inc. Idaho Falls, Idaho. **Field Measurements and Interpretation of TMI-2 Instrumentation: CF-1-PT4.** J.E. Jones; J.T. Smith; and M.V. Mathis. Springfield, VA: NTIS, November 1981. 42 p. (GEND-INF-017-v.2; DE82003587; CONTRACT AC07-761D01570)

An analysis of the function of Core Flood Tank 1B pressure monitor CF-1-PT3 is presented. Its output was compared with unit CF-1-PT4 and found to be similar. Therefore, it is believed that the unit is working properly.

*Core flood tank monitors; Instrument function

1145
U.S. Department of Energy. E, G, and G Idaho, Inc. Idaho Falls, Idaho. **Field Measurements and Interpretation of TMI-2 Instrumentation: CF-2-LT2.** J.E. Jones; J.T. Smith; and M.V. Mathis. Springfield, VA: NTIS, January 1982. 44 p. (GEND-INF-017-v.5; DE82010492; CONTRACT AC07-761D01570)

Evaluation of function of core flood tank 1A level monitor CF-2-LT2 is presented and evidence indicates that the monitor is working.

*Core flood tank monitors; Instrument function

1146
U.S. Department of Energy. E, G, and G Idaho, Inc. Idaho Falls, Idaho. **Field Measurements and Interpretation of TMI-2 Instrumentation: CF-2-LT4.** J.E. Jones; J.T. Smith; and M.V. Mathis. Springfield, VA: NTIS, January 1982. 44 p. (GEND-INF- 017-v.4; DE82010035; CONTRACT AC07-761D01570)

Evaluation of the function of core flood tank 1B level monitor CF-2-LT4 is presented. Evidence indicates that the monitor continues to work.

*Core flood tank monitors; Instrument function

1147
U.S. Department of Energy. E, G, and G Idaho, Inc. Idaho Falls, Idaho. **Field Measurements and Interpretation of TMI-2 Instrumentation: HP-R-211.** J.E. Jones; J.T. Smith; and M.V. Mathis. Springfield, VA: NTIS, January 1982. 179 p. (GEND-INF-017-v.3; DE82011046; CONTRACT AC07-761D01570)

The components of radiation monitor HP-R-211 are reported, its approximate location in the TMI-2 reactor given, and the speculative reasons for its failure to function are discussed.

*Radiation monitors; Instrument function; Instrument failure

1148
U.S. Department of Energy. E, G, and G, Idaho, Inc. Idaho Falls, Idaho. **Field Measurements and Interpretation of TMI-2 Instrumentation: HP-R-212.** J.E. Jones; J.T. Smith; and M.V. Mathis. Springfield, VA: NTIS January 1982. 63 p. (GEND-INF-017-v.8; DE82010679; CONTRACT AC07-761D01570)

Area radiation monitor HP-R-212 inside the containment at TMI-2 apparently failed to function properly and will be replaced as early as possible so that radiation levels inside the containment can be accurately monitored.

*Radiation monitors; Instrument function; Instrument failure

1149
U.S. Department of Energy. E, G, and G Idaho, Inc. Idaho Falls, Idaho. **Field Measurements and Interpretation of TMI-2 Instrumentation: HP-R-213.** J.E. Jones; J.T. Smith; and M.V. Mathis. Springfield, VA: NTIS, January 1982. 64 p. (GEND-INF-017-v.9; DE82010678; CONTRACT AC07-761D01570)

Area radiation monitor, HP-R-213, located inside the containment at TMI-2 apparently failed and is scheduled to be replaced as early as possible in order to monitor the radiation inside the containment.

*Radiation monitors; Instrument failure

1150
U.S. Department of Energy. E, G, and G Idaho, Inc. Idaho Falls, Idaho. **Field Measurements and Interpretation of TMI-2 Instrumentation: HP-R-214.** J.E. Jones; J.T. Smith; and M.V. Mathis. Springfield, VA: NTIS, April 1982. 62 p. (GEND-INF-017-v.10; DE82013780; CONTRACT AC07-761D01570)

A technical inquiry into the probable failure of HP-R-214 at TMI-2 is reported.

*Radiation monitors; Instrument failure

1151
U.S. Department of Energy. E, G, and G Idaho, Inc. Idaho Falls, Idaho. **Field Measurements and Interpretation of TMI-2 Instrumentation: IC-10-DPT.** J.E.

Jones; J.T. Smith; and M.V. Mathis. Springfield, VA: NTIS, January 1982. 44 p. (GEND-INF-017-v.6; DE82010037; CONTRACT AC07-761D01570)

Evaluation of control rod drive bypass flow IC-10-DPT indicates that although the status of the instrument is uncertain, it appears to be functioning.

*Instrument function

1152
U.S. Department of Energy. E, G, and G Idaho, Inc. Idaho Falls, Idaho. **Field Measurements and Interpretation of TMI-2 Instrumentation: NI-AMP-2.** J.E. Jones; J.T. Smith; and M.V. Mathis. Springfield, VA: NTIS, April 1982. 45 p. (GEND-INF-017-v.11; DE82013782; CONTRACT AC07-761D01570)

Reasons for the apparent failure of TMI-2 detector NI-AMP-2 are speculated.

*Instrument failure

1153
U.S. Department of Energy. E, G, and G Idaho, Inc. Idaho Falls, Idaho. **Field Measurements and Interpretation of TMI-2 Instrumentation: YM-AMP-7023 and YM-AMP-7025.** J.E. Jones; J.T. Smith; and M.V. Mathis. Springfield, VA: NTIS, January 1982). 74 p. (GEND-INF-017-v.7; DE82010493; CONTRACT AC07-761D01570)

This report communicates the make-up and function of loose Particle Monitor Channels at TMI-2.

*Instrument function; Instrument failure

1154
U.S. Department of Energy. E, G, and G Idaho, Inc. Idaho Falls, Idaho. **Final Analysis Data on TMI-2 Reactor Coolant System and Reactor Coolant Bleed-Tank Samples.** T.E. Cox; J.T. Horan; and R.L. Nitschke. Springfield, VA: NTIS, June 1983. 19 p. (GEND-INF-039; DE83013538; CONTRACT AC07-761D01570)

This report is a summary of analysis on four reactor coolant samples, four reactor coolant-bleed tank samples, one steam generator sample, and two reactor coolant gamma scans performed at four different laboratories.

*Coolant; Experimental data

1155
U.S. Department of Energy. E, G, and G Idaho, Inc. Idaho Falls, Idaho. **Fission Product Transfer in the TMI-2 Purification System.** T.E. Cox. Springfield, VA: NTIS, 1982. 4 p. American Nuclear Society Winter Meeting. Washington, D.C., November 14, 1982. (EGG-TIO-M-01682; CONF-821103-81; DE83005547; CONTRACT AC07-761D01570)

The filters and demineralizers of the TMI-2 reactor coolant make-up purification were examined to determine amount and type of materials retained therein.

*Coolant; Cooling system; Filtration

1156
U.S. Department of Energy. E, G, and G Idaho, Inc. Idaho Falls, Idaho. **GEND Planning Report.** Springfield, VA: NTIS, October 1980. 716 p. (GEND-001; DE81025533; CONTRACT AC07-761D01570)

This report summarizes planning for the TMI-2 cleanup program in all phases from instrument survival to disposal of radioactive waste.

*Cleanup; Waste handling; Waste disposal; Decontamination

1157
U.S. Department of Energy. E, G, and G Idaho, Inc. Idaho Falls, Idaho. **Immobilization of Three-Mile Island Core Debris.** J.M. Welch; R.L. Miller; and J.E. Finn. Springfield, VA: NTIS, 1983. 19 p. American Ceramic Society Annual Meeting. Chicago, Illinois, April 25, 1983. (EGG-M-26082; CONF-830451-17; DE83013540; CONTRACT AC07-761D01570)

The possible extraction of uranium fuel from the TMI-2 reactor core by first dissolving it in iron-enriched basalt (IEB) and later leaching it from the IEB is considered.

*Uranium; Fuel recovery; Fuel debris

1158
U.S. Department of Energy. E, G, and G Idaho, Inc. Idaho Falls, Idaho. **Initial SCDAP Predictions of the TMI-2 Event.** C.M. Allison; T.M. Howe; and G.P. Marino. Springfield, VA: NTIS, 1982. 9 p. Water Reactor Safety Research Conference. Gaithersburg, Maryland. October 12, 1982. (EGG-M-21682; CONF-821037-58; DE83003238; CONTRACT AC07-761D01570)

Severe Core Damage Analysis Package (SCDAP) computer code was developed to predict fuel rod and core behavior during an accident progression model. The results of this study will be applied to the prediction of TMI-2 fuel rod behavior.

*Computer programs; Computer simulation; Decision making, computer assisted

1159
U.S. Department of Energy. E, G, and G Idaho, Inc. Idaho Falls, Idaho. **Instrumentation and Electrical Program at the Three Mile Island Unit 2, Technical Integration Office.** L.A. Hecker. Springfield, VA: NTIS, 1982. 12 p. Society of Women Engineers Annual Meeting. Dearborn, Michigan, June 17, 1982. (EGG-TIO-M-00182; CONF-820689-1; DE82018895; CONTRACT AC07-761D01570)

The Technical Integration Office at Three Mile Island was established by the Department of Energy after the March 28, 1979, accident to coordinate a research program on the effects of the accident. One part of that program deals with the electrical system analysis, identification of parts failure, causes and longevity of those units that did survive the accident.

*Electrical equipment; Equipment function; Equipment failure; Research and development

1160
U.S. Department of Energy. E, G, and G Idaho, Inc. Idaho Falls, Idaho. **Instrumentation and Electrical Program at TMI-2.** W.F. Schwarz. Springfield, VA: NTIS, 1982. 3 p. American Nuclear Society Winter Meeting. Washington, D.C., November 14, 1982. (EGG-TIO-M-00982; CONF-821103-73; DE83005542; CONTRACT AC07-761D01570)

Through the instrumentation and electrical program at TMI-2, sample electrical instruments from the containment building were examined for functional capacity of the devices.

*Electrical equipment; Instrument function

1161
U.S. Department of Energy. E, G, and G Idaho, Inc. Idaho Falls, Idaho. **Interim Status Report of the TMI Personnel-Dosimetry Project.** B.L. Rich; J.L. Alvarez; and S.R. Adams. Springfield, VA: NTIS, 1981. 94 p. (GEND-004; DE81025672; CONTRACT AC07-761D01570)

Because the personnel dosimeter used at TMI-2 was inadequate for cleanup operations, investigation of alternative personnel dosimeters was made. This report reviews the positives and negatives of several commercial dosimeters as well as one developed by the Idaho National Engineering Laboratory and includes recommendations.

*Dosimeters; Cleanup; Recommendations

1162
U.S. Department of Energy. E, G, and G Idaho, Inc. Idaho Falls, Idaho. **Investigation of Acoustic Methods for Detecting TMI-2 Fuel Debris.** L.S. Beller. Springfield, VA: NTIS, August 1981. 45 p. (EGG-FM-5437; DE82001005/GAR; CONTRACT AC07-761D01570)

Ultrasound techniques for location of fuel debris and other reactor non-fuel debris in the piping system of the TMI-2 reactor is described, and experimental data interpreted. Recommendations for use of ultrasonic methods and current problems are presented.

*Fuel debris; Ultrasonic technique; Experimental data

1163
U.S. Department of Energy. E, G, and G Idaho, Inc. Idaho Falls, Idaho. **Lessons Learned from TMI-2.** H.J. Zeile. Springfield, VA: NTIS, 1981. 7 p. Department of Energy Nuclear Safety Conference. Augusta, Georgia, October 26, 1981. (EGG-M-09681; CONF-811805-6; DE82005647; CONTRACT AC07-761D01570)

Among the lessons learned at TMI-2 on which the nuclear industry will make efforts to improve are operator training, accident response, communication to the public, better technical support, management support, and independent operations assessment.

*Safety; Operators; Emergency preparedness; Communications; Lessons learned

1164
U.S. Department of Energy. E, G, and G Idaho, Inc. Idaho Falls, Idaho. **Metallurigcal Examination of, and Resin Transfer from, Three Mile Island Prefilter Liners.** J.W. McConnell, Jr. and H.W. Spaletta. Springfield, VA: NTIS, August 1984. 72 p. (GEND-036; DE84016086; CONTRACT AC07-761D01570)

This report looks at the findings from metallurigcal examination of the EPICOR II prefilter liners to evaluate the expected lifetime of the liners in interim storage at Idaho National Engineering Laboratory. The determination of a minimum lifetime governs the length of time in interim storage and the final disposal of such waste.

*EPICOR; Prefilter liners; Waste handling; Waste disposal

1165
U.S. Department of Energy. E, G, and G Idaho, Inc. Idaho Falls, Idaho. **Micro-Raman Analysis of Three Mile Island Samples.** T.E. Doyle and J.L. Alvarez. Springfield, VA: NTIS, 1983. 7 p. Annual Microbeam Analysis Society Conference. Phoenix, Arizona, August 8, 1983. (EGG-M-06383; CONF-830815-10; DE84001003; CONTRACT AC07-761D01570)

TMI-2 samples were analyzed using a Raman microprobe to learn the chemical elements in a nuclear accident, the release into the environment of these elements, and the progression of the accident. Chemical examination may provide clues to what happened at the TMI-2 core.

*Chemical releases; Reactor accident analysis; Atmospheric releases

1166
U.S. Department of Energy. E, G, and G Idaho, Inc. Idaho Falls, Idaho. **Nondestructive Techniques for Assaying Fuel Debris in Piping at Three Mile Island Unit 2.** K. Vinjamuri; C.V. McIsaac; L.S. Beller; L. Isaacson; and J.W. Mandler. Springfield, VA: NTIS, November 1981. 159 p. (GEND-018; DE82003039; CONTRACT AC07-761D01570)

In an effort to determine the amount of fuel in the piping system at TMI-2, four non-invasive techniques were considered. Passive gamma ray detection seems to be the most promising, but the technique needs further refinement.

*Fuel; Fuel debris; Fuel damage; Experimental data

1167
U.S. Department of Energy. E, G, and G Idaho, Inc. Idaho Falls, Idaho. **Overview of the TMI-2 Core-Examination Plan.** K.C. Sumpter; K.A. Trickett; E. Feinauer; D.E. Owen; and M.R. Martin. Springfield, VA: NTIS, 1982. 3 p. American Nuclear Society Winter Meeting. Washington, D.C., November 14, 1982. (EGG-M-18482; CONF-821103-72; DE83005280; CONTRACT AC07-761D01570)

The TMI-2 core examination plan is discussed relative to the defueling procedures.

*Defueling; Reactor core analysis

1168
U.S. Department of Energy. E, G, and G Idaho, Inc. Idaho Falls, Idaho. **Plan for Shipment, Storage, and Examination of TMI-2 Fuel.** G.J. Quinn; I.A. Engen; M.J. Tyacke; and H.W. Reno. Springfield, VA: NTIS, May 1984. 68 p. (EGG-TMI-0536; DE84013084; CONTRACT AC07-761D01570)

This report describes a two-part coordinated program between TMI and the Idaho National Engineering Laboratory (1) to package and transport fuel debris from TMI, and (2) to receive, analyze, and maintain the material at the Idaho National Engineering Laboratory.

*Debris, radioactive; Waste handling; Waste shipment; Reactor accident analysis; Fuel debris

1169
U.S. Department of Energy. E, G, and G Idaho, Inc. Idaho Falls, Idaho. **Pre-Decontamination Gamma-Ray Surface Scans in TMI-2 Containment Building 305' Elevation.** E.D. Barefoot; D.D. Cameron; J.S. Cline; J.A. Daniel; and D.G. Keefer. Springfield, VA: NTIS, February 1982. 18 p. (GEND-INF-009; DE82008184; CONTRACT AC07-761D01570)

As a result of gamma-ray surface scans conducted inside the TMI-2 containment building on December 16, 1981, it appears that contamination has decreased considerably from previous scans.

*Contamination; Reactor containment building

1170
U.S. Department of Energy. E, G, and G Idaho, Inc. Idaho Falls, Idaho. **Preliminary Characterization of EPICOR II Prefilter 16 Liner.** Springfield, VA: NTIS, November 1981. 35 p. (GEND-INF-015; DE82003588; CONTRACT AC07-761D01570) .

Evaluation of the EPICOR II prefilter liner 16 used in the TMI-2 water decontamination cleanup of the Auxiliary and Fuel Handling Building is discussed. Design and shipment, problems, ion exchange samples, gas analysis and liner temperatures are a few of the areas noted.

*EPICOR; Prefilter liners; Decontamination; Cleanup; Reactor auxiliary building; Ion exchange; Waste handling; Waste shipment

1171
U.S. Department of Energy. E, G, and G Idaho, Inc. Idaho Falls, Idaho. **Preliminary Report: Examination of H8 and B8 Leadscrews from Three Mile Island Unit 2 (TMI-2).** K. Vinjamuri; D.W. Akers; and R.R. Hobbins. Springfield, VA: NTIS, April 1985. 243 p. (EGG-TMI-6685; DE85010919/GAR; CONTRACT AC07-761D01570)

Leadscrews were obtained from the TMI-2 reactor control rods after head removal: H8 from the center of the core and B8 from near the outer edge. The results of studies on these leadscrews is reported.

*Leadscrews; Control rods

1172
U.S. Department of Energy. E, G, and G Idaho, Inc. Idaho Falls, Idaho. **Preliminary Report of TMI-2 In-Core Instrument Damage.** M.E. Yancey and N. Wilde. Springfield, VA: NTIS, June 1983. 29 p. (GEND-INF-031; DE83007281; CONTRACT AC07-761D01570)

Tests of the TMI-2 reactor core instruments were performed to determine the viability of the in-core thermocouples, self-powered neutron detectors, and background detectors. Indications were that the center of the core experienced the greatest change.

*Radiation detectors; Instrument function; Reactor core analysis; Reactor core damage

1173
U.S. Department of Energy. E, G, and G Idaho, Inc. Idaho Falls, Idaho. **Preliminary Results of the TMI-2 Radioactive Iodine Mass Balance Study.** C.A. Pelletier; T.E. Cox; D.L. Reeder; P.G. Vollique; and C.D. Thomas. Springfield, VA: NTIS, 1982. 6 p. American Nuclear Society Winter Meeting. Washington, D.C., November 14, 1982. (EGG-TIO-M-00782; CONF-821103-63; DE83005540; CONTRACT AC07-761D01570)

Measurement of radioactive iodine releases from the TMI-2 core into the containment building are presented.

*Iodine radioisotopes; Radioactive releases; Reactor containment building

1174
U.S. Department of Energy. E, G, and G Idaho, Inc. Idaho Falls, Idaho. **Preparations to Ship the Damaged TMI-2 Reactor Core.** R.C. Schmitt and G.J. Quinn. Springfield, VA: NTIS, 1985. 20 p. American Nuclear Society Winter Meeting. San Francisco, California, November 10, 1985. (EGG-M-17985; CONF-851115-45; DE86005315/GAR; CONTRACT AC07-761D01570)

This paper reviews the progress made in developing a rail cask for removal and transport of radioactive core debris from TMI-2 to the Idaho National Engineering Laboratory for examination and disposal.

*Cannister, high integrity; Waste shipment; Waste disposal

1175
U.S. Department of Energy. E, G, and G Idaho, Inc. Idaho Falls, Idaho. **Proceedings of the First International Information Meeting on the TMI-2 Accident.** S. Langer and W.R. Young. Springfield, VA: NTIS, 1985. 211 p. International Information Meeting on the TMI-2 Accident. Washington, D.C., October 21, 1985. (CONF-8510166; DE86010354/GAR; CONTRACT AC07-761D01570)

A collection of fourteen papers pertaining to the TMI accident comprise this report.

*Reactor accident analysis

1176
U.S. Department of Energy. E, G, and G Idaho, Inc. Idaho Falls, Idaho. **Processing and Removal of the Three Mile Island Makeup and Purification System Resins.** J.K. Reilly; T.W. McIntosh; L.M. Northey; J.J. Gatanto; and T.R. Osterhoudt. Springfield, VA: NTIS, 1985. 6 p. Waste Management '85. Tucson, Arizona, March 24, 1985. (EGG-TIO-M-00485; CONF-850314-29; DE85008928/GAR; CONTRACT AC07-761D01570)

This is a report on the progress of the removal of radioactive cesium and fuel from the TMI-2 demineralizers. Inventory of the demineralizers with regard to escaped fuel and fission products post accident and subsequent chemical separation of radioactive isotopes began in September 1984.

*Cesium radioisotopes; Fuel damage; Fission product release; Cleanup

1177
U.S. Department of Energy. E, G, and G Idaho, Inc. Idaho Falls, Idaho. **Program Plan for Shipment, Receipt, and Storage of the TMI-2 Core, Revision 1.** G.J. Quinn; H.W. Reno; and R.C. Schmitt. Springfield, VA: NTIS, January 1985. 56 p. (EGG-TMI-6536-rev.1; DE85008018/GAR; CONTRACT AC07-761D01570)

A revision of the previous "plan for shipment, storage, and examination of TMI-2 fuel," this document describes the preparation and shipment of TMI-2 fuel debris to the Idaho National Engineering Laboratory.

*Fuel debris; Waste handling; Waste shipment

1178
U.S. Department of Energy. E, G, and G Idaho, Inc. Idaho Falls, Idaho. **Quick Look Report, Entry 3: Three Mile Island Unit 2.** Springfield VA: NTIS, July 1981. 34 p. (GEND-INF-003; DE81028758)

This report summarizes the findings of, and tasks performed by, the Quick Look Entry team into TMI-2 on October 16, 1980. Radiation surveys and maintenance work were conducted during the one and one-half hour entry.

*Quick look entry

1179
U.S. Department of Energy. E, G, and G Idaho, Inc. Idaho Falls, Idaho. **Quick Look Report, Entry 4: Three Mile Island Unit 2.** G.E. Eidam. Springfield, VA: NTIS, June 1981. 32 p. November 13. 1980. (GEND-004; DE80023966; CONTRACT AC07-761D01570)

This report relates the activities performed on Entry 4 into the TMI-2 reactor building on November 13, 1980. Photography and videotapes were taken in addition to radiation surveys and equipment tests.

*Quick look entry; Photography; Radiation monitoring; Equipment function

1180
U.S. Department of Energy. E, G, and G Idaho, Inc. Idaho Falls, Idaho. **Radiological Impacts of Transporting Three Mile Island Core Debris.** N.D. Cox. Springfield, VA: NTIS, January 1985. 26 p. (EGG-TMI-7108; DE86007652/GAR; CONTRACT AC07-761D01570)

The shipment of radioactive core debris from Three Mile Island is discussed with special reference to radiation exposure to persons along the shipping routes. The container cask design is reported and allowable radiation dose exposure limits.

*Debris, radioactive; Fuel debris; Cannister, high integrity; Population exposure; Radiation dosage

1181
U.S. Department of Energy. E, G, and G Idaho, Inc. Idaho Falls, Idaho. **Radionuclide Distribution on TMI-2 Reactor Building Basement Liquids and Solids.** J.T. Horan; C.V. McIsaac; and D.G. Keefer. Springfield, VA: NTIS, 1984. 12 p. Topical Meeting on Fission Product Behavior and Source Term Research. Snowbird, Utah, July 15, 1984. (EGG-TIO-M-0284; CONF-840701-22; DE84016216/GAR; CONTRACT AC07-761D01570)

The path of the contaminated water release from TMI-2 reactor was traced from the damaged core through the reactor coolant system, various valves and tanks to the reactor building basement. Data from radiochemical analyses of the fission products in basement water and solids are presented.

*Contaminated water; Chemical water pollution; Water pollutants, radioactive; Fission products release; Reactor building

1182
U.S. Department of Energy. E, G, and G, Idaho, Inc. Idaho Falls, Idaho. **Raman and Luminescence Spectroscopy of Zirconium Oxide with the Use of the MOLE Microprobe.** T.E. Doyle and J.L. Alvarez. Springfield, VA: NTIS, 1984. 13 p. (EGG-M-08484; DE84017234/GAR; CONTRACT AC07-761D01570)

Examination of samples of zirconium oxide from the fuel-rod cladding from TMI-2 provide information about temperatures and hydrogen formation in the reactor core.

*Zirconium oxide; Fuel rods; Hydrogen; Reactor core temperature; Molecular optical laser examiner

1183
U.S. Department of Energy. E, G, and G Idaho, Inc. Idaho Falls, Idaho. **Reactor-Building-Basement Radionuclide and Source Distribution Studies. Volume 3.** T.E.

Cox; J.T. Horan; and G. Worku. Springfield, VA: NTIS, June 1983. 27 p. (GEND-INF-11-v.3; DE83014875; CONTRACT AC07-761D01570)

Analysis of the liquid and solid samples obtained from the TMI-2 reactor building basement are reported in addition to radiation measurements gathered from thermoluminescent dosimeters.

*Radiation monitoring; Radiation monitors; Reactor building

1184
U.S. Department of Energy. E, G, and G Idaho, Inc. Idaho Falls, Idaho. **RELAP5/MOD2 Overview and Development Assessment Results from TMI-1 Plant Transient Analysis.** J.C. Lin; C.C. Tsai; V.H. Ransom; and G.W. Johnsen. Springfield, VA: NTIS, 1984. 8 p. International Thermal Hydraulics and Plant Operations Topical Meeting. Taipei, Taiwan, October 22, 1984. (EGG-M-12284; CONF-841075-5; DE85003549/GAR; CONTRACT AC07-761D01570)

Analyses on TMI-1 of new version of the RELAP5 thermal- hydraulic computer code, RELAP5/MOD2, indicates that it continues to function well after the introduction of highly subcooled water into a high pressure primary coolant system.

*Decision making, computer assisted; Cooling system; Transients

1185
U.S. Department of Energy. E, G, and G Idaho, Inc. Idaho Falls, Idaho. **Requirements for Transporting the TMI-2 Core.** D.E. Wilkins. Springfield, VA: NTIS, August 1983. 18 p. (EGG-TMI-6272; DE84000859; CONTRACT AC07-761D01570)

The shipment requirements for portions of the TMI-2 core to be transported to the Idaho National Engineering Laboratory are discussed.

*Reactor core; Waste handling; Waste shipment

1186
U.S. Department of Energy. E, G, and G Idaho, Inc. Idaho Falls, Idaho. **Response of the SPND Measurement System to Temperature During the Three Mile Island Unit 2 Accident.** N. Wilde and J.L. Morrison, Jr. Springfield, VA: NTIS, December 1981. 88 p. (GEND-017; DE82004663; CONTRACT AC07-761D01570)

The response/non-response of the Self Powered Neutron Detector (SPND) Measuring System in providing temperatures of fuel rods at TMI-2 during the accident is presented. It is noted that the temperature range was too great for the system to function effectively.

*Reactor core temperature; Fuel rods; Instrument function

1187
U.S. Department of Energy. E, G, and G Idaho, Inc. Idaho Falls, Idaho. **Results of Analyses Performed on Concrete Cores Removed from Floors and D-Ring Walls of the TMI-2 Reactor Building.** C.V. McIsaac; C.M. Davis; J.T. Horan; and D.G. Keefer. Springfield, VA: NTIS, June 1984. 44 p. (GEND-INF-054; DE85001455/GAR; CONTRACT AC07-761D01570)

A study of the effectiveness of surface protectors (several coats of nuclear grade paint) on the walls and floor of the TMI reactor building was evaluated by samples. Where protective coatings were undamaged, radionuclides remained within a few millimeters of the surface; where damaged, particularly floors, penetration of radiocesium occurred to a depth of several centimeters.

*Cesium radioisotopes; Decontamination; Cleanup; Reactor building

1188
U.S. Department of Energy. E, G, and G Idaho, Inc. Idaho Falls, Idaho. **Results of Surface Activity and Radiation Field Measurements Made During Surface Decontamination Experiments Conducted at TMI-2.** C.V. McIsaac; C.M. Davis; J.T. Horan; and D.G. Keefer. Springfield, VA: NTIS, 1984. 16 p. Topical Meeting on Fission Product Behavior and Source Term Research. Snowbird, Utah, July 15, 1984.

(EGG-TIO-M-0384; CONF-840701-24; DE84016217/GAR; CONTRACT AC07-761D01570)

Surface decontamination investigations using low pressure water, water lances, high pressure water spray, strippable coatings or mechanical scrubbing were conducted on different surfaces at TMI-2 reactor building, February and March 1982.

*Decontamination; Reactor building

1189
U.S. Department of Energy. E, G, and G, Idaho, Inc. Idaho Falls, Idaho. **Review of TMI-2 Resistance Temperature Detectors Accident Data and In-Situ Testing.** J.W. Mock. Springfield, VA: NTIS, November 1982. 29 p. (GEND-INF-024; DE83004265; CONTRACT AC07-761D01570)

This report reviews the function of the sixteen resistance temperature detectors in the TMI-2 reactor building air handling system.

*Temperature detectors; Reactor building; Instrument function

1190
U.S. Department of Energy. E, G, and G, Idaho, Inc. Idaho Falls, Idaho. **SCDAP/MOD1 Analysis of the Progression of Core Damage During the TMI-2 Accident.** C.M. Allison; S.T. Polkinghorne; and M.S. Sohal. Springfield, VA: NTIS, November 1985. 170 p. (EGG-SAR-7104; DE86005283/GAR; CONTRACT AC07-761D01570)

Damage to the TMI-2 reactor core and fission products originating from the accident are examined relative to the SCDAP/MOD1 computer code (Severe Core Damage Analysis Package - Major Modification 1).

*Computer simulation; Fission product release; Reactor core damage

1191
U.S. Department of Energy. E, G, and G, Idaho, Inc. Idaho Falls, Idaho. **Solidification of EPICOR II Resin Waste Forms.** R.M. Neilson, Jr. and J.W. McConnell, Jr. Springfield, VA: NTIS, August 1984. 40 p. (GEND-INF-055; DE84017410/GAR; CONTRACT AC07-761D01570)

Experimental immobilization of radioactive resin EPICOR II prefilter wastes from TMI-2 using Portland type I-II cement and a vinyl-ester-styrene formulation are evaluated.

*EPICOR; Resins; Waste handling

1192
U.S. Department of Energy. E, G, and G, Idaho, Inc. Idaho Falls, Idaho. **Special Handling Requirements for Highly Loaded Organic Resins.** R.E. Ogle; J.M. Bower; and T.E. Rekart. Springfield, VA: NTIS, 1982. 4 p. American Nuclear Society Winter Meeting. Washington, D.C., November 14, 1982. (EGG-TIO-M-01382; CONF-821103-69; DE83005545; CONTRACT AC07-761D01570)

This report examines the reasons for selection of EPICOR II prefilter liners, PF-16, (which were used in the water filtration process to remove cesium and strontium) for study.

*EPICOR; Prefilter liners; Filtration; Cesium radioisotopes; Strontium radioisotopes

1193
U.S. Department of Energy. E, G, and G Idaho, Inc. Idaho Falls, Idaho. **Status of the TMI-2 Core: A Review of Damage Assessments.** D.W. Croucher. Springfield, VA: NTIS, 1981. 15 p. Topical Meeting on Reactor Safety Aspects of Fuel Behavior. Sun Valley, Idaho, August 2, 1981. (CONF-810803-18; DE82005251; CONTRACT AC07-761D01570)

A review of probable core damage at TMI-2 as a result of the accident is presented by reconstructing the event. Thermal-hydraulic sequence hydrogen gas production and radionuclide fission products are all considered.

*Reactor core damage; Fission products release; Hydrogen; Thermal hydraulics

1194
U.S. Department of Energy. E, G, and G Idaho, Inc. Idaho Falls, Idaho. **Status of TMI-2 Instruments and Electrical Components.** H.J. Helbert. Springfield, VA: NTIS, August 1982. 17 p. (GEND-INF-022; DE82021877; CONTRACT AC07-761D01570)

Data on the status and survivability of the TMI-2 instrumentation and electrical components is presented as it is called for in the Planning Report of the Instrumentation and Electrical Equipment Survivability Planning Group.

*Electrical equipment; Equipment function; Equipment failure

1195
U.S. Department of Energy. E, G, and G Idaho, Inc. Idaho Falls, Idaho. **Submerged Demineralizer System Processing of TMI-2 Accident Waste Water.** H.F. Sanchez and G.J. Quinn. Springfield, VA: NTIS, Feburary 1983. 65 p. (GEND-031; DE83009223; CONTRACT AC07-761D01570)

The decontamination processing of water from the TMI-2 reactor coolant bleed tanks, reactor building basement, and reactor coolant system by the Submerged Demineralizer System (SDS) was completed in August 1982. Other wastes not able to be processed by this system were processed through EPICOR II. This report is primarily on the development and operation of the SDS.

*Contaminated water; Water pollutants, radioactive; Decontamination; Submerged Demineralizer System; Coolant; Cooling system; Reactor building

1196
U.S. Department of Energy. E, G, and G, Idaho, Inc. Idaho Falls, Idaho. **Submerged Demineralizer System Vessel Shipment Report.** G.J. Quinn; J.O. Harris; and J. Greenborg. Springfield, VA: NTIS, June 1984. 81 p. (GEND-035; DE84013866; CONTRACT AC07-761D01570)

The processing of contaminated water by the Submerged Demineralizer System produced some waste material that was found to generate radiolytic hydrogen and oxygen gases. The potential for explosion and burning precluded shipment in this form; to alleviate the problem, a catalyst was added to stop gas generation.

*Contaminated water; Radiolytic gas; Hydrogen; Oxygen; Submerged Demineralizer System; Combustion potential

1197
U.S. Department of Energy. E, G, and G, Idaho, Inc. Idaho Falls, Idaho. **Surface Activity and Radiation Field Measurements of the TMI-2 Reactor Building Gross Decontamination Experiment.** C.V. McIsaac. Springfield, VA: NTIS, October 1983. 122 p. (GEND-037; DE84003469; CONTRACT AC07-761D01570)

Samples of the concrete and metal surfaces of walls and floors of the TMI-2 reactor building were obtained on December 15 and 17, 1981. The reactor building was decontaminated and samples were again collected on March 25 and 26, 1982. Comparison of the two samples indicated that the decontamination operations were successful in reducing cesium and strontium surface activity.

*Decontamination; Reactor building; Cesium radioisotopes; Strontium radioisotopes

1198
U.S. Department of Energy. E, G, and G, Idaho, Inc. Idaho Falls, Idaho. **Surface Deposition Measurement of the TMI-2 Gross Decontamination Experiment.** C.V. McIsaac and D.C. Hetzer. Springfield, VA: NTIS, 1982. American Nuclear Society Winter Meeting. Washington, D.C., November 14, 1982. (EGG-TIO-M-1782; CONF-821103-67; DE83005387; CONTRACT AC07-761D01570)

The equipment necessary and process by which surface samples were collected from the TMI-2 reactor building are described. Samples were accumulated before and after gross decontamination.

*Decontamination; Reactor building

1199
U.S. Department of Energy. E, G, and G Idaho, Inc. Idaho Falls, Idaho. **Technical Assistance and Advisory Group, Report Number 5. Volume 5.** W.H. Hamilton. Springfield, VA: NTIS, December 1983. 182 p. (GEND-INF-037-v.5; DE84004467; CONTRACT AC07-761D01570)

Nine separate subjects ranging from core samples to radiation exposure are outlined relative to the TMI-2 accident.

*Reactor accident analysis; Technical assessment

1200
U.S. Department of Energy. E, G, and G Idaho, Inc. Idaho Falls, Idaho. **Technology Transfer at Three Mile Island Unit 2.** H.M. Burton and W.W. Bixby. Springfield, VA: NTIS, 1982. 6 p. International Conference on Nuclear Technology Transfer. Buenos Aires, Argentina, November 1, 1982. (EGG-TIO-M-00282; CONF-821104-2; DE83005278; CONTRACT AC07-761D01570)

The Department of Energy's TMI-2 Information and Examination Program is reported and its application for future uses in specifications, licensure, operations, etc., of nuclear power plants discussed.

*Nuclear power plants; Licensing; Recommendations

1201
U.S. Department of Energy. E, G, and G Idaho, Inc. Idaho Falls, Idaho. **Tellurium Behavior During and After the TMI-2 Accident.** K. Vinjamuri; D.J. Osetek; and R.R. Hobbins. Springfield, VA: NTIS, 1984. 20 p. International Meeting on Thermal Nuclear Reactor Safety. Karlsruhe, F.R. Germany, September 10, 1984. (EGG-M-05384; CONF-840914-22; DE85003429/GAR; CONTRACT AC07-761D01570)

Tellurium behavior at TMI-2 was examined based on available measurements and calculated release. Studies suggest that zircaloy cladding holds tellurium until oxidation occurs.

*Tellurium

1202
U.S. Department of Energy. E, G, and G, Idaho, Inc. Idaho Falls, Idaho. **Tellurium Chemistry, Tellurium Release and Deposition During the TMI-2 Accident.** K. Vinjamuri; D.W. Akers; R.A. Sallach; D.J. Osetek; and R.R. Hobbins. Springfield, VA: NTIS, August 1985. 70 p. (39 refs.) (EGG-TMI-6894; DE85017500/GAR; CONTRACT AC07-761D01570)

Tellurium behavior during and after the TMI accident, as well as estimated release, is presented with the conclusion that probably very little escaped because of the cladding and other materials surrounding the core.

*Tellurium; Radioactive releases

1203
U.S. Department of Energy. E, G, and G, Idaho, Inc. Idaho Falls, Idaho. **Tellurium Chemistry, Tellurium Release and Deposition During the TMI-2 Accident.** K. Vinjamuri; R.A. Sallach; D.J. Osetek; R.R. Hobbins; and D.W. Akers. Springfield, VA: NTIS, 1985. 23 p. 37 references. Water Reactor Safety Research Information Meeting. Gaithersburg, Maryland, October 22, 1985. (EGG-M-23485; CONF-8510173-31; DE86003224/GAR; CONTRACT AC07-761D01570)

The study of tellurium chemistry and release concludes that very little was released and transported from the TMI-2 core as a result of cladding and other structural materials.

*Tellurium; Radioactive releases; Accident investigations

1204
U.S. Department of Energy. E, G, and G Idaho, Inc. Idaho Falls, Idaho. **Tellurium Release and Deposition During the TMI-2 Accident.** K. Vinjamuri; D.J. Osetek; R.R.

Hobbins; and J.S. Jessup. Springfield, VA: NTIS, September 1984. 56 p. (EGG-TMI-6701; DE85006169/GAR; CONTRACT AC07-761D01570)

Tellurium behavior during and after the TMI-2 accident is presented with approximations of its location throughout the reactor. The report suggests that the zircaloy cladding holds tellurium until significant oxidation occurs.

*Tellurium; Radioactive releases

1205
U.S. Department of Energy. E, G, and G Idaho, Inc. Idaho Falls, Idaho. **Three Mile Island Ambient-Air Temperature Sensor Measurements.** M.O. Fryer. Springfield, VA: NTIS, 1983. 4 p. International Symposium of the Instrument Society of America. Albuquerque, New Mexico, May 2, 1983. (EGG-M-27782; CONF-830518-8; DE83013517; CONTRACT AC07-761D01570)

Data from ambient-air-temperature-sensors were examined to determine the length of time the hydrogen burn continued.

*Hydrogen burn; Ambient-air-temperature sensors

1206
U.S. Department of Energy. E, G, and G Idaho, Inc. Idaho Falls, Idaho. **Three Mile Island Technical Information and Examination Program. Instrumentation and Electrical Summary Report.** R.D. Meininger. Springfield, VA: NTIS, July 1985. 21 p. (GEND-050; DE85015387/GAR; CONTRACT AC07-761D01570)

A summary of previous reports, instruments and electrical systems affected by the TMI-2 accident is scrutinized.

*Electrical equipment; Instrument function; Reactor accident analysis

1207
U.S. Department of Energy. E, G, and G Idaho, Inc. Idaho Falls, Idaho. **Three Mile Island Unit-2 Core Status: A Basis for Tool Development for Reactor Disassembly and Defueling.** D.W. Croucher. Springfield, VA: NTIS, May 1981. 73 p. (GEND-007; CONTRACT AC07-761D01570)

Several indirect methods to estimate core damage from the TMI-2 accident have been employed. In this report, thermal-hydraulic, hydrogen generation and fission products released are all correlated in an attempt to calculate the minimum and maximum bounds of core damage.

*Reactor core damage; Reactor core analysis; Thermal hydraulics; Radiolytic gas; Fission product release; Experimental data

1208
U.S. Department of Energy. E, G, and G Idaho, Inc. Idaho Falls, Idaho. **TMI Abnormal Waste Project Plan.** A.L. Ayers, Jr. Springfield, VA: NTIS, June 1984. 36 p. (EGG-TMI-6588; DE84014715; CONTRACT AC07-761D01570)

Plans for abnormal waste disposition, i.e. that from EPICOR system, are discussed including packaging, shipment, storage, processing, and final disposal.

*Radioactive waste; EPICOR; Waste handling; Waste shipment; Waste disposal

1209
U.S. Departmentof Energy. E, G, and G Idaho, Inc. Idaho Falls, Idaho. **TMI Abnormal Wastes Disposal Options.** A.L. Ayers, Jr. Springfield, VA: NTIS, May 1984. 6 p. Waste Management '84. Tucson, Arizona, March 11, 1984 (EGG-M-19983; CONF-840307-40; DE84011746; CONTRACT AC07-761D01570)

Options are explored for storing highly radioactive wastes from the TMI-2 accident that cannot be disposed of because of current regulations.

*Radioactive waste; Waste disposal; Regulations

1210
U.S. Department of Energy. E, G, and G Idaho, Inc. Idaho Falls, Idaho. **TMI Cable Tracer Operation and Maintenance Manual for Assembly 417910.** R.L. Sumstine.

Springfield, VA: NTIS, November 1983. 53 p. (EGG-TMI-6443; DE84004945; CONTRACT AC07-761D01570)

The cable tracer system at TMI-2 was developed so that cables could be traced for testing and examination. This document is a technical instruction manual for using the cable tracer equipment.

*Cables; Equipment function

1211
U.S. Department of Energy. E, G and G Idaho, Inc. Idaho Falls, Idaho. **TMI-2 Accident Evaluation Program.** E.L. Tolman; J.M. Broughton; R.K. McCardell; S. Langer; and R.R. Hobbins. Springfield, VA: NTIS, February 1986. 132 p. (EGG-TMI-7048; DE86007614/GAR; CONTRACT AC07-761D01570)

The TMI-2 accident is reviewed, including core damage, progression of core damage, and fission product behavior. Unresolved technical issues are addressed.

*Reactor core damage; Fission product release; Reactor core analysis; Technical assessment

1212
U.S. Department of Energy. E, G, and G Idaho, Inc. Idaho Falls, Idaho. **TMI-2 Accident Evaluation Program Sample Acquisition and Examination Plan. Executive Summary.** M.L. Russell; R.K. McCardell; and J.M. Broughton. Springfield, VA: NTIS, December 1985. 33 p. (EGG-TNI-7121; DE86007659/GAR; CONTRACT AC07-761D01570)

The TMI-2 inspection plan covers the TMI-2 equipment that may have been damaged in the accident and the core fission summary by sample acquisition measurements and in-place measurements.

*Equipment function; Reactor core analysis; Reactor core damage

1213
U.S. Department of Energy. E, G, and G Idaho, Inc. Idaho Falls, Idaho. **TMI-2 Cable/Connection Program: A Look at In-Situ Test Data.** L.A. Kecker and H.J. Helbert. Springfield, VA: NTIS, December 1983. 57 p. (GEND-INF-042; DE84004114; CONTRACT AC07-761D01570)

Evaluation of the cable/connector systems at TMI-2 after the severe abuses they received during the accident was made in an effort to develop guidelines for the production of a more durable system.

*Cables; Equipment function

1214
U.S. Department of Energy. E, G, and G Idaho, Inc. Idaho Falls, Idaho. **TMI-2 Cable/Connections Program: FY-84 Status Report.** H.J. Helbert; N.S. Cannon; D.R. Chick; M.R. Donaldson; and R.D. Meininger. Springfield, VA: NTIS, September 1984. 106 p. (GEND-INF-056; DE85005506/GAR; CONTRACT AC07-761D01570)

Two hundred and thirty-three electrical cable channels of the 567 channels tested at TMI-2 are the subject of this report. It is a part of an ongoing Department of Energy effort to determine the condition of the electric channels in the TMI-2 Reactor Building.

*Cables; Electrical equipment; Equipment function

1215
U.S. Department of Energy. E, G, and G Idaho, Inc. Idaho Falls, Idaho. **TMI-2 Core Debris Analytical Methods and Results.** D.W. Akers and B.A. Cook. Springfield, VA: NTIS, 1984. 11 p. Topical Meeting on Fission Product Behavior and Source Term Research. Snowbird, Utah, July 15, 1984. (EGG-M-11984; CONF-840701-19; DE84016108; CONTRACT AC07-761D01570)

Analysis plans for six grab samples extracted from the TMI-2 core in September 1983 are detailed.

*Debris, radioactive; Fuel debris; Reactor core damage

1216
U.S. Department of Energy. E, G, and G Idaho, Inc. Idaho Falls, Idaho. **TMI-2 Core Debris Bed Coolability.** P Kuan. Springfield, VA: NTIS, March 1986. 41 p. (EGG-TMI-7150; DE86009091/GAR; CONTRACT AC07-761D01570)

In five sections, this report looks at the temperature moderation that occurs in the TMI-2 core debris that is not within the range of cooling water.

*Debris, radioactive; Reactor core damage; Reactor core temperature

1217
U.S. Department of Energy. E, G, and G Idaho, Inc. Idaho Falls, Idaho. **TMI-2 Core Debris - Cesium Release/Settling Test. Draft Report.** D.W. Akers and D.A. Johnson. Springfield, VA: NTIS, September 1984. 26 p. (EGG-TMI-6697-Dr; DE85003401/GAR; CONTRACT AC07-761D01570)

Cesium release was studied on TMI-2 core debris in both undisturbed particles and disturbed (or crushed) particles. Crushing increased soluability. Releases into air occurred during evaporation and were highest in magnitude just before dryout.

*Debris, radioactive; Fuel debris; Cesium radioisotopes; Experimental data

1218
U.S. Department of Energy. E, G, and G Idaho, Inc. Idaho Falls, Idaho. **TMI-2 Core Debris - Cesium Release/Settling Test. Volume 3.** D.W. Akers and D.A. Johnson. Springfield, VA: NTIS, December 1984. 34 p. (GEND-INF-60-v.3; DE85008020/GAR; CONTRACT AC07-761D01570)

Tests on cesium release from the TMI-2 core were conducted on as-is specimens and on crushed (simulating core breakup) specimens to aid in predicting reactor environment during the defueling process.

*Debris, radioactive; Fuel debris; Cesium radioisotopes; Defueling

1219
U.S. Department of Energy. E, G, and G Idaho, Inc. Idaho Falls, Idaho. **TMI-2 Core Debris Grab Sample Quick Look Report.** D.W. Akers and R.L. Nitschke. Springfield, VA: NTIS, March 1984. 42 p. (EGG-TMI-6531-rev.1; DE84011032; CONTRACT AC07-761D01570)

Six core samples were obtained for examination in the fall of 1983. Eight particles selected from the samples are the subject of this report. Analysis revealed that these particles consisted of fuel pellets, cladding, previously melted materials, and coated or glazed fuel pellets.

*Quick look entry; Reactor core analysis

1220
U.S. Department of Energy. E, G, and G Idaho, Inc. Idaho Falls, Idaho. **TMI-2 Core Examination Plan.** D.E. Owen; P.E. MacDonald; R.R. Hobbins; and S.A. Ploggr. Springfield, VA: NTIS, 1982. 11 p. International Meeting on Thermal Nuclear Reactor Safety. Chicago, Illinois, August 29, 1982. (EGG-M-05582;CONF-820802-39; DE83000565; CONTRACT AC07-761D01570)

The TMI-2 core examination plan is presented including the proposed stages preparatory to eventual offsite inspection and analysis. Evaluation of core damage from the accident may aid in development of safer operating procedures.

*Reactor core damage; Reactor core analysis; Safety

1221
U.S. Department of Energy. E, G, and G Idaho, Inc. Idaho Falls, Idaho. **TMI-2 Core Examination Plan.** J.O. Carlson. Springfield, VA: NTIS, July 1984. 117 p. (EGG-TMI-6169-rev.; DE84016303/GAR; CONTRACT AC07-761D01570)

The TMI-2 core examination plan conducted by E,G, and G Idaho covers nuclear safety issues, recommendations for on-and off-site activities, and data needs and data recovery.

*Safety; Reactor core analysis; Recommendations

1222
U.S. Department of Energy. E, G, and G Idaho, Inc. Idaho Falls, Idaho. **TMI-2 Core Sample Acquisition and Examination Project Status.** K.C. Sumpter. Springfield, VA: NTIS, 1984. 3 p. Water Reactor Safety Research Information Meeting. Gaithersburg, Maryland, October 23, 1984. (EGG-M-24384; CONF-8410142-81; DE85006378; CONTRACT AC07-761D01570)

This report reviews the examination of the TMI-2 core to date and the development of additional areas of inquiry as the reactor recovery program continues.

*Reactor core analysis; Cleanup

1223
U.S. Department of Energy. E, G, and G Idaho, Inc. Idaho Falls, Idaho. **TMI-2 Fuel Canister Interface Requirements for Idaho National Engineering Laboratory. Revision 1.** D.E. Wilkins; D.E. Martz; and H.W. Reno. Springfield, VA: NTIS, June 1984. 90 p. (EGG-TMI-6156-Rev.1; DE84014716; CONTRACT AC07-761D01570)

Criteria for design of TMI-2 fuel cannisters are discussed relative to handling and storage at the Idaho National Engineering Laboratory.

*Cannister, high integrity; Waste handling

1224
U.S. Department of Energy. E, G, and G Idaho, Inc. Idaho Falls, Idaho. **TMI-2 Fuel-Recovery Plant. Feasibility Study.** D.L. Evans. Springfield, VA: NTIS, December 1982. 129 p. (EGG-TMI-6130; DE83004852; CONTRACT AC07-761D01570)

This report summarizes the feasibility of constructing a fuel recovery plant at the Idaho National Engineering Laboratory for the recovery of uranium fuel from the TMI-2 core debris for the disposal of the waste products.

*Reactor core damage; Fuel recovery; Uranium; Debris, radioactive; Waste disposal

1225
U.S. Department of Energy. E, G, and G Idaho, Inc. Idaho Falls, Idaho. **TMI-2 In-Core Instrument Damage - An Update. Volume II.** M.E. Yancey; R.D. Meininger; and L.A. Hecker. Springfield, VA: NTIS, April 1984. 78 p. (GEND-INF-031-v.2; DE84015257; CONTRACT AC-07-761D01570)

Further tests on TMI-2 throughout July and August 1982 to determine the extent of damage to the reactor core indicate significant disruption of the central portion of the lower core.

*Reactor core damage

1226
U.S. Department of Energy. E, G, and G Idaho, Inc. Idaho Falls, Idaho. **TMI-2 Isotopic Inventory Calculations.** B.G. Schnitzler and J.B. Briggs. Springfield, VA: NTIS, August 1985. 80 p. (EGG-PBS-6798; DE86004110/GAR; CONTRACT AC07-761D01570)

Fission products and heavy metal inventories of the TMI-2 core are presented along with a full-core inventory calculation. Isotopic ratio data allow estimates of fission product release correlated to core location of fuel debris.

*Reactor core damage; Fission product release; Fuel debris

1227
U.S. Department of Energy. E, G, and G Idaho, Inc. Idaho Falls, Idaho. **TMI-2 Leadscrew Radionuclide Deposition and Characterization.** K. Vinjamuri; D.W. Akers; and R.R. Hobbins. Springfield, VA: NTIS, June 1, 1984. 14 p. Topical Meeting on Fission Product Behavior and Source Item Research. Snowbird, Utah, July 15, 1984. (EGG-M-10684; CONF-840701-18; DE84016117; CONTRACT AC07-761D01570)

Two leadscrews from the TMI-2 reactor core -- one from near the center and one from the outer edge -- are examined by metallography, hardness measurements, radiologic and chemical analyses to infer radionuclide activity and location.
*Leadscrews; Reactor core analysis

1228
U.S. Department of Energy. E, G, and G Idaho, Inc. Idaho Falls, Idaho. **TMI-2 Plant Demineralizer Sample Analysis.** J.D. Thompson and J.B. Knauer. Springfield, VA: NTIS, 1984. 4 p. Joint Meeting of the American Nuclear Society and the Atomic Industrial Forum. Washington, D.C., November 11, 1984. (EGG-TIO-M-00784; CONF-841105-46; DE85003263/GAR; CONTRACT AC07-761D01570)
Analysis of the submerged demineralizer system (SDS) in the TMI-2 plant indicates that the system will successfully wash out cesium.
*Submerged Demineralizer System; Cesium radioisotopes; Filtration

1229
U.S. Department of Energy. E, G, and G Idaho, Inc. Idaho Falls, Idaho. **TMI-2 Pressure Transmitter Examination and Evaluation of CF-1-PT1, CF-2-LT1, and CF-2-LT2. Volume 2.** M.E. Yancey and R.C. Strahm. Springfield, VA: NTIS, April 1984. 25 p. (GEND-INF-029-v.2; DE84011033; CONTRACT AC07-761D01570)
Three transmitters were removed from TMI-2 in order to examine them and to determine the reasons for failure of two of the three.
*Pressure transmitters; Instrument function; Instrument failure

1230
U.S. Department of Energy. E, G, and G Idaho, Inc. Idaho Falls, Idaho. **TMI-2 Pressure Transmitter Examination Program Year-End Report: Examination and Evaluation of Pressure Transmitters CF-1-PT3 and CF-2-LT3.** R.C. Strahm and M.E. Yancey. Springfield, VA: NTIS, February 1983 42 p. (GEND-INF-029; DE83006479; CONTRACT AC07-761D01570)
Two pressure transmitters removed from TMI-2 were examined and the results reported. The first was functional and had survived the accident while the second had failed and was too badly damaged to determine the cause.
*Pressure transmitters; Instrument function; Instrument failure

1231
U.S. Department of Energy. E, G, and G, Idaho, Inc. Idaho Falls, Idaho. **TMI-2 Purification Demineralizer Resin Study.** J.D. Thompson and T.R. Osterhoudt. Springfield, VA: NTIS, May 1984. 30 p. (GEND-INF-013; DE84014736; CONTRACT AC07-761D01570)
TMI-2 resin purification studies indicate that fuel quantities in the reactor vessel are low and the radioactive cesium can be chemically processed out.
*Resins; Cesium radioisotopes; Decontamination

1232
U.S. Department of Energy. E, G, and G Idaho, Inc. Idaho Falls, Idaho. **TMI-2 Pyrophoricity Studies.** V.F. Baston; W.E. Austin; K.J. Hoffstetter; and D.E. Owen. Springfield, VA: NTIS, November 1984. 35 p. (GEND-043; DE85003777/GAR; CONTRACT AC07-761D01570)
Because of the presence of zirconium, zirconium alloys and possibly oxidizable zirconium, the potential for spontaneous ignition of the TMI-2 core was considered. This report is a summary of the experimental evaluation and a literature review of that potential.
*Combustion potential; Experimental data

1233
U.S. Department of Energy. E, G, and G, Idaho, Inc. Idaho Falls, Idaho. **TMI-2 Quick Look Examination.** W.A. Frantz; R.L. Rider; W.A. Austin; and N. Cole. Springfield, VA: NTIS, 1982. 4 p. American Nuclear Society Winter Meeting.

Washington, D.C. November 14, 1982. (EGG-TIO-M-01182; CONF-821103-68; DE83005544; CONTRACT AC07-761D01570)

This report describes various possible techniques for Quick Look examination of TMI-2 to determine core and plenum assembly conditions.

*Quick look; Reactor core analysis; Plenum

1234
U.S. Department of Energy. E, G, and G Idaho, Inc. Idaho Falls, Idaho. **TMI-2 Reactor Building Source Term Measurements. Surfaces and Basement Water and Sediment.** C.V. McIsaac and D.G. Keefer. Springfield, VA: NTIS, October 1984. 88 p. (GEND-042; DE85003888/GAR; CONTRACT AC07-761D01570)

Examination of external surfaces of the TMI-2 reactor building, as well as the water and sediment in the containment building, were performed on samples collected from August 1979 to December 1983. Radionuclide dispersal was largely found to be in the water and the basement sediment.

*Reactor building; Reactor containment building; Radiation analysis

1235
U.S. Department of Energy. E, G, and G Idaho, Inc. Idaho Falls, Idaho. **TMI-2 Reactor Vessel Head Removal and Damaged-Core-Removal Planning.** J.A. Logan; C.W. Hultman; and T.J. Lewis. Springfield, VA: NTIS, 1982. 4 p. American Nuclear Society Winter Meeting. Washington, D.C., November 14, 1982. (EGG-TIO-M-02282; CONF-821103-76; DE83005393; CONTRACT AC07-761D01570)

Examination of the reactor core and damage debris will be extremely valuable in tracing the effect of the accident on various materials and components. To conduct this examination efficiently, the reactor head and core will be removed and this report describes procedures for the operation.

*Reactor core damage; Reactor core analysis; Reactor head removal

1236
U.S. Department of Energy. E, G, and G Idaho, Inc. Idaho Falls, Idaho. **TMI-2 Technical Information and Examination Program. 1982 Annual Report.** Springfield, VA: NTIS, April 1983. 97 p. (GEND-032; DE83010177; CONTRACT AC07-761D01570)

This report is a summary of the work accomplished by the Technical Information and Examination Program for TMI-2 for the year 1982. The information acquired by this program will be applied to the nuclear power industry in general.

*Reactor accident analysis; Nuclear power plants

1237
U.S. Department of Energy. E, G, and G Idaho, Inc. Idaho Falls, Idaho. **TMI-2 Technical Information and Examination Program. 1983 Annual Report.** D.E. Scardena. Springfield, VA: NTIS, April 1984. 67 p. (GEND-039; DE84010392; CONTRACT AC07-761D01570)

Research over the past four years on the TMI accident has centered on waste immobilization, reactor evaluation, data acquisition, information and industry coordination, core activities and EPICOR II waste and disposition. Problem solving at TMI-2 and development of nuclear safety are also pursued.

*Research and development; Reactor accident analysis; EPICOR; Safety

1238
U.S. Department of Energy. E, G, and G Idaho, Inc. Idaho Falls, Idaho. **TMI-2 Technical Information and Examination Program. 1984 Annual Report.** C.J. Hess. Springfield, VA: NTIS, April 1985. 54 p. (GEND-049; DE85010614/GAR; CONTRACT AC07-761D01570)

Research over the past five years on the TMI accident has centered on waste immobilization, reactor evaluation, data acquisition, information and industry coordination, core activities and EPICOR II waste and disposition. Problem solving at TMI-2 and development of nuclear safety are also pursued.

*Research and development; Reactor accident analysis; EPICOR; Safety

1239
U.S. Department of Energy. E, G, and G Idaho, Inc. Idaho Falls, Idaho. **TMI (Three Mile Island) Unit-2 Technical Information and Examination Program Update.** Springfield, VA: NTIS, November 20, 1981. 12 p. (EGG-B-094-1130-2-5M; DE82007358; CONTRACT AC07-761D01570)

This report encompasses a number of technical facets of the TMI-2 examination plan such as information on the submerged demineralizer system and shipment of EPICOR II wastes.

*Submerged Demineralizer System; Waste shipment; EPICOR; Technical assessment

1240
U.S. Department of Energy. E, G, and G, Idaho, Inc. Idaho Falls, Idaho. **TMI-2 (Three Mile Island Unit 2) Core Damage: A Summary of Present Knowledge.** D.E. Owen; R.E. Mason; R.D. Meininger; and W.A. Franz. Springfield, VA: NTIS, 1983. 8 p. International Meeting on Light-Water Reactor Severe Accident Evaluation. Cambridge, Massachusetts, August 28, 1983. (EGG-M-08483; CONF-830816-41; DE84000892)

This report presents a summary of the core damage in the TMI-2 reactor gathered to date. It includes information on core fragmentatioin, core segments relocated, fuel assemblies at the bottom of the reactor vessel, and the interaction of core materials due to temperature extremes within the reactor vessel as evidenced by leadscrew examination.

*Reactor core damage; Fuel damage; Fuel assemblies; Reactor core temperature; Temperature variation; Leadscrews

1241
U.S. Department of Energy. E, G, and G Idaho, Inc. Idaho Falls. Idaho. **TMI-2 (Three Mile Island Unit 2) Core Examination.** R.R. Hobbins; P.E. MacDonald; and D.E. Owen. Springfield, VA: NTIS, 1983. 9 p. International Meeting on Light-Water Reactor Severe Accident Evaluation. Cambridge, Massachusetts, August 28, 1983. (EGG-M-16383; CONF-830816-39; DE84000888; CONTRACT AC07-761D01570)

Evaluation of the damaged core at TMI-2 is discussed as it relates to safety and examination techniques, e.g. radiation releases, containment integrity, prevention of criticality, transport, etc.

*Reactor core damage; Reactor core analysis; Safety

1242
U.S. Department of Energy. E, G, and G Idaho, Inc. Idaho Falls, Idaho. **Ultrasonic Mapping of the Postaccident TMI-2 Core Configuration.** L.S. Beller and M.R. Martin. Springfield, VA: NTIS, 1984. 5 p. Annual Meeting of the American Nuclear Society. New Orleans, Louisiana, June 3, 1984. (EGG-M-02684; CONF-840614-93; DE84014952; CONTRACT AC07-761D01570)

Ultrasonic techniques are described as they have been applied to the TMI-2 core in an effort to obtain a clearer picture of the core measurements and damage.

*Reactor core damage; Reactor core analysis; Ultrasonic technique

1243
U.S. Department of Energy. E, G, and G Idaho, Inc. Idaho Falls, Idaho. **Understanding the TMI-2 Accident. An Overview.** R.R. Hobbins; J.M. Broughton; and C.M. Allison. Springfield, VA: NTIS, 1986. 7 p. Thermal Reactor Safety Meeting. San Diego, California, February 2, 1986. (EGG-M-19585; CONF-860204-27; DE86005319/GAR; CONTRACT AC07-761D01570)

An overview of the accident at TMI-2 is presented including core damage, radioactive releases, temperature in core debris, and fission products distribution in the reactor.

*Reactor accidents; Reactor core damage; Radioactive releases; Reactor core temperature; Fission product release; Overview

1244
U.S. Department of Energy. E, G, and G Idaho, Inc. Idaho Falls, Idaho. **Uranium Recovery from a Nuclear Fuel.** R.L. Miller; J.M. Welch; and J.E. Finn. Springfield, VA: NTIS, Feburary 1984. 18 p. (EGG-MS-6522; DE84011363; CONTRACT AC07-761D01570)

Methods to recover the uranium fuel from TMI-2 debris were examined. Using a man-made simulation of debris containing uranium, leaching with three different solutions was attempted. Results were considered successful.

*Fuel recovery; Uranium; Experimental data

1245
U.S. Department of Energy. E, G, and G, Idaho, Inc. Idaho Falls, Idaho. **U.S. Department of Energy Three Mile Island Research and Development Program. 1985 Annual Report.** G.R. Brown. Springfield, VA: NTIS, April 1986. 41 p. (GEND-055; DE86009454; CONTRACT AC07-761D01570)

Three major aspects of the TMI program are focused upon: (1) fuel and waste handling, (2) reactor evaluation, and (3) accident evaluation. It is thought that the technology learned from TMI-2 will be applied to the safe operation of other nuclear facilities.

*Waste handling; Reactor accident analysis; Safety; Research and development

1246
U.S. Department of Energy. E, G, and G Idaho, Inc. Idaho Falls, Idaho. **Venting Krypton-85 from the Three Mile Island Unit 2 Reactor Building.** H.M. Burton. Springfield, VA: NTIS, 1981. 7 p. Seminar on the Management of Radioactive Waste from Nuclear Power Plants. Karlsruhe, F.R. Germany, October 5, 1981. (EGG-M-08281; CONF-811056-3; DE82007348/GAR; CONTRACT AC07-761D01570)

This report details the reasons for the necessity of krypton-85 venting from TMI-2 and details the approval process pursued prior to the venting by General Public Utilities. Over seven months were expended in review of the proposed venting before the actual event took place on June 28, 1980.

*Krypton; Venting; Atmospheric releases

1247
U.S. Department of Energy. E, G, and G Idaho, Inc. Idaho Falls, Idaho. Babcock and Wilcox Company. New York, New York. **In-Vessel Inspection Before Head Removal. TMI II: Phase I. (Conceptual Development).** N.E. Calloway; D.W. Greenlee; G.R. Lawrence; A.L. Paglia; and T.D. Piatt. Springfield, VA: NTIS, August 1981. 90 p. (GEND-010-v.1; DE81029492; CONTRACT AC07-761D01570)

The importance of in-vessel inspection to determine the extent of core damage prior to the planning of reactor head removal and core dismantling is discussed.

*Reactor core damage; Reactor head; Reactor core analysis

1248
U.S. Department of Energy. E, G, and G Idaho, Inc. Idaho Falls, Idaho. Babcock and Wilcox Company. Lynchburg, Virginia. **In-Vessel Inspection Before Head Removal. TMI II: Phase II. (Tooling and Systems Design).** D.W. Greenlee. Springfield, VA: NTIS, July 1981. 122 p. (GEND-010-v.2; DE81026957; CONTRACT AC07-761D01570)

This report summarizes the inspection of the TMI-2 reactor core via videocamera through a control rod drive nozzle prior to reactor head removal.

*Videocamera use; Photographs; Reactor core analysis; Reactor head

1249
U.S. Department of Energy. E, G, and G Idaho, Inc. Idaho Falls, Idaho. Allied General Nuclear Services. Barnwell, South Carolina. **Cannister-Design Considerations for Packaging of TMI Unit 2 Damaged Fuel Debris.** G.A. Townes. Springfield, VA: NTIS, September 1981. 91 p. (GEND-011; DE82000885; CONTRACT AC07-761D01570)

Cannister designs for TMI-2 fuel debris containment are discussed. Fuel debris will first be stored on-site and later shipped to off-site locations either for storage or reprocessing. Eventually, the debris will be buried. Multi-application cannisters, to support all of the various technical processing chores, are considered.

*Cannister, high integrity; Fuel debris; Debris, radioactive; Waste shipment; Waste disposal

1250
U.S. Department of Energy. E, G, and G, Idaho, Inc. Middletown, Pennsylvania. **Technology Transfer in the Department of Energy Three Mile Island Research Programs.** F.L. Meltzer and B.A. Ettinger. Springfield, VA: NTIS, 1982. 4 p. American Society of Mechanical Engineers Annual Winter Meeting. Phoenix, Arizona, November 14, 1982. (EGG-TIO-M-00382; CONF-821101-18; DE83005396; CONTRACT AC07-761D01570)

The importance of sharing technical information relating to the accident at TMI-2 is emphasized and the joint Technical Information and Examination Program is described.

*Reactor accident analysis; Technical assessment; Research and development

1251
U.S. Department of Energy. Electric Power Research Institute. Palo Alto, California. Nuclear Safety Analysis Center. **Analysis of Three Mile Island - Unit 2 Accident.** Springfield, VA: NTIS, March 1980. 527 p. (EPRI-NSAC-80-1-rev.; DE82901520)

The accident at TMI-2 was studied by the Nuclear Safety Analysis Center at EPRI. This report contains a detailed analysis of the accident and includes references and appendices.

*Reactor accident analysis

1252
U.S. Department of Energy. Electric Power Research Institute. Palo Alto, California. Nuclear Power Division. **Joint TMI-2 Information and Examination Program: EPRI Participation and Support, 1982 Progress and 1983 Program Plan.** Springfield, VA: NTIS, February 1983. 95 p. (EPRI-NP-2907-SR; DE83901866)

This report describes the contributions to the TMI-2 cleanup program made through the projects of the Electric Power Research Institute.

*Cleanup; Decontamination

1253
U.S. Department of Energy. Electric Power Research Institute. Palo Alto, California. Nuclear Safety Analysis Center. **Indexed Bibliography of TMI-2 Documents in NSAC Working File.** Springfield, VA: NTIS, January 1980. 472 p. (EPRI-NSAC-Bib4; DE82901521)

The bibliography contains 1212 documents accumulated by NSAC/EPRI largely pertaining to the accident at TMI-2 and its consequences.

*Bibliographies

1254
U.S. Department of Energy. Electric Power Research Institute. Palo Alto, California. Nuclear Safety Analysis Center. **Mitigation of Small-Break LOCAs in Pressurized Water Reactor Systems.** Springfield, VA: NTIS, March 1980. 54 p. (EPRI-NSAC-2; DE82901454)

This report explores the events that would have been necessary for the TMI-2 reactor core to breach containment.

*Loss of coolant accident; Pressurized water reactors; Experimental data

1255
U.S. Department of Energy. Electric Power Research Institute. Palo Alto, California. Nuclear Safety Analysis Center. **TMI-2 Accident: Core Heat-Up Analysis.** K.H. Ardron and D.G. Cain. Springfield, VA: NTIS, January 1981. 92 p. (EPRI-NSAC-24; DE82901453)

This report reviews the available data both direct and indirect pertaining to the TMI-2 reactor core during the first two hours immediately following the accident.

*Reactor core temperature; Reactor accident analysis

1256
U.S. Department of Energy. General Public Utilities Service Corporation. Parsippany, New Jersey. **Control of Radiolytic Gases in Liners of Radioactive Zeolites.** J. Greenberg; J.O. Henrie; and G.J. Quinn. Springfield, VA: NTIS, 1982. 5 p. American Nuclear Society Winter Meeting. Washington, D.C., November 14, 1982. (EGG-TIO-M-01482; CONF-821103-79; DE83005546; CONTRACT AC07-761D01570)

The production of radiolytic gases is a recognized hazard as the liners from the TMI-2 Submerged Demineralizer System decompose. This report describes the use of a catalyst to control the gases.

*Radiolytic gas; EPICOR; Filters; Submerged Demineralizer System

1257
U.S. Department of Energy. General Public Utilities Nuclear Corporation. Middletown, Pennsylvania. **Preparations to Ship EPICOR Liners.** S.P. Queen. Springfield, VA: NTIS, June 1983. 59 p. (GEND-029; DE83015920; CONTRACT AC07-761D01570)

This report discusses the radiolytic hydrogen generation from the EPICOR II prefilters used in decontamintation of the water at TMI-2, gas sampling tools and methods for inactivating gas production.

*Radiolytic gas; Hydrogen; EPICOR; Prefilter liners; Decontamination; Cleanup

1258
U.S. Department of Energy. General Public Utilities Nuclear Corporation. Middletown, Pennsylvania. **TMI-2 Reactor Vessel Plenum Final Lift.** D.C. Wilson. Springfield, VA: NTIS, January 1986. 107 p. (GEND-054; DE86005859/GAR; CONTRACT AC07-761D01570)

A description of the preparation for, and final removal of, the plenum assembly is presented as it was accomplished on May 15, 1985. The removal of the plenum was necessary to the defueling process of the damaged reactor at TMI-2.

*Plenum; Defueling

1259
U.S. Department of Energy. GPU Nuclear Corporation. Middletown, Pennsylvania. **Research and Development Activities on Three Mile Island Unit Two. Annual Report of 1985.** Springfield, VA: NTIS, April 1986. 96 p. (DOE/NE/34109-T1; DE86011045/GAR; CONTRACT GI01-84NE34109)

This report summarizes the TMI-2 cleanup activities for the year 1985. Remarkable among these operations were the plenum lift revealing access to the reactor for removal of damaged fuel. A number of other topics were also covered: fuel shipping casks, waste management, decontamination, etc.

*Cleanup; Plenum; Fuel damage; Defueling; Waste handling; Decontamination; Research and development

1260
U.S. Department of Energy. GPU Nuclear Corporation. Middletown, Pennsylvania. **TMI-2 Reactor Vessel Head Removal.** P.R. Bengel; M.D. Smith; and G.A. Estabrook. Springfield, VA: NTIS, December 1984. 99 p. (GEND-INF-062; DE85005498/GAR; CONTRACT AC07-761D01570)

This report described the removal and storage of the TMI-2 reactor head in order to gain access to the core.

*Reactor head removal

1261
U.S. Department of Energy. GPU Nuclear Corporation. Middletown, Pennsylvania. **TMI-2 Reactor Vessel Head Removal.** P.R. Bengel; M.D. Smith; and G.A. Estabrook. Springfield, VA: NTIS, September 1985. 118 p. (GEND-044; DE85017613/GAR; CONTRACT AC07-761D01570)

This report discusses the removal and storage of the TMI-2 reactor head preparatory to the removal of the reactor core. Data obtained from the activity will aid in the fuel removal operations.

*Reactor head removal; Defueling

1262
U.S. Department of Energy. Hanford Engineering Development Laboratory. Richland, Washington. **Advances in Continuous Gamma-Ray Spectrometry and Applications.** R. Gold; J.P. McNeece; and B.J. Kaiser. Springfield, VA: NTIS, July 1984. 16 p. American Society for Testing Materials. Euratom Symposium on Reactor Dosimetry. Geesthacht, F.R. Germany, September 24, 1984. (HEDL-SA-3125; CONF-840902-17; DE85003018/GAR; CONTRACT AC06-76FF02170)

Uses of continuous gamma-ray spectrometry are presented with emphasis on applications in several specific areas, notably the assessment of fuel distributions for TMI-2 reactor recovery.

*Spectrometry; Fuel; Fuel recovery; Experimental data

1263
U.S. Department of Energy. Hanford Engineering Development Laboratory. Richland, Washington. **Applications of Solid State Track Recorder Neutron Dosimetry for Fuel Debris Location in the Three Mile Island Unit 2 Reactor Coolant System.** F.H. Ruddy; J.H. Roberts; R. Gold; and C.C. Preston. Springfield, VA: NTIS, 1983. 10 p. International Conference on Solid State Nuclear Track Detectors. Acapulco, Mexico, September 4, 1983. (HEDL-SA-2861; CONF-830969-10; DE84006351; CONTRACT AC06-76FF02170)

The usefulness of solid state track recorder in the location and recovery of TMI-2 fuel debris is discussed and estimated measurement of fuel in Demineralizer A is presented.

*Fuel debris; Fuel; Fuel recovery; Dosimetry, neutron

1264
U.S. Department of Energy. Hanford Engineering Development Laboratory. Richland, Washington. **Characterization of Fuel Distributions in the Three Mile Island Unit 2 (TMI-2) Reactor System by Neutron and Gamma-Ray Dosimetry.** R. Gold; J.H. Roberts; F.H. Ruddy; C.C. Preston; and J.P. McNeece. Springfield, VA: NTIS, April 1984. 5 p. American Society for Testing and Materials. Euratom Symposium on Reactor Dosimetry. Geesthacht, F.R. Germany, September 24, 1984. (HEDL-SA-3063A; CONF-840902-13; DE85000292/GAR; CONTRACT AC06-76FF02170)

Assessment studies of fuel and waste and their disposal in the TMI-2 reactor are considered as they relate to reactor recovery operations. The importance of the technical issues is noted.

*Waste disposal; Fuel recovery; Reactor accident analysis; Dosimetry, neutron; Dosimetry, gamma-ray

1265
U.S. Department of Energy. Hanford Engineering Development Laboratory. Richland, Washington. **Effect of Three Mile Island LOCA on Cables, Connector Components.** N.S. Cannon; H.J. Helbert; R. Trjillo; L.A. Hecker; and C.P. Cannon. Springfield, VA: NTIS, June 1983. 7 p. American Nuclear Society Winter Meeting. San Francisco, California, October 30, 1983. (HEDL-SA-2980; CONF-831047-136; DE84006363; CONTRACT AC06-76FF02170)

Tests on the TMI-2 cables and connectors that were subjected to a wide variety of adverse conditions during the accident were performed both on location and in the laboratory to determine the effects of the various circumstances.

*Loss of coolant accident; Cables; Electrical equipment

1266
U.S. Department of Energy. Hanford Engineering Development Laboratory. Richland, Washington. **Fuel Debris Assessment for Three Mile Island Unit 2 (TMI-2) Reactor Recovery by Gamma-Ray and Neutron Dosimetry.** R. Gold; J.H. Roberts; J.P. McNeece; B.J. Kaiser; and F.H. Ruddy. Springfield, VA: NTIS, August 1983. 17 p. International Conference on Nondestructive Evaluation in the Nuclear Industry. Zurich, Switzerland, November 27, 1983. (HEDL-SA-2867-FP; CONF-831175-1; DE84003950; CONTRACT AC06-76FF02170)

The use of solid state track recorder neutron dosimetry and gamma-ray dosimetry is described as applied to the study of TMI-2 fuel debris. The results of the two methods appear to be in harmony.

*Fuel debris; Dosimetry, neutron; Dosimetry, gamma-ray; Experimental data

1267
U.S. Department of Energy. Hanford Engineering Development Laboratory. Richland, Washington. **Neutron Dosimetry in the Three Mile Island Unit 2 Reactor Cavity with Solid State Track Recorders.** R. Gold; J.H. Roberts; F.H. Ruddy; C.C. Preston; and W.N. McElroy. Springfield, VA: NTIS, 1985. 5 p. International Conference on Solid State Nuclear Tract Detectors. Rome Italy, September 23, 1985. (HEDL-SA-3293; CONF-850933-1; DE85018370/GAR; CONTRACT AC06-76FF02170)

Solid state track recorder (SSTR) neutron dosimetry was conducted in the TMI-2 reactor cavity to determine the amount of fuel debris lying at the bottom. Recorded intensity indicates that probably four fuel assemblies lie at the base of the vessel.

*Reactor core analysis; Fuel debris; Fuel assemblies; Dosimetry, neutron

1268
U.S. Department of Energy. Hanford Engineering Development Laboratory. Richland, Washington. **Planning Study, Resin and Debris Removal System. Three Mile Island Nuclear Station Unit 2 Make-Up and Purification Demineralizers.** E.J. Rankey and W.W. Jenkins. Springfield, VA: NTIS, June 1983. 33 p. (HEDL-7377; DE83017009; CONTRACT AC06-76FF02170)

Methods for the removal of debris, cesium and complex compounds, and resins from the purification demineralizers at TMI-2 are discussed. Keeping the system as free of contamination as possible during the operations and eventual disposal of the waste removed are the goals.

*Decontamination; Demineralizers; Waste disposal; Debris, radioactive; Cesium radioisotopes; Resins

1269
U.S. Department of Energy. Hanford Engineering Development Laboratory. Richland, Washington. **Solid State Track Recorder Neutron Dosimetry in the Three Mile Island Unit-2 Reactor Cavity.** R. Gold; J.H. Roberts; F.H. Ruddy; C.C. Preston; and W.N. McElroy. Springfield, VA: NTIS, April 1985. 29 p. (HEDL-7484; DE85011360/GAR; CONTRACT AC06-76FF02170)

Solid state track recorder neutron dosimetry measurements indicate that a significant amount of fuel debris is on the floor of the reactor vessel, approximately four fuel assemblies.

*Reactor core analysis; Fuel debris; Fuel assemblies; Dosimetry, neutron

1270
U.S. Department of Energy. Hanford Engineering Development Laboratory. Richland, Washington. **Solid State Track Recorder Neutron Dosimetry Measurements for Fuel Debris Assessment of TMI-2 Demineralizer A.** F.H. Ruddy; J.H. Roberts; R. Gold; C.C. Preston; and J.A. Ulseth. Springfield, VA: NTIS, December 1982. 45 p. (HEDL-SA-2834-FP; DE83017004; CONTRACT AC06-76FF02170)

The use of solid state track recorder neutron dosimetry to evaluate the amount of fuel debris in TMI-2 is presented and compared with data from Compton recoil gamma-ray spectrometry. The two correlate well.

*Fuel debris; Dosimetry, neutron; Experimental data

1271
U.S. Department of Energy. Idaho National Engineering Laboratory. Idaho Falls, Idaho. **Development of In Situ Test Procedures for TMI-2 Axial Power Shaping Rod-Drive Mechanisms.** J.A. Gannon. Springfield, VA: NTIS, November 1982. 48 p. (GEND-INF-028; DE83004395; CONTRACT AC07-761D01570)

The development of Axial Power Shaping Rod-Drive (APSR) mechanisms for assessment of the reactor core at TMI-2 is described.

*Axial power shaping rod; Reactor core analysis

1272
U.S. Department of Energy. Idaho National Engineering Laboratory. Idaho Falls, Idaho. **EPICOR II Research and Disposition Program at the INEL (Idaho National Engineering Laboratory).** H.W. Reno and R.L. Dodge. Springfield, VA: NTIS, September 13, 1982. 15 p. American Nuclear Society Workshop on LLW Packing and Shipping. San Diego, California, September 12, 1982. (EGG-M-19482; CONF-820919-3; DE83001352; CONTRACT AC07-761D01570)

Cleanup activities at TMI-2 have produced 50 EPICOR II prefilter liners which were used in filtration of contaminated water. This report discusses their handling through the use of high integrity containers to disposal at a remote commercial waste disposal site.

*EPICOR; Prefilter liners; Filtration; Cannister, high integrity; Waste disposal; Research and development

1273
U.S. Department of Energy. Idaho National Engineering Laboratory. Idaho Falls, Idaho. **EPICOR II Resin/Liner Research Plan.** J.W. McConnell. Springfield, VA: NTIS, March 1983. 34 p. (EGG-TMI-6198; DE83009196; CONTRACT AC07-761D01570)

A description of the present EPICOR II prefilter study from the TMI-2 accident is presented, and several possible areas for expansion of the study explored.

*EPICOR; Prefilter liners; Research and development

1274
U.S. Department of Energy. Idaho National Engineering Laboratory. Idaho Falls, Idaho. **Interim Report on the TMI-2 Purification Filter Examination.** R.E. Mason. Springfield, VA: NTIS, February 1983. 137 p. (EGG-TMI-6181; DE83009199; CONTRACT AC07-761D01570)

This interim report discusses the filters from TMI-2 purification/make-up system, including the nature and size of debris removed from them as well as the radionuclide content.

*Filters; Filtration; Debris, radioactive; Radioactive waste

1275
U.S. Department of Energy. Idaho National Engineering Laboratory. Idaho Falls, Idaho. **Passive Surveillance: A Technique to Characterize the Condition of Power and Control Circuits in a Nuclear Plant.** R.D. Meininger and M.R. Dinsel. Springfield, VA: NTIS, 1985. 6 p. IEEE Nuclear Science Symposium. San Francisco, California, October 23, 1985. (EGG-M-27785; CONF-851009-34; DE86004153/GAR; CONTRACT AC07-761D01570)

This short paper details the progress made in the examination of electrical circuits in the TMI-2 unit after the accident.

*Electrical equipment; Nuclear power plants; Reactor accidents

1276
U.S. Department of Energy. Idaho National Engineering Laboratory. Idaho Falls, Idaho. **Task Plan for the U.S. Department of Energy TMI-2 Programs.** Springfield, VA: NTIS, October 1982. 36 p. (GEND-INF-036; DE83002815; CONTRACT AC07-761D01570)

The Task Plan presented for the D.O.E. TMI-2 Program includes three major areas: (l) Data acquisition, (2) Waste immobilization, and (3) Reactor evaluation.

*Reactor accident analysis; Waste handling

1277
U.S. Department of Energy. Idaho National Engineering Laboratory. Idaho Falls, Idaho. **Technical Assessment of Radiation Overexposures at Three Mile Island from August 28, 1979, Entry into the Unit 2 Fuel-Handling Building Makeup Valve Room.** B.L. Rich and S.R. Adams. Springfield, VA: NTIS, June 1980. 28 p. (EGG-SD-5159; DE83005518; CONTRACT AC07-761D01570)

An entry into the TMI-2 Fuel Handling Building Makeup Valve Room on August 28, 1979 resulted in radiation overexposure to six personnel. The overexposure was investigated and a report by Metropolitan Edison Comapny was submitted to the Nuclear Regulatory Commission, who, in turn, asked an independent contractor to evaluate the doses.

*Radiation exposure, occupational; Nuclear workers; Radiation dosage

1278
U.S. Department of Energy. Idaho National Engineering Laboratory. Idaho Falls, Idaho. **Testing and Examination of TMI-2 Electrical Components and Discrete Devices.** F.T. Soberano. Springfield, VA: NTIS, November 1982. 81 p. (GEND-INF-034; DE83000810; CONTRACT AC07-761D01570)

Electrical material at the TMI-2 reactor was examined after the accident, and tests performed while it was in place are described.

*Electrical equipment; Reactor accidents; Equipment function

1279
U.S. Department of Energy. Idaho National Engineering Laboratory. Idaho Falls, Idaho. **Thermal Hydraulic Features of the TMI Accident.** B. Tolman. Springfield, VA: NTIS, 1985. 28 p. American Chemical Society Meeting. Miami, Florida, May 1, 1985 (EGG-M-12085; CONF-8505179-1; DE85014569; CONTRACT AC07-761D01570)

Based upon TMI-2 core damage data, a scenario of the accident -- with severe fuel damage -- is proposed. Better understanding of critical core damage conditions provides information upon which safety measures can be predicated.

*Reactor core damage; Thermal hydraulics; Reactor accident analysis; Simulation

1280
U.S. Department of Energy. Idaho National Engineering Laboratory. Idaho Falls, Idaho. **TMI-2 Core Examination Program INEL Facilities - Readiness Study.** T.B. McLaughlin. Springfield, VA: NTIS, September 1982. 91 p. (EGG-TMI-5966; DE83000679; CONTRACT AC07-761D01570)

Evaluation of the Idaho National Engineering Laboratories facilities for handling of TMI-2 offsite core samples is presented.

*Reactor core analysis

1281
U.S. Department of Energy. Idaho National Engineering Laboratory. Idaho Falls, Idaho. **TMI-2 Core-Examination Program: INEL Facilities Readiness Study.** T.B. McLaughlin. Springfield, VA: NTIS, February 1983. 83 p. (EGG-TMI-5966-rev.1; DE83009200; CONTRACT AC07-761D01570)

A description of the Idaho National Engineering Laboratories facilities readiness for receiving, examining, and storing TMI-2 core samples is presented.

*Reactor core analysis

1282
U.S. Department of Energy. Idaho National Engineering Laboratory. Idaho Falls, Idaho. **TMI-2 Fuel/Canister Interface Requirements for INEL.** D.E. Wilkins. Springfield, VA: NTIS, January 1983. 102 p. (EGG-TMI-6156; DE83005746; CONTRACT AC07-761D01570)

This document discusses procedures for receipt and storage of the TMI-2 core debris cannisters at the Idaho National Engineering Laboratory.

*Cannister, high integrity; Debris, radioactive; Fuel debris

1283
U.S. Department of Energy. International Atomic Energy Agency. Vienna, Austria. **International Experience in the Implementation of the Lessons Learned from the Three Mile Island Incident.** Springfield, VA: NTIS, August 1983. 271 p. (IAEA-TECDOC-294; DE84700763)

A three-part document consisting of TMI-2 lessons learned, specific responses in ten identified areas and individual country examples of use of TMI-2 lessons, this report should be valuable to any country with existing nuclear power generation programs or contemplating such a program.

*Accident investigations; International raction; Recommendations; Lessons learned

1284
U.S. Department of Energy. International Energy Associates, Limited. Washington, D.C. **Examination of Mechanical Snubbers from TMI-2. Final Report.** F.L. Wadsworth; R.S. Walker; and E.P. Kurdziel. Springfield, VA: NTIS, March 1983. 116 p. (EPRI-NP-2966; DE83901978)

Five mechanical snubbers were the first components removed from TMI-2 for examination. Four were found to be severely damaged and non-operative from heat; one remained undamaged.

*Mechanical snubbers; Instrument function; Instrument failure

1285
U.S. Department of Energy. International Energy Associates, Limited. Washington, D.C. **TMI-2 Mechanical-Component Information and Examination Program and Progress Report.** E.P. Kurdziel; C.A. Negin; F.L. Wadsworth; R.S. Walker; and G. Worku. Springfield, VA: NTIS, September 1982. 187 p. (EPRI-NP-2625; DE83900153)

A collection and examination program for TMI-2 mechanical components exposed to the accident is reported. It is the intention of the program to apply the information obtained to future safety measures and better design of mechanical equipment. The most important, and expensive, mechanical device discussed is the polar crane.

*Equipment function; Information; Reactor accidents; Safety; Polar crane

1286
U.S. Department of Energy. International Nuclear Law Association. Brussels, Belgium. **Public Attitudes Toward Nuclear Power and the TMI Accident.** B.K. Kanga. Springfield, VA: NTIS, 1983. 6 p. Nuclear Inter Jura '83. San Francisco, California, September 11, 1983. (INIS-mf-9549; CONF-8309278-22; DE85780998/GAR)

Drawing attention to the public attitudes on the TMI-accident and nuclear power in general, the importance of a safe nuclear power generation track record is noted if there is to be public understanding and acceptance for the future.

*Public opinion; Attitude to nuclear power; Attitude to TMI; Nuclear industry; Nuclear power plants

1287
U.S. Department of Energy. Japan Atomic Energy Research Institute. Tokyo, Japan. **Examination of TMI Accident Simulation Test at LSTF (Large Scale Test Facility) in ROSA-IV Program.** M. Tanaka; K. Tasaka; H. Ito; T. Owada; and A. Owada. Springfield, VA: NTIS, July 1983. 65 p. (JAERI-M-83-105; DE84703557/GAR)

A simulation of the TMI accident using a LSTF is examined.

*Computer simulations; Reactor accident analysis

1288
U.S. Department of Energy. Japan Atomic Energy Research Institute. Tokyo, Japan. **Simulation of a TMI-2 Type Scenario at the ROSA-IV Program's Large Scale Test Facility: A First Look.** R.R. Schultz; Y. Kukita; and K. Tasaka. Springfield VA: NTIS, September 1984. 74 p. (JAERI-M-84-176; DE86701003/GAR)

Computer simulation analysis of the TMI-2 accident was studied with the ROSA-IV program.

*Computer simulations; Reactor accident analysis; Experimental data

1289
U.S. Department of Energy. Kernenergie-Studiengesellschaft-Studiengesellschaft zur Foerderung der Kernenergieverwertung in Schiffbau ud Schuffahrt e.V. Hamburg, Germany, F.R. **Three Mile Island 2. The Accident of March 28, 1979, as Seen from the Point of View of Radiation Protection Experts.** W. Mueller. Springfield, VA: NTIS, 1980. 28 p. Text of a paper presented November 13, 1980. (INIS-mf-8506; DE83703592)

A paper presented at a conference in Hamburg, Germany, this document reviews the accident at TMI-2 from the perspective of radiation protection personnel and judges the effects to be comparatively mild.

*Reactor accident analysis; Radiation protection; Health effects

1290
U.S. Department of Energy. Lawrence Livermore National Laboratory. Livermore, California. **Assessment of Extent and Degree of Thermal Damage to Polymeric Materials in the Three Mile Island Unit 2 Reactor Building. Volume VI.** N.J. Alvares. Springfield, VA: NTIS, February 1984. 59 p. (GEND-INF-023-v.6; DE84010031; CONTRACT AC07-761D01570)

A survey of the burn damage to the TMI-2 Reactor Building revealed non-uniform burn patterns. In some areas where like materials were adjacent, one sustained damage while the other did not. Tests indicate that the reactor building atmosphere consisted of about 8% hydrogen, a factor consistent with thermal damage.

*Hydrogen burn; Reactor building; Thermal damage

1291
U.S. Department of Energy. Lawrence Livermore National Laboratory. Livermore, California. **Investigation of Hydrogen-Burn Damage in the TMI-2 Reactor Building.** N.J. Alvares; D.G. Beason; G.R. Eidam; and D.L. Reeder. Springfield, VA: NTIS, 1982. 5 p. American Nuclear Society Winter Meeting. Washington, D.C.,

November 14, 1982. (EGG-TIO-M-01982; CONF-821103-80; DE83005548; CONTRACT AC07-761D01570)

This report is a joint analysis of the hydrogen burn damage by the Fire Science Group of the Hazards Control Department of Lawrence Livermore National Laboratory and E, G, and G, the Nuclear Regulatory Commission, and the Department of Energy through the use of photographs obtained from the previous 15 entries into the TMI-2 reactor building.

*Hydrogen burn; Reactor building; Photographs; Thermal damage

1292
U.S. Department of Energy. Lawrence Livermore National Laboratory. Livermore, California, and E, G, and G, Inc. Idaho Falls, Idaho. **Investigation of Hydrogen-Burn Damage in the Three Mile Island Unit 2 Reactor Building.** N.J. Alvares; D.G. Beason; and G.R. Eidam. Springfield, VA: NTIS, June 1982. 67 p. (GEND-INF-023-v.1; DE82018639; CONTRACT AC07-761D01570)

Photographic data obtained from the first fifteen entries into the TMI-2 reactor building have permitted the determination of possible damage caused by the hydrogen burn which occurred on the day of the accident.

*Hydrogen burn; Photographs; Reactor building; Thermal damage

1293
U.S. Department of Energy. Los Alamos National Laboratory. Los Alamos, New Mexico. **Fuel Models and Results from the TRAC-PF1/MIMAS TMI-2 Accident Calculation.** E.C. Schwegler and P.J. Maudlin. Springfield, VA: NTIS, 1983. 4 p. International Meeting on Light-Water Reactor Severe Accident Evaluation. Cambridge, Massachusetts, August 28, 1983. (LA-UR-83-963; CONF-830816-10; DE83015240; CONTRACT W-7405-ENG-36)

Calculations by TRAC-PF1/MIMAS of fuel rod behavior at TMI-2 are presented and analyzed.

*Computer simulations; Fuel rods; Models

1294
U.S. Department of Energy. Los Alamos National Laboratory. Los Alamos, New Mexico. **NDA (Nondestructive Analysis) Measurement of the Demineralizers at TMI-2.** J.R. Phillips; J.K. Halbig; P.M. Rinard; and G.E. Bosier. Springfield, VA: NTIS, August 1983. 34 p. (LA-9795-MS; DE84002295; CONTRACT W-7405-ENG-36)

Using gross gamma-ray measurements, the amount of fuel in the two TMI-2 demineralizers was estimated at be 2 to 7 kg in Demineralizer A and 0.7 kg in Demineralizer B.

*Fuel recovery; Filtration; Demineralizers; Experimental data

1295
U.S. Department of Energy. Los Alamos National Laboratory. Los Alamos, New Mexico. **TMI-2 Fission-Product Elemental and Isotopic Inventories.** T.R. England and W.B. Wilson. Springfield, VA: NTIS, January 1983. 5 p. (LA-9622-MS; DE83007121; CONTRACT W-7405-ENG-36)

A catalog of TMI-2 generated nuclides is reported along with the radiation production calculation.

*Fission products; Radionuclides; Radiation analysis

1296
U.S. Department of Energy. Mound Facility. Miamisburg, Ohio. **Tritium Waste Control: October 1981 - March 1982.** P.H. Lamberger and M.L. Rogers. Springfield, VA: NTIS, August 1982. 21 p. (MLM-2969; DE82021902; CONTRACT AC04-76DP00053

Tritium waste control processes are utilizing a number of methods. Of particular relevance is the decontamination of tritium contaminated water at TMI-2.

*Tritium; Contaminated water; Waste handling

1297
U.S. Department of Energy. New York University Medical Center. New York, New York. Institute of Environmental Medicine. **Gamma-Ray Spectrometer System for High Radiation Fields. Final Report.** G.R. Lauer; T. Rahon; and M. Eisenbud. Springfield, VA: NTIS, November 1984. 38 p. (EPRI-NP-3804; DE85005800/GAR; CONTRACT AC02-78EV04957)

The availability of a new gamma-ray spectrometer system much improved from the one used at TMI-2 is noted. The spectrometer measures fission products and fuel debris particles.

*Fuel debris; Fission products; Spectrometry

1298
U.S. Department of Energy. North Carolina University at Chapel Hill. Chapel Hill, North Carolina. School of Law. **TMI-2 Information and Examination Program 1981. Annual Report.** Springfield, VA: NTIS, April 1982. 94 p. (GEND-022; DE82012570; CONTRACT AC07-761D01570)

The annual report for 1981 on the TMI-2 Technical Information and Examination Program notes the areas of concentration of study. This program is important in that the knowledge gained is shared among other nuclear facilities.

*Technical assessment; Reactor accident analysis

1299
U.S. Department of Energy. Nuclear Associates International Corporation. Rockville, Maryland. **TMI-2 Accident: Core Heat-Up Analysis. A Supplement.** Springfield, VA: NTIS, June 1981. 236 p. (EPRI-NSAC-25; DE82901519)

This report analyzes the time chain of the TMI-2 accident relating to core heat-up. This was accomplished in three parts: by single pin analysis; by thermal-hydraulic core analysis; and by TMI-2 materials design and characteristics.

*Reactor core temperature; Reactor core analysis; Temperature variation

1300
U.S. Department of Energy. NUS Corporation. Gaithersburg, Maryland. **Radionuclide Mass Balance for the TMI-2 Accident: Data-Base System and Preliminary Mass Balance. Volume 1.** M.I. Goldman; R.J. Davis; J.F. Strahl; W.C. Arcieri; and D.W. Tonkay. Springfield, VA: NTIS, April 1983. 95 p. (GEND-INF-032-v.1; DE83012090; CONTRACT AC07-761DO1570)

An accounting of the radionuclide species generated by the TMI-2 accident is presented. The construction of a computerized data base to support future study is followed.

*Radionuclides; Reactor accidents; Computer programs

1301
U.S. Department of Energy. NUS Corporation. Gaithersburg, Maryland. **Radionlucide Mass Balance for the TMI-2 Accident: Data-Base System and Preliminary Mass Balance. Volume 2.** M.I. Goldman; R.J. Davis; J.F. Strahl; W.C. Arcieri; and D.W. Tonkay. Springfield, VA: NTIS, April 1983. 169 p. (GEND-INF-032-v.2; DE83012091; CONTRACT AC07-761D01570)

The second volume of a two-volume report, this document contains tables of radionuclide levels from TMI-2.

*Radionuclides; Reactor accidents

1302
U.S. Department of Energy. NUS Corporation. Gaithersburg, Maryland. **Radionuclide Mass Balance for the TMI-2 Accident: Data Through 1979 and Preliminary Assessment of Uncertainties.** R.J. Davis; D.W. Tonkay; E.A. Vissing; T.D. Nguyen; and L.W. Shawn. Springfield, VA: NTIS, November 1984. 137 p. (GEND-INF-047; DE85004314/GAR; CONTRACT AC07-761D01570)

Disposition of escaped radionuclides (tritium, cesium, iodine) from the TMI-2 accident are presented. Information has been assembled into a systematic data base.

*Radionuclides; Fission product release; Tritium; Iodine radioisotopes; Cesium radioisotopes

1303
U.S. Department of Energy. Oak Ridge Associated Universities, Inc. Oak Ridge, Tennessee. **Review of Light Water Reactor Safety Through the Three Mile Island Accident.** D.L. Phung. Springfield, VA: NTIS, May 1984. 130 p. (ORAU/IEA-84-2(M); DE84012891; CONTRACT AC05-76OR00033)

A pre-TMI accident reactor safety assessment is examined in light of the ensuing events, and it is concluded that "basic flaws" in technology and safety philosophy existed.

*Safety; Light water reactors; Reactor accident analysis

1304
U.S. Department of Energy. Oak Ridge National Laboratory. Oak Ridge, Tennessee. **Accident at Three Mile Island and Its Aftermath.** A.P. Malinauskas. Springfield, VA: NTIS, April 18, 1983. 35 p. (DOE/NBM-3012666; DE83012066; CONTRACT W-7405-ENG-26)

The TMI-2 reactor accident, activity after, and costs of, are described.

*Reactor accident analysis; Economic effects; Costs

1305
U.S. Department of Energy. Oak Ridge National Laboratory. Oak Ridge, Tennessee. **Cleanup of TMI-2 Demineralizer Resins.** W.D. Bond; L.J. King; J.B. Knauer; K.J. Hofstetter; and J.D. Thompson. Springfield, VA: NTIS, 1985. 27 p. National Meeting of the American Chemical Society. Miami, Florida, April 28, 1985. (CONF-850417-18; DE85011431/GAR; CONTRACT AC05-840R21400)

The process of removal of the radiocesium from TMI-2 demineralizer resins is discussed. Seventeen stages are required and, in this report, fifteen stages have been completed with 750 curies of cesium removed from demineralizer A and 3300 curies from demineralizer B.

*Cleanup; Decontamination; Demineralizers; Resins; Cesium radioisotopes

1306
U.S. Department of Energy. Oak Ridge National Laboratory. Oak Ridge, Tennessee. **Criticality Analysis for TMI-2 Defueling.** R.M. Westfall. Springfield, VA: NTIS, 1985. 24 p. Topical Meeting of Criticality Safety and the Storage of Fissile Material. Jackson, Wyoming, September 9, 1985. (CONF-850917-7; DE86008974/GAR; CONTRACT AC05-84OR21400)

A strategy outlining the optimum soluable boron level in the reactor core for the safe subcritical defueling operation at TMI-2 is presented.

*Defueling; Criticality; Boron

1307
U.S. Department of Energy. Oak Ridge National Laboratory. Oak Ridge, Tennessee. **Development of a Promising Filtration Method for Liquid Clarification in Nuclear Facilities.** E.D. Collins; J.B. Knauer; L.A. Byrd; R.G. Ross; and H.C. Savage. Springfield, VA: NTIS, 1982. 22 p. Annual Meeting of Heat Transfer and Energy Conversion of the American Institute of Chemical Engineers. Los Angeles, California, November 14, 1982. (CONF-821106-34; DE84004387; CONTRACT W-7405-ENG-26)

Diatomaceous earth was proposed as a filter for contaminated TMI-2 water since, as an inorganic compound, diatomaceous earth is resistant to radiation decomposition.

*Filters; Filtration; Contaminated water

1308
U.S. Department of Energy. Oak Ridge National Laboratory. Oak Ridge, Tennessee. **Development of the Flowsheet Used for Decontaminating High-Activity-Level Water at TMI-2. Draft.** E.D. Collins; D.O. Campbell; L.J. King; J.B. Knauer; and R.M. Wallace. Springfield, VA: NTIS, 1985. 31 p. National Meeting of the American Chemical Society. Miami, Florida, April 28, 1985. (CONF-850417-23; DE85012782/GAR; CONTRACT AC05-84OR21400)

Devising a flowsheet using small samples of water from TMI-2 contaminated with cesium, antimony, strontium, and ruthenium is detailed.

*Contaminated water; Cesium radioisotopes; Strontium radioisotopes; Ruthenium; Antimony

1309
U.S. Department of Energy. Oak Ridge National Laboratory. Oak Ridge, Tennessee. **Effect of Boron and Gadolinium Concentration on the Calculated Neutron Multiplication Factor of U(3)O sub 2 Fuel Pins in Optimum Geometries.** J.T. Thomas. Springfield, VA: NTIS, October 1984. 74 p. (ORNL/CSD/TM-218; DE85003793/GAR; CONTRACT AC05-84OR21400)

Studies on the KENO-Va improved Monte Carlo program at the TMI-2 reactor indicate usefulness in the defueling process.

*Computer programs; Decision making, computer assisted; Defueling; Boron; Gadolinium

1310
U.S. Department of Energy. Oak Ridge National Laboratory. Oak Ridge, Tennessee. **Estimated Amount of Exp 85 Kr Available for Release from Intact Fuel Rods in the Three Mile Island Unit 2 Nuclear Power Station.** R.A. Lorenz. Springfield, VA: NTIS, October 1983. 28 p. (ORNL/TM-8730; DE84001140; CONTRACT W-7405-ENG-26)

Because of the lack of definite knowledge concerning the number of intact fuel rods, the temperatures to which they were exposed, and the amount of krypton produced both before and after the accident at TMI-2, calculations were attempted to try to determine the amount of free krypton in the reactor core. Based on the available data, it was estimated that 30 curies of Krypton might be present but a possible range of 0 to 100 curies was cited.

*Reactor core analysis; Krypton; Fuel rods; Temperature variation; Experimental data

1311
U.S. Department of Energy. Oak Ridge National Laboratory. Oak Ridge, Tennessee. **Evaluation of the Use of Zeolite Mixtures in the Submerged Demineralizer System (SDS) Flowsheet for Decontamination of High-Activity-Level Water at the Three Mile Island Nuclear Power Station Unit 2.** L.J. King; D.O. Campbell; E.D. Collins; J.B. Knauer; and R.M. Wallace. Springfield, VA: NTIS, 1983. 4 p. International Zeolite Conference. Reno, Nevada, July 10, 1983. (CONF-830756-8; DE83015700; CONTRACT W-7405-ENG-26)

Evaluation of the use of Linde Ionsiv IE-96 and Linde Ionsiv A-51 for the decontamination of high activity level water in the submerged demineralizer system of TMI-2 is reported.

*Zeolites; Submerged Demineralizer System; Decontamination; Contaminated water

1312
U.S. Department of Energy. Oak Ridge National Laboratory. Oak Ridge, Tennessee. **Evaluation of Zeolite Mixtures for Decontamination of High-Activity-Level Water in the Submerged Demineralizer System (SDS) Flowsheet at the Three Mile Island Nuclear Power Station, Unit 2.** L.J. King; D.O. Campbell; E.D. Collins; J.B. Knauer; and R.M. Wallace. Springfield, VA: NTIS, 1983. 16 p. International

Zeolite Conference. Reno, Nevada, July 10, 1983. (DPST-83-248; CONF-830756-4; DE83012203; CONTRACT AC09-76SR00001)

Various mixtures of zeolites for use in the Submerged Demineralizer System (SDS) were tested for the most efficient combination to remove both cesium and strontium at the same time from the contaminated water at TMI-2.

*Zeolites; Submerged Demineralizer System; Cesium radioisotopes; Strontium radioisotopes; Contaminated water

1313
U.S. Department of Energy. Oak Ridge National Laboratory. Oak Ridge, Tennessee. **Nuclear-Criticality-Safety Studies on Interest to TMI-2 Recovery Operations.** J.T. Thomas. Springfield, VA: NTIS, October 1982. 22 p. (ORNL/CSD/TM-193; DE83000423; CONTRACT W-7405-ENG-26)

Criticality studies were conducted and applied to possible situations that might be encountered in the TMI-2 cleanup operation.

*Criticality; Reactor core analysis; Cleanup

1314
U.S. Department of Energy. Oak Ridge National Laboratory. Oak Ridge, Tennessee. **Population Dose Estimation from a Hypothetical Release of 2.4 x 10 exp Curies of Noble Gases and 1 x 10 exp 4 Curies of exp I131 at the Three Mile Island Nuclear Station, Unit 2.** C.D. Berger; B.H. Lance; S.J. Cotter; C.W. Miller; and S.R. Glandon. Springfield, VA: NTIS, September 1981. 30 p. (ORNL/TM-7980; DE81030701; CONTRACT W-7405-END-26)

This report presents estimated population doses of noble gases and I131 received from approximations of TMI-2 releases.

*Population exposure; Radiation dosage; Noble gas releases; Atmospheric releases; Iodine radioisotopes

1315
U.S. Department of Energy. Oak Ridge National Laboratory. Oak Ridge, Tennessee. **Process Improvement Studies for the Submerged Demineralizer System (SDS) at the Three Mile Island Nuclear Power Station, Unit 2.** D.O. Campbell; E.D. Collins; L.J. King; and J.B. Knauer. Springfield, VA: NTIS, May 1982. 30 p. (ORNL/TM-7756; DE82014568; CONTRACT W-7405-ENG-26)

Studies were conducted on the Submerged Demineralizer System used at TMI-2 for decontamination of water in an effort to improve its performance. It was concluded in this report that "aging" the water in zeolite columns was efficient in removing appreciable amounts of cesium and strontium prior to further filtration through the systems.

*Submerged Demineralizer System; Zeolites; Cesium radioisotopes; Strontium radioisotopes; Filtration; Contaminated water

1316
U.S. Department of Energy. Oak Ridge National Laboratory. Oak Ridge, Tennessee. **Restarting TMI Unit One. Social and Psychological Impacts.** J. Sorensen; J. Soderstrom; R. Bolin; E. Copenhaver; and S. Carnes. Springfield, VA: NTIS, December 1983. 249 p. (ORNL-5891; DE84004107; CONTRACT W-7405-ENG-26)

The social and psychological impact of restarting TMI-1 on the community is defined as expressed by literature review, focus group discussion, community profiling and surveys. It is concluded that TMI-1 restart would generate adverse effects.

*Restart; Health effects - psychological; Social effects

1317
U.S. Department of Energy. Oak Ridge National Laboratory. Oak Ridge, Tennessee. **Three Mile Island A Year Later.** D.B. Trauger. Springfield, VA: NTIS, 1980. 8 p. (DOE/TIC-11172; CONTRACT W-7405-ENG-26)

The TMI-2 accident and subsequent events are reported for the year post-accident.

*Accident investigations; TMI - anniversary (1)

1318
U.S. Department of Energy. Oak Ridge National Laboratory. Oak Ridge, Tennessee. **TMI-2 Criticality Studies. Lower Vessel Rubble and Analytical Benchmarking.** R.M. Westfall; J.R. Knight; P.B. Fox; O.W. Herman; and J.C. Turner. Springfield, VA: NTIS, December 1985. 52 p. tables. (ORNL/CSD/TM-222; DE86004639/GAR; CONTRACT AC05-84OR21400)

A "safe" boron level for the entire TMI-2 reactor is determined so that the reactor will not gain criticality during the defueling operations.

*Criticality; Boron; Defueling; Reactor core damage; Reactor core analysis

1319
U.S. Department of Energy. Oak Ridge National Laboratory. Oak Ridge, Tennessee. **Water-Decontamination Process-Improvement Tests and Considerations.** E.D. Collins; D.O. Campbell; L.J. King; and J.B. Knauer. Springfield, VA: NTIS, 1981. 33 p. National Meeting of the American Institute of Chemical Engineers. Detroit Michigan, August 16, 1981. (CONF-810814-7; DE81028064; CONTRACT W-7405-ENG-26)

Although the Submerged Demineralizer System is effective in removing a large amount of the cesium and strontium from the contaminated water at TMI-2, it was not sufficiently decontaminated for interim storage. This report examines the changes that would be necessary to improve radionuclide removal.

*Submerged Demineralizer System; Contaminated water; Cesium radioisotopes; Strontium radioisotopes

1320
U.S. Department of Energy. Oak Ridge National Laboratory. Oak Ridge, Tennessee. **Chemical Technology Division. Progress Report. April 1, 1979 - March 31, 1981.** Springfield, VA: NTIS, November 1981. 207 p. (ORNL-5757; DE82003522; CONTRACT W-7405-ENG-260

Three Mile Island support makes up one section of this report.

*Reactor accidents; Accident investigations

1321
U.S. Department of Energy. Oak Ridge National Laboratory. Oak Ridge, Tennessee. **Chemical Technology Division. Progress Report. April 1, 1981 - March 31, 1983.** Springfield, VA: NTIS, September 1983. 178 p. (ORNL-5933; DE84000647; CONTRACT W-7405-ENG-26)

Mention is made of TMI support studies as a portion of the Progress Report for this period.

*Reactor accidents: Accident investigations

1322
U.S. Department of Energy. Pennsylvania State University. University Park, Pennsylvania. Department of Nuclear Engineering. **Develop a Practical Means to Monitor the Criticality of the TMI-2 Core.** S.S. Kim; S.H. Levine; and G. Imel. Springfield, VA: NTIS, June 1984. 189 p. (DOE/ER/75088-1; DE85000059/GAR; CONTRACT FG02-83ER75088)

An Asymmetric Multiple Position Neutron Source (AMPNS) method of determining/monitoring the subcriticality level of the degraded TMI-2 core is described.

*Criticality; Reactor core damage

1323
U.S. Department of Energy. Quadrex. Richland, Washington. **Comparison of Decontamination Techniques for Reactor Coolant System Applications. Final**

**Report.** H.R. Gardner; L.M. Polentz; R.P. Allen; and W.E. Skiens. Springfield, VA: NTIS, December 1982. 58 p. (EPRI-NP-2777; DE83901122)

The most effective techniques for cleanup of the TMI-2 reactor coolant system are discussed. Both "in-place" and "out-of-system" methods are presented.

*Cooling system; Decontamination; Cleanup

1324
U.S. Department of Energy. Quadrex. Richland, Washington. **High-Pressure Water Systems for Flushing Loose Fuel Debris from the Reactor Underhead Area at TMI-2.** H.R. Garnder; L.M. Polentz; and D.B. Bateman. Springfield, VA: NTIS, June 1984. 82 p. (EPRI-NP-3510; DE84015211)

This report evaluates the effectiveness of two different remote controlled high-pressure water systems inserted through control rod openings into the reactor vessel. The purpose of the operation would be the flushing of radioactive debris from the underhead and plenum covers of the TMI-2 reactor.

*Debris, radioactive; Fuel debris

1325
U.S. Department of Energy. Quadrex. Richland, Washington. **Use of Pressurized Water to Decontaminate TMI-2 Leadscrew Sections.** H.R. Gardner and L.M. Polentz. Springfield, VA: NTIS, June 1984. 85 p. (EPRI-NP-3509; DE84015209)

This report evaluates the use of high pressure water at various pressures and distances, to remove debris and radioactive cesium from the leadscrews obtained from the TMI-2 reactor.

*Leadscrews; debris, radioactive; Cesium radioisotopes; Decontamination

1326
U.S. Department of Energy. Rand Corporation. Santa Monica, California. **Some Implications of the Three Mile Island Accident for LMFBR Safety and Licensing. The Design Basis Issue.** K.A. Solomon. Springfield, VA: NTIS, August 1980. 48 p. (RAND/N-1559-DOE; DE82009276; CONTRACT AS03-78ET37300)

The TMI-2 accident has generated a number of safety studies concerning light water reactors. One of the factors considered is the design basis accident, i.e. separating those accidents which are plausible from those that are unusual or improbable. Changes in design and operation will effect not only light water reactors but also liquid metal fast breeder reactors (LMFBR).

*Light water reactors; Nuclear reactor design; Operators; Liquid metal fast breeder reactors

1327
U.S. Department of Energy. Rockwell International. Richland, Washington. **Proposed Methods for Defueling the TMI-2 Reactor Core.** J.O. Henrie. Springfield, VA: NTIS, May 1984. 27 p. (GEND-INF-027; DE84010879; CONTRACT AC07-761D01570)

A consensus report of recommendations for removal of fuel debris from the TMI-2 reactor is presented including information on the core configuration and equipment necessary to complete the task.

*Fuel debris; Reactor core; Recommendations

1328
U.S. Department of Energy. Rockwell International. Richland, Washington. Energy Systems Group. **Analysis of Three Mile Island Unit 2 Reactor Cooling System Transients. Volume 5.** J.O. Henrie and A.K. Postma. Springfield, VA: NTIS, August 1983. 35 p. (GEND-INF-023-v.5; DE83017034; CONTRACT AC07-761D01570)

The cause of the release of transients early on the day of the TMI-2 accident was thought to be because of rapid energy releases from the cooling system. This study denies that possibility and proposes instead a reactimeter power source malfunction.

*Cooling system; Transients

1329
U.S. Department of Energy. Rockwell International. Richland, Washington. Rockwell Hanford Operations. **Evaluation of Special Safety Issues Associated with Handling the Three Mile Island Unit 2 Core Debris.** J.O. Henrie and J.N. Appel. Springfield, VA: NTIS, June 1985. 45 p. (GEND-051; DE85013763; CONTRACT AC07-761D01570)

This document reviews safety features and concerns connected with the movement of TMI-2 core debris in cannisters for containing the shipment.

*Cannister, high integrity; Fuel debris; Debris, radioactive; Waste shipment; Safety

1330
U.S. Department of Energy. Rockwell International. Richland, Washington. Rockwell Hanford Operations. **Hydrogen Control in the Handling, Shipping, and Storage of Wet Radioactive Waste.** J.O. Henrie; D.J. Flesher; G.J. Quinn; and J. Greenborg. Springfield, VA: NTIS, February 1986. 55 p. (GEND-052; DE86007447; CONTRACT AC07-761D01570)

The handling and shipment of wet radioactive wastes such as those from TMI-2 are discussed and the methods of controlling the generation of flammable gas (hydrogen) reported.

*Waste handling; Waste shipment; Radiolytic gas; Hydrogen; Combustion potential

1331
U.S. Department of Energy. Sandia National Laboratories. Albuquerque, New Mexico. **Analysis and Testing of the HP-R-214 Dome Monitor Cable from Three Mile Island Unit 2.** E.H. Richards and V.J. Dandini. Springfield, VA: NTIS, March 1986. 34 p. (GEND-INF-072; DE86009093/GAR; CONTRACT AC07-761D01570)

Two sections of cable connected to the TMI-2 dome monitor were removed for examination. One was directly exposed to the forces of the accident while a like cable also exposed was encased in a conduit. Both sections were subjected to various tests and were compared with an unused control sample of cable.

*Cables; Electrical equipment; Equipment function; Reactor accidents

1332
U.S. Department of Energy. Sandia National Laboratories. Albuquerque, New Mexico. **Examination Results of the Three Mile Island Radiation Detector HP-R-211.** M.B. Murphy; G.M. Mueller; and F.V. Thome. Springfield, VA: NTIS, September 1981. 133 p. (SAND-81-0725; DE82000939; CONTRACT AC04-76DP00789)

Reasons for failure of area radiation detector HP-R-211 are explored. HP-R-211 was removed from TMI-2 on August 15, 1980.

*Radiation detectors; Instrument failure

1333
U.S. Department of Energy. Sandia National Laboratories. Albuquerque, New Mexico. **Examination Results on TMI-2 LPM Charge Converters YM-AMP-7023 and YM-AMP-7025.** M.B. Murphy and R.E. Heintzelman. Springfield, VA: NTIS, September 1982. 28 p. (SAND-82-0980; DE83001661; CONTRACT AC04-76DP00789)

Two TMI-2 charge converters YM-AMP-7023 and YM-AMP-7025 were examined and found to be severely degraded by radiation. On the basis of these findings, recommendations for changes in the U.S. Nuclear Regulatory Guide are made.

*Charge converters; Instrument function; Instrument failure

1334
U.S. Department of Energy. Sandia National Laboratories. Albuquerque, New Mexico. **Examination Results on TMI-2 LPM Charge Converters YM-AMP-7023 and YM-AMP-7025.** M.B. Murphy and R.E. Heintzelman. Springfield, VA: NTIS, November 1982. 29 p. (GEND-020; DE83002893; CONTRACT AC07-761D01570)

Two Endevco charge converters removed from the TMI-2 coolant systems during loose parts survey were examined and evaluated. These units were found to be severely damaged by radiation and because of these studies, recommendations will be made to the Nuclear Regulatory Commission.

*Charge converters; Instrument function; Instrument failure; Recommendations

1335
U.S. Department of Energy. Sandia National Laboratories. Albuquerque, New Mexico. **Examination Results of the Three Mile Island Radiation Detector HP-R-211.** M.B. Murphy; G.M. Mueller; and F.V. Thome. Springfield, VA: NTIS, October 1981. 132 p. (GEND-014; DE82004023; CONTRACT AC07-761D01570)

Reasons for failure of area radiation detector HP-R-211 are discussed. The detector was removed from TMI on August 15, 1980.

*Radiation detectors; Instrument failure

1336
U.S. Department of Energy. Sandia National Laboratories. Albuquerque, New Mexico. **Examination Results of the Three Mile Island Radiation Detector HP-R-212.** G.M. Mueller. Springfield, VA: NTIS, December 1983. 45 p. (SAND-83-1930; DE84004121; CONTRACT AC04-76DP00789)

Radiation detector HP-R-212 was removed from the TMI-2 containment building on November 13, 1981 because it had apparently begun to fail about November 1979. This report discusses the failure.

*Radiation detectors; Instrument failure; Reactor containment building

1337
U.S. Department of Energy. Sandia National Laboratories. Albuquerque, New Mexico. **Examination Results of the Three Mile Island Radiation Detector HP-R-212.** G.M. Mueller. Springfield, VA: NTIS, January 1984. 42 p. (GEND-INF-049; DE84007354; CONTRACT AC07-761D01570)

Radiation detector HP-R-212 was removed from the TMI-2 containment building on November 13, 1981 because it had ceased to function. This paper discusses the failure.

*Radiation detectors; Instrument failure; Reactor containment building

1338
U.S. Department of Energy. Sandia National Laboratories. Albuquerque, New Mexico. **Examination Results of the Three Mile Island Radiation Detector HP-R-213.** G.M. Mueller. Springfield, VA: NTIS, August 1982. 33 p. (SAND-82-1173; DE82022153; CONTRACT AC04-76DP00789)

Reasons for the failure of TMI-2 radiation detector HP-R-213 are reported with particular reference to the effect of the hydrogen burn.

*Radiation detectors; Instrument failure; Hydrogen burn

1339
U.S. Department of Energy. Sandia National Laboratories. Albuquerque, New Mexico. **Examination Results of the Three Mile Island Radiation Detector HP-R-213.** G.M. Mueller. Springfield, VA: NTIS, November 1982. 32 p. (GEND-019; DE83002765; CONTRACT AC07-761D01570)

The cause of failure of TMI-2 radiation detector HP-R-213 is reported with special reference to the effect of the hydrogen burn.

*Radiation detectors; Instrument failure: Hydrogen burn

1340
U.S. Department of Energy. Sandia National Laboratories. Albuquerque, New Mexico. **TMI-2 Instrument Analysis Results.** M.B. Murphy and F.V. Thome. Springfield, VA: NTIS, 1981. 12 p. American Nuclear Society Winter Meeting. San Francisco, California, November 29, 1981. (SAND-81-1736C; CONF-811103-18; DE81027099; CONTRACT AC04-77DP00789)

The TMI-2 experience provides an opportunity to examine electrical equipment in reactor operations with regard to its function during and after a loss of coolant accident. Instrument capability of several specific parts were reported.

*Electrical equipment; Instrument function; Loss of coolant accident

1341
U.S. Department of Energy. Science Applications, Inc. Idaho Falls, Idaho. **Estimated Source Terms for Radionuclides and Suspended Particulates During TMI-2 Defueling Operations Phase II. Volume II.** P.G. Voilleque; J.R. Noyce; and C.A. Pelletier. Springfield, VA: NTIS, May 1983. 60 p. (GEND-INF-019-v.2; DE83011877; CONTRACT AC07-761D01570)

Particulates, radionuclides and other solids in suspension in the reactor coolant at TMI-2 will affect the defueling process. This report reviews the parameters.

*Coolant; Radionuclides; Contamination; Defueling

1342
U.S. Department of Energy. Science Applications, Inc. Rockville, Maryland. Nuclear Environmental Services Division. **Characterization of Contaminants in TMI-2 Systems.** J.A. Daniel; T.L McVey; E.A. Schlomer; and D.G. Keefer. Springfield, VA: NTIS, March 1983. 309 p. (EPRI-NP-2922; DE83901942)

The contaminants in the TMI-2 reactor coolant system are discussed with regard to nature and location, and cleanup efforts for various components mentioned.

*Coolant; Cooling system; Contamination; Cleanup; Decontamination

1343
U.S. Department of Energy. Science Applications, Inc. Rockville, Maryland. **Iodine-131 Behavior During the TMI-2 Accident. Final Report.** C.A. Pelletier; C.D. Thomas, Jr.; R.L. Ritzman; and F. Trooper. Springfield, VA: NTIS, September 1981. 81 p. (EPRI-NSAC-30; DE82901463)

Iodine is recognized to be an isotope of major health consequence and its release from TMI was monitored closely. According to the report, only small amounts excaped into the surrounding environment, most having been absorbed in the water in the containment areas.

*Iodine radioisotopes; Health effects; Atmospheric releases

1344
U.S. Department of Energy. Science Applications, Inc. Rockville, Maryland. **Preliminary Radioiodine Source-Term and Inventory Assessment for TMI-2.** C.A. Pelletier; P.G. Voilleque; C.D. Thomas; J.A. Daniel; and E.A. Schlomer. Springfield, VA: NTIS, March 1983. 65 p. (GEND-028; DE83009735; CONTRACT AC07-761 D01570)

The behavior of radioactive iodine both during and after the accident is described. Estimates, according to this report, are that about 1% of iodine from the core escaped to the air at the time of the accident.

*Iodine radioisotopes; Atmospheric releases

1345
U.S. Department of Energy. Secretary for Nuclear Energy. Washington, D.C. **Evaluation of Increased Cesium Loading on Submerged Demineralizer System (SDS) Zeolite Beds.** Springfield VA: NTIS, May 1981. 127 p. (DOE/NE-0012)

The Department of Energy examined the Submerged Demineralizer System to determine whether it could safely handle higher levels of cesium than were previously planned. The study concluded that a cesium load of up to 60,000 curies could be disposed safely.

*Submerged Demineralizer System; Cesium radioisotopes; Cleanup; Zeolites

1346
U.S. Department of Energy. SRI International. Menlo Park, California. **Two Phase Natural Circulation Experiments in a Test Facility Modeled after Three Mile Island Unit 2. Final Report.** R.L. Kiang. Springfield, VA: NTIS, October 1981. 81 p. (EPRI-NP-2069; DE82900971)

This report explores the possibility of conducting the TMI-2 reactor core heat away by natural circulation. Experiments performed at a test facility are presented.

*Reactor core temperature; Cooling system; Experimental data

1347
U.S. Department of Energy. Studsvik Energitenik A.B. Nykoeping. Sweden. **Appraisal of Power Reactor Status Following Unplanned Shutdown by Using Dynamic Signal Analysis. Stage 1.** F. Aakerhielm. Springfield, VA: NTIS, March 1981. 39 p. (TUDSVIK- K5-81,SKI-B-18-80; DE82702109)

The development of a noise analysis system to assess unplanned reactor shutdown for Swedish BWR and LWR is presented and conclusions applied to the TMI-2 accident.

*Shutdown; Nuclear reactors; Sweden

1348
U.S. Department of Energy. Stuttgart University. Germany, F.R. Institute fuer Kernenergetik und Energiesysteme. **Theoretical Investigations on Fuel Rod Behavior During the TMI-2 Accident and Similar Calculations for a German PWR of Type Biblis.** W. Tuerk; F. Schmidt; and R. Bisanz. Springfield, VA: NTIS, May 1981. 96 p. (IKE-4-102; DE82750341)

Calculations of core survival at TMI were made using three different computer programs. One predicted a completely destroyed core while the other two predicted the beginning of a melting process.

*Fuel rods; Reactor core analysis; Computer programs; Computer simulations; West Germany

1349
U.S. Department of Energy. Technology for Energy Corporation. Knoxville, Tennessee. **Recommendations for TMI-2 Instrumentation Surveillance Program.** J.E. Jones and M.V. Mathis. Springfield, VA: NTIS, April 1982. 88 p. (GEND-INF-017-v.12; DE82012878; CONTRACT AC07-761D01570)

Because a number of instruments, detectors, monitors, etc., failed after the TMI-2 accident, the Technical Integration Office undertook to design a program to monitor devices in adverse environments or under radiation conditions. This report lists a number of proposed instruments to be surveyed.

*Instrument function; Instrument failure; Reactor accidents; Recommendations

1350
U.S. Department of Energy. Technology for Energy Corporation. Knoxville, Tennessee. **Review of the Population Radiation Exposure at TMI-2. Final Report.** P.F. Knight; J.T. Robinson; F.J. Slagle; and P.M. Garrett. Springfield, VA: NTIS, August 1981. 191 p. (EPRI-NSAC-26; DE82901462)

This final report collected and compared various investigators' findings on radiation exposure to the population from the TMI-2 accident.

*Radiation dosage; Population exposure

1351
U.S. Department of Energy. Temple, Barker, and Sloane, Inc. Lexington, Maine. **Preliminary Analysis of the Effects of the Accident at Three Mile Island on**

**Metropolitan Edison Company.** Springfield, VA: NTIS, August 1981. 46 p. (DOE/RG/06380-T2; DE83006020; CONTRACT AC01-87RG06380)

This is an analysis of the financial aspect of cleanup operations at TMI-2 by the Metropolitan Edison Company.

*Financing; Cleanup; Metropolitan Edison

1352
U.S. Department of Energy. United Engineers and Constructors, Inc. Philadelphia, Pennsylvania. **Axial-Power-Shaping Rod-Insertion Test.** K. Parlee; J. Weissburg; W. Austin; and G. Carter. Springfield, VA: NTIS, 1981. 5 p. American Nuclear Society Winter Meeting. Washington, D.C., November 14, 1982. (EGG-TIO-M-0182; CONF-821103-70; DE83005543; CONTRACT AC07-761D01570)

This document describes the test insertion of eight axial power shaping rods to determine motion capability preparatory to uncoupling leadscrews and reactor head removal.

*Axial power shaping rod; Leadscrews; Reactor head romoval

1353
U.S. Department of Energy. United Engineers and Constructors, Inc. Philadelphia, Pennsylvania. **Evaluation Results of TMI-2 Solenoids AH-V6 and AH-V74.** F.T. Soberano. Springfield, VA: NTIS, January 1984. 37 p. (GEND-INF-045; DE84006085; CONTRACT AC07-761D01570)

Two solenoids from the TMI-2 reactor building were evaluated and found to be functional post-accident.

*Solenoids; Instrument function

1354
U.S. Department of Energy. United Engineers and Constructors, Inc. Philadelphia, Pennsylvania. **Final Report on the In Situ Testing of Electrical Components and Devices at TMI-2.** F.T. Soberano. Springfield, VA: NTIS, June 1984. 108 p. (GEND-040; DE84014163; CONTRACT AC07-761D01570)

Evaluation of 88 electrical components and devices at the TMI reactor was conducted post-accident. Of these, eleven failed and twenty-one were impaired. Reasons for failure are discussed and detailed where known.

*Electrical equipment; Equipment failure

1355
U.S. Department of Energy. United Engineers and Constructors, Inc. Philadelphia, Pennsylvania. **Status In Situ Test of the Axial Power Shaping Rod and Shim Safety Control Rod Mechanisms.** F.T. Soberano; J.A. Gannon; and K.A. Parlee. Springfield, VA: NTIS, August 1982. 20 p. (GEND-INF-026; DE82022186; CONTRACT AC07-761D01570)

Tests were performed on eight axial power shaping rods and three shim safety control rods to help assess any damage that may have occurred to the TMI-2 reactor core.

*Axial power shaping rod; Reactor core damage

1356
U.S. Department of Energy. Westinghouse Electric Corporation. Madison, Pennsylvania. Waste Technology Services Division. **TMI-2 Defueling System Design Description.** D.E. Falk and C.E. Swenson. Springfield, VA: NTIS, March 1985. 92 p. (GEND-INF-065; DE85014611/GAR; CONTRACT AC07-761D01570)

Design for the construction of a platform and support structure over the reactor vessel are described in preparation for defueling operations at TMI-2.

*Defueling

1357
U.S. Department of Energy. Westinghouse Electric Corporation. Pittsburgh, Pennsylvania. Nuclear Services Division. **Evaluation of Abrasive Grit-High-Pressure**

**Water Decontamination. Final Report.** J. Sejvar and P.H. Dawson. Springfield, VA: NTIS, October 1982. 62 p. (EPRI-NP-2691; DE83900613)

The decontamination of the TMI-2 steam generator channel heads by an abrasive grit-high pressure water technique is reported. Guidelines, etc., are presented for the decontamination method.

*Steam generator; Decontamination

# 16

# Federal Documents: Miscellaneous Reports

1358
Atomic Energy of Canada Limited. Sheridan Park (Ontario). Power Projects. **Analysis of the Three Mile Island Accident.** G.L. Brooks and E. Siddal. Springfield, VA: NTIS, September 1980. 9 p. (AECL-7065)

This Canadian analysis of the TMI-2 accident is applied to the Canadian reactor safety program.

*Reactor accident analysis; Safety; Canada

1359
Atomic Industrial Forum, Inc. New York. **Lessons of Three Mile Island - Public Acceptance of Nuclear Power.** T.R. Jordan. Springfield, VA: NTIS, 1979. 16 p. International Conference on Financing Nuclear Power. Copenhagen, Denmark, September 23, 1979. (CONF-7909121-1; INIS-mf-5327)

This conference reviewed the effect that the accident at TMI-2 had on the public's attitude toward nuclear power generation and the effect any change in attitude will have on future directions.

*Attitude to nuclear power; Public opinion; Financing; Lessons learned

1360
Babcock and Wilcox Company. Lynchburg, Virginia. **TMI-2 H8A Core Debris Sample Examination. Final Report.** G.O. Hayner. Springfield, VA: NTIS, May 1985. 62 p. (4 refs.) 5 appendices. (GEND-INF-060; CONTRACT AC07-761D01570)

This document reports on the TMI-2 core debris sample composition. The sample studied was obtained at a depth of three inches below the top of the rubble from position H8.

*Debris, radioactive; Fuel debris; Reactor core analysis

1361
Babcock and Wilcox Company. Lynchburg, Virginia. Nuclear Power Generation Division. **Development of Sensors and Instrumentation for the TMI-2 OSTG Tube Vibration Measurements Program.** J.M. Anthony and J.S. Olszewski. Springfield, VA: NTIS, June 1981. 168 p. (EPRI-NP-1875)

Babcock and Wilcox Company developed sensors to monitor the once through steam generator tube vibrations at normal and operating conditions.

*Once through steam generator tubes; Vibration; Equipment function; Babcock & Wilcox

1362
Battelle Pacific Northwest Laboratories. Richland, Washington. **Decontamination Demonstration Facility (D.D.F.) Modularization/Mobility Study.** V.F. FitzPatrick; H.L. Butts; R.G. Moles; and R.A. Lundgren. Springfield, VA: NTIS, November 1980. 36 p. (CONTRACT AC06-76RL01830; PNL-3628)

The TMI-2 Decontamination Demonstration Facility is of value to other nuclear facilities in areas of waste management, decommissioning of nuclear power plants, and reaction to accidents.

*Decontamination; Waste handling; Research and development

1363
Battelle Pacific Northwest Laboratories. Richland, Washington. **Demonstration of Alternative Decontamination Techniques at Three Mile Island.** H.W. Arrowsmith and R.P. Allen. Springfield, VA: NTIS, November 1979. 41 p. DOE/EPRI Symposium on Three Mile Island Reactor Problems. Harrisburg, Pennsylvania, November 11, 1979. (CONF-7911104-1; PNL-SA-8143; CONTRACT AC06-76RL01830)

Several different decontamination techniques, e.g. adsorption, electroplating, etc., are considered for applicability to the TMI-2 cleanup.

*Decontamination; Cleanup; Research and development

1364
Billings Energy Corporation. Provo, Utah. **Experiments on Hydrogen for Three Mile Island.** R.L. Wooley; J.H. Ruckman; G.L. Kimball; A.L. Ayers, Jr.; and J.L. Liebenthal. Springfield, VA: NTIS, 1980. 11 p. American Nuclear Society Thermal Reactor Safety Meeting. Knoxville, Tennessee, April 8, 1980. (CONF-800403-7; CONTRACT EY-76-C-07-1570)

This is a report of the investigation of the hydrogen in the TMI-2 coolant water, the ways in which the hydrogen could be removed, and the best method of depressurization. Less hydrogen remained in the system than was initially thought.

*Hydrogen; Coolant; Cleanup; Research and development

1365
Brookhaven National Laboratory. Upton, New York. **Critique of Source Term and Environmental Measurement at Three Mile Island.** A.P. Hull. Springfield, VA: NTIS, 1979. 20 p. IEEE Nuclear Science Symposium. San Francisco, California, October 17, 1979. (CONF-791037-14; BNL-26970; CONTRACT EY-76C-02-0016)

Because radiation monitoring equipment at TMI was unreliable after the accident, air and other environmental samples were collected and analyzed by the Pennsylvania Bureau of Radiologic Health and mobile facilities of the Department of Energy. Some samples were transported to remote laboratories for examination.

*Radiation monitoring; Radiation analysis

1366
Brookhaven National Laboratory. Upton, New York. **OSTG (Once Through Steam Generator) Modeling for the Analysis of the TMI Incident.** C.J. Hsu; W.G. Shier; and M.M. Levine. Springfield, VA: NTIS, 1980. 5 p. American Nuclear Society Annual Meeting. Las Vegas, Nevada, June 8, 1980. (CONF-800607-23; BNL-NUREG- 27185; CONTRACT EY-76-C-02-0016)

A new modified code for analysis of the TMI accident through use of the OSTG model for pressurized water reactors is presented.

*Once through steam generator tubes; Pressurized water reactors; Models

1367
Budensministerium des Innern. Bonn (Germany F.R.). Referat Oeffentlichkeitsarbeit. Harrisburg Report. **Evaluation of the Accident at Harrisburg Nuclear Power Plant. 2nd Interim Report of the Committee on Internal Affairs of the German Federal Parliament.** Springfield, VA: NTIS, June 1, 1979. 110 p. (INIS- mf-5590)

The German evaluation and report of the TMI-2 accident is described and implications for German nuclear power plant safety checks commented upon.

*Safety; Nuclear power plants; Reactor accident analysis; West Germany

1368
California University. Berkley, California. Lawrence Berkley Laboratory. **Health Effects of the Nuclear Accident at Three Mile Island.** J.I. Fabrikant. Springfield, VA: NTIS, 1980. 24 p. Conference on Enviromental Regulation of the Nuclear Industry: A New Decade. San Francisco, California, May 18, 1980. (CONF-8005125-1; LBL-11297; CONTRACT W-7405-ENG-48)

This report examines the health effects of the TMI-2 accident. It concludes that the radiation doses received were negligible and will constitute minimal, if any, detectable problem. The most critical health side-effect cited was stress.

*Health effects; Radiation dosage; Stress

1369
California University. Livermore, California. Lawrence Livermore Laboratory. **Archiving the Paper Fallout from Three Mile Island.** H.L. Fisher and V.D. Caswell. Springfield, VA: NTIS, August 10, 1979. 7 p. (UCID-18243; CONTRACT W-7405-ENG-48)

The work of the Atmospheric Release Advisory Capability Center (ARAC) in monitoring the TMI radioactive releases has accumulated voluminous data. This report discusses the organization of this material so that it will be accessible for future use.

*Atmospheric releases; Radioactive releases; Radiation monitoring; Information

1370
California University. Livermore, California. Lawrence Livermore Laboratory. **Atmospheric Release Advisory Capability (ARAC) Response to the Three Mile Island Accident.** M.H. Dickerson and P.H. Gudikson. Springfield, VA: NTIS, October 1979. 5 p. IEEE Nuclear Science Symposium. San Francisco, California, October 17, 1979. (CONF-791037-3; URCL-83489; CONTRACT W-7405-ENG-48)

The role of the ARAC is presented as it related to the TMI-2 accident during March and April of 1979.

*Atmospheric releases; Radiation monitoring

1371
California University. Livermore, California. Lawrence Livermore Laboratory. **Operator/Instrumentation Interactions During the Three Mile Island Incident.** G.E. Cummings. Springfield, VA: NTIS, October 1979. 4 p. IEEE Symposium on Nuclear Power Systems. San Jose, California, October 19, 1979. (CONF-791060-1; UCRL-82227; CONTRACT W-7405-ENG-48)

A report of the operator-instrument interaction during the TMI accident during the first sixteen hours is presented and six specific areas of concern discussed.

*Operators; Human factors; Human error

1372
California University. Livermore, California. Lawrence Livermore Laboratory. **Utilization of the Atmospheric Release Advisory Capability (ARAC) Services During and After the Three Mile Island Accident.** J.B. Know; M.H. Dickerson; G.D. Greenly; P.H. Gudiksen; and T.J. Sullivan. Springfield, VA: NTIS, July 1, 1980. 37 p. (UCRL-52959; CONTRACT W-7405-ENG-48)

ARAC (Atmospheric Release Advisory Capability) monitored the krypton and xenon releases from the TMI-2 accident, and this report details the population dose calculations.

*Atmospheric releases; Krypton; Xenon radioisotopes; Population exposure; Radiation dosage

1373
Central Electric Generating Board. London, England. **Three Mile Mile Island Nuclear Accident. A Selective Bibliography.** Springfield, VA: NTIS, 1979. J.L. Watson. 1980. 12 p. (PB81-100885; CE-BIB-275)

A bibliography of the TMI-2 accident is selectively listed.
*Bibliographies

1374
E, G & G Energy Measurements. **Airborne Cloud Tracking Measurements During Three Mile Island Nuclear Station Accident. March - June 1979.** R.H. Beers; Z.G. Burson; T.C. Maguire; and G.R. Shipman. Springfield, VA: NTIS, December 1984. 75 p. (13 refs.) (EGG-10282-1009; CONTRACT DE-AC08-83NV-10282)

This report details measurements gathered in aerial monitoring missions of isotopes, plume direction, and maximum radiation levels in the plume up to one mile high after the TMI-2 accident.

*Radiation monitoring; Radioactive releases; Atmospheric releases

1375
E, G & G Idaho, Inc. Idaho Falls, Idaho. **Commercial Disposal of High Integrity Containers (HICS) Containing EPICOR II Prefilters from Three Mile Island.** J.W. McConnell; R.J. Lynch; and M.J. Tyacke. Springfield, VA: NTIS, September 1985. 33 p. (13 refs.) (GEND-048; CONTRACT AC07-761D01570)

This report describes the loading, transporting, and commercial disposal of 45 HICS containing EPICOR II prefilters generated during cleanup of the TMI Unit 2 Auxiliary and Fuel Handling Building.

*Cannister, high integrity; Waste shipment; Waste disposal; Cleanup; Reactor auxiliary building

1376
E, G & G Idaho, Inc. Idaho Falls, Idaho. **Three Mile Island Unit-2 Core Status Summary: A Basis for Tool Development for Reactor Disassembly and Defueling.** D.W. Croucher. Springfield, VA: NTIS, May 1981. 73 p. (GEND-007; CONTRACT AC07-761D01570)

Several indirect methods to estimate core damage from the TMI-2 accident have been employed. In this report, thermal-hydraulic, hydrogen generation, and fission products released are all correlated in an attempt to calculate the minimum and maximum bounds of core damage.

*Reactor core damage; Reactor core analysis; Hydrogen; Fission product release; Thermal hydraulics

1377
E, G & G Idaho, Inc. Idaho Falls, Idaho. **TMI-2 Information and Examination Program. Technical Integration Office. Annual Report.** Springfield, VA: NTIS, 1981. 56 p. (EGG/GEND-003; CONTRACT AC07-761D01570)

This annual report of the TMI-2 Information and Examination Program for 1980 includes all activities of the program since September 1980. Reactor accident assessments of equipment, core and fuel damage and radioactive releases are mentioned.

*Reactor accident analysis; Reactor core damage; Fuel damage; Equipment function

1378
E, G & G Idaho National Engineering Laboratory. Idaho Falls, Idaho. **Simulation of the Three Mile Island Transient in Semiscale.** T.K. Larson; G.G. Loomis; and R.W. Shumway. Springfield, VA: NTIS, July 1979. 94 p. (EGG/SEMI-TR-010; CONTRACT EY-76-C-07-1570)

Using a Semiscale Mod-3 System, eight simulations of the sequence of the TMI-2 accident were performed and analyzed. Simulations behaved in a like manner to the actual accident.

*Computer simulation; Reactor accident analysis

1379
Energy, Inc. Idaho Falls, Idaho. **TMI-2 Reactor Building Purge: Kr-85 Venting.** L.J. Kripps. Springfield, VA: NTIS, March 1981. 176 p. (GEND-013; CONTRACT AC07-761D01570)

A summary of the planned Kr-85 venting at TMI-2 in connection with the cleanup program, this document describes the venting itself, equipment used, how it was controlled, and the off-site atmospheric monitoring.

*Krypton; Venting; Atmospheric releases; Radiation monitoring

1380
Environmental Monitoring Systems Laboratory. Las Vegas, Nevada. **Environmental Radioactivity at the TMI (Three Mile Island) Venting Phase.** E. Bretthauer; F. Grossman; W. Efurd; G. Douglas; and A. Smith. Springfield, VA: NTIS, June 1985. 15 p. (EPA/600/4- 85/042; PB85-227502/GAR)

This report describes the monitoring of krypton-85 venting from TMI-2, June 28 through July 12, 1979, and concludes that the maximum whole body dose received was 0.0071 millirem.

*Krypton; Venting; Radiation dosage; Radiation monitoring

1381
Environmental Monitoring Systems Laboratory. Las Vegas, Nevada. **Three Mile Island Nuclear Reactor Accident of March 1979. Environmental Radiation Data. Volume 1. A Report of the Presidents Commission on the Accident at Three Mile Island.** E.W. Bretthauer; R.F. Grossman; D.J. Thome; and A.E. Smith. Springfield, VA: NTIS, March 1981. 762 p. (PB81-175374; EPA-600/4-81-013B)

Volume 1 of a six-volume report on environmental radiation data collected at TMI by four government agencies, the Commonwealth of Pennsylvania and/or Bethlehem Steel, contains the first five tables of the report. Tables include information on milk and air contamination, environmental data and agency comparisons.

*Environmental pollutants; Radiation monitoring; Radiation analysis

1382
Environmental Monitoring Systems Laboratory. Las Vegas, Nevada. **Three Mile Island Nuclear Reactor Accident of March 1979. Environmental Radiation Data. Volume 2. A Report of the President's Commission on the Accident at Three Mile Island.** E.W. Bretthauer; R.F. Grossman; D.J. Thome; and A.E. Smith. Springfield, VA: NTIS, March 1981. 578 p. (PB81-175382; EPA- 600/4-81-013C)

Volume 2 of a six-volume report on environmental radiation data collected at TMI by four government agencies, the Commonwealth of Pennsylvania and/or Bethlehem Steel, contains tables 6, 7, and 8 of the report. These include sample analysis procedures of various agencies.

*Environmental pollutants; Radiation analysis

1383
Environmental Monitoring Systems Laboratory. Las Vegas, Nevada. **Three Mile Island Nuclear Reactor Accident of March 1979. Environmental Radiation Data. Volume 3. A Report of the President's Commission on the Accident at Three Mile Island.** E.W. Bretthauer; R. F. Grossman; D.J. Thome; and A.E. Smith. Springfield, VA: NTIS, March 1981. 1347 p. (PB81-175390; EPA-600/4-81-013D)

Volume 3 of a six-volume report on environmental radiation data collected at TMI by four government agencies, the Commonwealth of Pennsylvania and/or Bethlehem Steel, contains table 9 of the report. This table is the environmental data collected by the Nuclear Regulatory Commission.

*Environmental pollutants; Radiation analysis

1384
Environmental Monitoring Systems Laboratory. Las Vegas, Nevada. **Three Mile Island Nuclear Reactor Accident of March 1979. Environmental Radiation Data.**

**Volume 4. A Report of the President's Commission on the Accident at Three Mile Island.** E.W. Bretthauer; R.F. Grossman; D.J. Thome; and A.E. Smith. Springfield, VA: NTIS, March 1981. 385 p. (PB81-175408; EPA-600/4-81-013E)

Volume 4 of a six-volume report on environmental radiation data at TMI by four government agencies, the Commonwealth of Pennsylvania, and/or Bethlehem Steel, contains six tables of the report. These tables include sample and analysis procedures for various agencies, the Commonwealth of Pennsylvania and the State of New Jersey.

*Environmental pollutants; Radiation analysis

1385
Environmental Monitoring Systems Laboratory. Las Vegas, Nevada. **Three Mile Island Nuclear Reactor Accident of March 1979. Environmental Radiation Data. Volume 5. A Report of the President's Commission on the Accident at Three Mile Island.** E.W. Bretthauer; R.F. Grossman; D.J. Thome; and A.E. Smith. Springfield, VA: NTIS, March 1981. 771 p. (PB81-175416; EPA-600/4-81-013F)

Volume 5 of a six-volume report on environmental radiation data collected at TMI by four government agencies, the Commonwealth of Pennsylvania and/or Bethlehem Steel, contains two tables of the report. These include Metropolitan Edison's sample and analysis procedures.

*Environmental pollutants; Radiation analysis

1386
Environmental Monitoring Systems Laboratory. Las Vegas, Nevada. **Three Mile Island Nuclear Reactor Accident of March 1979. Environmental Radiation Data. Volume 6. A Report of the President's Commission on the Accident at Three Mile Island.** E.W. Bretthauer; R.F. Grossman; D.J. Thome; and A.E. Smith. Springfield, VA: NTIS, March 1981. 158 p. (PB81-175424; EPA-600/4-81-013G)

Volume 6 of a six-volume report on environmental radiation data collected at TMI by four government agencies, the Commonwealth of Pennsylvania and/or Bethlehem Steel, contains the appendix of this report.

*Environmental pollutants; Radiation analysis

1387
Environmental Monitoring Systems Laboratory. Las Vegas, Nevada. **Three Mile Island Nuclear Reactor Accident of March 1979. Environmental Radiation Data: Update. A Report of the President's Commission on Three Mile Island.** E.W. Bretthauer; R.F. Grossman; D.J. Thome; and A.E. Smith. Springfield, VA: NTIS, March 1981. 333 p. (PB81-175366; EPA-600/4-81-013A)

This update of a previous report includes data on radiation counting in TMI workers and citizen volunteers in the area surrounding the power plant. No radionuclides traceable to the accident were found in this study.

*Population exposure; Nuclear workers; Radiation dosage; Radiation exposure, occupational

1388
General Public Utilities Nuclear Corporation. Middletown, Pennsylvania. **TMI-2 Reactor Vessel Plenum Final Lift.** D.C. Wilson. Springfield, VA: NTIS, January 1986. 64 p. (54 refs.) (5 appendices) (GEND-054; CONTRACT AC07-17ID01570)

Removal of the plenum was necessary to gain access to the core region of the TMI-2 reactor for defueling. The polar crane and three specially designed pendant assemblies were used. The plenum was transferred to the flooded deep end of the refueling canal and lowered to storage stand. The actual plenum lift was accomplished on May 15, 1985 in approximatley three hours.

*Plenum; Defueling; Polar crane

1389
Hanford Engineering Development Laboratory. Richland, Washington. **Accountability Study for TMI-2 Fuel.** P. Goris and D.D. Scott. Springfield, VA: NTIS, May 1981. 145 p. (GEND-016; CONTRACT AC07-761D01570)

This study proposes to measure the amount of fuel removed from TMI-2 during the cleanup operation and coordinate it with the amount known to be in the reactor pre-accident.

*Fuel recovery; Cleanup

1390
Human Sciences Research. McLean, Virginia. **Evacuation Planning in the TMI Accident. Final Report.** W.W. Chenault; G.D. Hilbert; and S.D. Reichlin. Springfield, VA: NTIS, January 1980. 205 p. (AD-A080 104/3; HSR-RS-2-8-34; CONTRACT DCPA01-78-C-0193)

This report reviews the crucial elements that entered into the mass evacuation planning associated with the TMI accident and examines the implications for the future.

*Evacuation; Emergency planning; Emergency preparedness

1391
Israel Nuclear Society. Yavne, Israel. **Accident at the Harrisburg Nuclear Reactor - Interim Conclusions.** S. Yiftah. Springfield, VA: NTIS, July 1979. 38 p. (INIS-mf-5203)

Upon review of the accident at TMI-2, the occurrence and the follow-up, the area in which most emphasis has been felt is in safety precautions.

*Safety; Reactor accident analysis

1392
Japan Atomic Energy Research Institute. Tokyo, Japan. **Prompt Technical Assessments of the TMI-2 Accident by JAERI Task Force.** Springfield, VA: NTIS, May 1979. 143 p. (JAERI-M-8373)

Based on early reports of the TMI-2 accident, the Japanese Atomic Energy Research Institute analyzed the accident and assessed the technical aspects such as fuel failure, hydrogen generation, and thermal hydraulic core and coolant activity.

*Reactor accident analysis; Technical assessment

1393
Japan Atomic Energy Research Institute. Tokyo, Japan. **Thermal-Hydraulic Analysis of the TMI Accident.** F. Tanabe; T. Shimooke; K. Yoshida; and K. Matsumoto. Springfield, VA: NTIS, January 1980. 89 p. (JAERI-M-8653)

Analysis of thermal-hydraulic and fuel heat-up behavior was made by comparing TMI-2 records for the first three hours of the accident and several computer code calculational models.

*Thermal hydraulics; Reactor core temperature; Computer models; Reactor accident analysis

1394
Los Alamos National Laboratory. Los Alamos, New Mexico. **Controlling Hydrogen Behavior in Light Water Reactors.** H.S. Cullingford and F.J. Edeskuty. Springfield, VA: NTIS, 1981. 15 p. Workshop on the Impact of Hydrogen on Water Reactor Safety. Albuquerque, New Mexico, January 26, 1981. (CONF-810120-8; LA-UR-81-681; CONTRACT W-7405-ENG-36)

Because of the hydrogen generation from the core damage at the TMI-2 nuclear station, an engineering study concerned with the safe handling of hydrogen from light water reactors was undertaken. A six-level defense strategy is proposed.

*Hydrogen; Reactor core damage; Radiolytic gas; Light water reactors

1395
Los Alamos Scientific Laboratory. Los Alamos, New Mexico. **Analysis of Early Core Damage at Three Mile Island.** P.K. Mast; T.R. Wehner; and J.R. Ireland. Springfield, VA: NTIS, 1980. 11 p. American Nuclear Society Thermal Reactor Safety Meeting. Knoxville, Tennessee, April 8, 1980. (CONF-800403-10; LA-UR-79-2942; CONTRACT W-7405-ENG-36

Analysis of the probable core damage in light of the thermal-hydraulic event at TMI-2 is reported.

*Reactor core damage; Thermal hydraulics

1396
Los Alamos Scientific Laboratory. Los Alamos, New Mexico. **Evaluation of the Three Mile Island Accident in the Context of WASH-1400.** R.D. Burns, III. Springfield, VA: NTIS, 1980. 6 p. American Nuclear Society Thermal Reactor Safety Meeting. Knoxville, Tennessee, April 8, 1980. (CONF-800403-3; LA-UR-79-2946; CONTRACT W-7405-ENG-36)

Accident probabilities for nuclear reactor operation as reported by the Reactor Safety Study (WASH-1400) were evaluated against the TMI accident and found to closely approximate. Health effects were predicted to be small and it is felt that this will be born out.

*Health effects; Reactor Safety Study (WASH 1400); Risk

1397
Los Alamos Scientific Laboratory. Los Alamos, New Mexico. **Fuel Damage Estimates for the TMI-2 Reactor.** W.L. Kirchner; J.R. Ireland; and P.K. Mast. Springfield, VA: NTIS, 1980. 8 p. American Nuclear Society Thermal Reactor Safety Meeting. Knoxville, TN, April 8, 1980. (CONF-800403-42; LA-UR-80-1069; CONTRACT W-7405-ENG-36)

Using the TRAC computer code, estimates indicate that fuel melting was negligible, but clad melting and dislocation did take place in the central core.

*Fuel melting; Reactor core damage; Computer programs; Computer simulation

1398
Los Alamos Scientific Laboratory. Los Alamos, New Mexico. **System Calculations Related to the Accident at Three Mile Island Using TRAC (Transient Reactor Analysis Code).** J.R. Ireland. Springfield, VA: NTIS, 1980. 11 p. National Heat Transfer Conference. Orlando, Florida, July 27, 1980. (CONF-800723-12; LA-UR-79-3301; CONTRACT W-7405-ENG-36)

A TRAC-P1A simulation of the TMI-2 accident was performed and evaluated while instituting single aspect hypothetical conditions in order to gather data for comparisons.

*Computer simulation; Reactor accident analysis

1399
Los Alamos Scientific Laboratory. Los Alamos, New Mexico. **TMI-2 Decay Power: LASL Fission-Product and Actinide Decay Power Calculations for the President's Commission at Three Mile Island.** T.R. England and W.B. Wilson. Springfield, VA: NTIS, October 1979. 217 p. (LA-8041-MS; CONTRACT W-7405-ENG-36)

Data on fission-product and actinide decay, gas content and curies that were supplied to the President's Commission are reported here.

*Fission products; Radiolytic gas; Radiation analysis

1400
Los Alamos Scientific Laboratory. Los Alamos, New Mexico. **TMI-2 Decay Power: LASL Fission-Product and Actinide Decay Power Calculations for the President's Commission on the Accident at Three Mile Island.** T.R. England and W.B. Wilson. Springfield, VA: NTIS, March 1980. 228 p. (LA-8041-MS (rev); CONTRACT W-7405-ENG-36)

Data on fission product and actinide decay, gas content and curies that were supplied to the President's Commission is reported on this revision of an earlier document.

*Fission products; Radiolytic gas; Radiation analysis

1401
Lovelace Biomedical and Environmental Research Institute. Albuquerque, New Mexico. **Characterization of an Aerosol Sample from Three Mile Island Reactor Auxiliary Building.** G.M. Kanapilly; J.A. Stanley; G.J. Newton; B.A. Wong; and P.B. DeNee. Springfield, VA: NTIS, January 1981. 13 p. (LMF-70; CONTRACT AC04-76EV01013)

Composition of air samples obtained from one week of continuous filtration from the TMI-2 auxiliary building were analyzed. Cesium and strontium were the major radionuclides found.

*Air pollution, radioactive; Filtration; Reactor auxilliary building; Cesium radioisotopes; Strontium radioisotopes

1402
Mideast Center for Radiological Physics. Pittsburgh, Pennsylvania. **Known Effects of Low-Level Radiation Exposure: Health Implications of the TMI (Three Mile Island) Accident.** P.K. Shrivastava. Springfield, VA: NTIS, April 1979. 158 p. Proceedings of a Conference held at Pittsburgh, Pennsylvania, on April 25, 1979. (HRP-0902889/5; DHEW/PUB/NIH-80-2087)

A conference on the known effects of low-level radiation exposure and probable health implications of TMI was held in April 1979. This conference was conducted to prepare health professionals to answer questions stemming from the radiation releases subsequent to TMI and provide the most up-to-date information on the topic.

*Health effects; Radiation, low level; Radiation effects; Radioactive releases

1403
National Academy of Public Administration. **Major Alternatives for Government Policies, Organizational Structures and Actions in Civilian Nuclear Reactor Emergency Management in the United States: A Report to M. Rogovin for the Nuclear Regulatory Commission Three Mile Island Special Inquiry.** Springfield, VA: NTIS, January 1980. 180 p. (NUREG-CR-1225; CONTRACT NRC-04-79-210)

This report evaluated alternatives to the currently existing nuclear emergency response system. It includes analysis of issues, review of present state of nuclear reactor emergency planning and response, and conclusions.

*Emergency planning; Emergency preparedness; Reactor accidents

1404
NUS Corporation. Gaithersburg, Maryland. **Summary Report: Development and Application of Technology Transfer Techniques for TMI-2 R & D Information.** S. Levine; F.T. Stetson; D.D. Orvis; R.R. Sherry; and E.R. Wiot. Springfield, VA: NTIS, 1983. 8 sections. (ll refs.) (GEND INF-033; CONTRACT AC07-761D01570)

"This report describes efforts undertaken by NUS Corporation to assist the Department of Energy and their on-site contractor, E G & G, in technology transfer from DOE TMI-2 R & D Program to the nuclear community."

*Communications; Information exchange; Research and development

1405
Oak Ridge National Laboratory. Oak Ridge, Tennessee. **Emergency Actions Concerned with Effluent Controlled at TMI.** R.E. Brooksbank; D.O. Campbell; J.W. Snyder; and O.O. Yarbro. Springfield, VA: NTIS, 1980. 34 p. American Nuclear Society Annual Meeting. Las Vegas, Nevada, June 8, 1980. (CONF-800607-60; CONTRACT W-7405-ENG-26)

Water treatment and control of gaseous releases at TMI-2 are discussed relative to the emergency requirements of the accident.

*Contaminated water; Water pollutants, radioactive; Noble gas releases

1406
Oak Ridge National Laboratory. Oak Ridge, Tennessee. **Estimates of Dose Due to Noble Gas Releases from the Three Mile Island Incident Using the AIRDOS-EPA Computer Code.** S.J. Cotter; C.W. Miller; R.E. Moore; and C.A. Little. Springfield, VA: NTIS, December 1980. 57 p. (ORNL-5649; CONTRACT W-7405-ENG-26)

Taking into account the total radionuclide release from TMI-2 and wind dispersion factors, it was estimated in this report that the highest radiation dose received from the accident was 1% of annual expected background.

*Noble gas releases; Air pollution, radioactive; Radiation dosage; Computer programs

1407
Oak Ridge National Laboratory. Oak Ridge, Tennessee. **Estimates of Dose to the Population within Fifty Miles to Noble Gas Releases from the Three Mile Island Incident.** C.W. Miller; S.J. Cotter; R.E. Moore; and C.A. Little. Springfield, VA: NTIS, 1980. 8 p. American Nuclear Society Thermal Reactor Meeting. Knoxville, Tennessee, April 8, 1980. (CONF-800403-13; CONTRACT W-7405-ENG-260)

This report of the Task Group on Health Physics and Dosimetry of the President's Commission on the Accident at TMI includes estimates of population exposure received from release of noble gases during the period March 28, through April 15, 1979.

*Noble gas releases; Population exposure

1408
Oak Ridge National Laboratory. Oak Ridge, Tennessee. **Flowsheet Development Studies for the Decontamination of High-Activity-Level Water at Three Mile Island Unit 2.** E.D. Collins; J.E. Bigelow; D.O. Campbell; L.J. King; and J.B. Knauer. Springfield, VA: NTIS, 1980. 33 p. American Institute of Chemical Engineers Annual Meeting. Portland, Oregon, August 17, 1980. (CONF-800-802-13; CONTRACT W-7405-ENG-26)

Using samples of contaminated water from TMI-2, a zeolite exchange process was tested for removal of cesium and strontium while an ion exchange process was evaluated for the lower-level radiation removal.

*Contaminated water; Ion exchange; Strontium radioisotopes; Cesium radioisotopes; Zeolites; Decontamination; Cleanup

1409
Oak Ridge National Laboratory. Oak Ridge, Tennessee. **Involvement of the ORNL Chemical Technology Division in Contaminated Air and Water Handling at the Three Mile Island Nuclear Power Station.** R.E. Brooksbank and L.J. King. Springfield, VA: NTIS, August 1979. 61 p. (ORNL/TM-7044; CONTRACT W-7405-ENG-26)

The Chemical Technology division of the Oak Ridge National Laboratory is active in the TMI-2 decontamination program in the areas of instrument diagnostics and removal of radionuclides from air and water at the power station.

*Decontamination; Contaminated water; Instrument function; Equipment function

1410
Oak Ridge National Laboratory. Oak Ridge, Tennessee. **Plenary Lecture. A Special Radioelement Problem: ORNL Assistance to Three Mile Island in Handling Contaminated Air and Water.** R.E. Brooksbank. Springfield, VA: NTIS, 1979. 44 p. Conference on Analytical Chemistry in Energy Technology. Gatlinburg, Tennessee, October 9, 1979. (CONF-791049-26; CONTRACT W-7405-ENG-26).

The Oak Ridge National Laboratory's assistance in the decontamination and waste handling, analytical service and flowsheet development, at TMI-2 are presented. Specifically addressed are efforts in the areas of contaminated air and water handling.

*Decontamination; Waste handling; Radiation analysis; Contaminated water; Air pollution, radioactive; Oak Ridge National Laboratory

1411
Oak Ridge National Laboratory. Oak Ridge, Tennessee. **Post-Accident Cleanup of Radioactivity at the Three Mile Island Station.** R.E. Brooksbank; W.J. Armento; D.O. Campbell; E.D. Collins; F.E. Harrington; L.J. King; J.R. Parrott, Sr.; W.A. Shannon; and J.W. Snider. Springfield, VA: NTIS, February 1980. 61 p. (15 refs.) (ORNL/TM-7081; CONTRACT W-7405-ENG-26)

This report is a compilation of data, in summarized form, assembled by the ORNL staff. It includes TMI on-site assistance and accomplishments post-accident, and flowsheet development for the TMI recovery and cleanup.

*Cleanup; Decontamination

1412
Oak Ridge National Laboratory. Oak Ridge, Tennessee. **Post-Accident Cleanup of Radioactivity at Three Mile Island.** R.E. Brooksbank. Springfield, VA: NTIS, 1979. 27 p. (DOE/TIC/-11023; CONTRACT W-7405-ENG-26)

Radioactivity levels at TMI-2 and processing of radioactive water are discussed.

*Water pollutants, radioactive; Cleanup

1413
Oak Ridge National Laboratory. Oak Ridge, Tennessee. **Post-Accident Examination of Platinum Resistance Thermometers Installed in the TMI-2 Reactor.** R.M. Carrol and R.L. Shepard. Springfield, VA: NTIS, September 1985. 82 p. (12 refs.) 3 appendices. (GEND INF-064; CONTRACT AC07-761D01570)

Laboratory tests on one resistance thermometer showed that neither calibration not time response was adversely affected by the accident or post-accident conditions to which it was exposed.

*Resistance thermometers; Instrument function

1414
Oak Ridge National Laboratory. Oak Ridge, Tennessee. **Radiation Release and Health Effects Lessons from the Three Mile Island Accident: Application to Emergency Response Training.** F.R. Mynatt and C.D. Berger. Springfield, VA: NTIS, November 1980. 5 p. (DOE TIC 11356; CONTRACT W-7405-eng-26)

This document reviews the emergency response planning especially related to the Nuclear Regulatory Commission Emergency Action guidlines.

*Health effects; Radioactive releases; Emergency planning

1415
Oak Ridge National Laboratory. Oak Ridge, Tennessee. **Radiation Release and Health Effects Lessons from the Three Mile Island Incident: Assessment of Objective Risks for Emergency Preparedness Planning.** C.D. Berger and F.R. Mynatt. Springfield, VA: NTIS, 1980. 5 p. Kentucky Special Advisory Commission on Nuclear Issues. Northern Kentucky University Campus, Highland Heights, Kentucky. November 21, 1980. (DOE TIC 11360; CONTRACT W-7405-eng-26)

An emergency response plan for potential "accidents" at commercial nuclear facilities like Three Mile Island is presented.

*Health effects; Radioactive releases; Emergency planning

1416
Oak Ridge National Laboratory. Oak Ridge, Tennessee. **Recent Chemical Engineering Requirements as the Result of TMI On-Site Experience.** R.E. Brooksbank. Springfield, VA: NTIS, 1980. 24 p. American Institute of Chemical Engineers. Portland, Oregon, August 17, 1980. (CONF-800802-10; CONTRACT W-7405-ENG-26)

As a result of the TMI-2 accident, it is speculated that the role of, and demands upon, chemical engineers will increase.
*Chemical engineering; Cleanup

1417
Oak Ridge National Laboratory. Oak Ridge, Tennessee. **Three Mile Island in Perspective.** D.B. Trauger. Springfield, VA: NTIS, 1979. 12 p. A Joint Meeting of American Institute of Chemical Engineers, ASME, and IEEE. Paducah, Kentucky, November 13, 1979. (CONF-791131-1; CONTRACT W-7405-ENG-26)
The TMI accident, its possible causes, and the implications for the future are discussed.
*Reactor accident analysis; Accident - causes; Lessons learned

1418
Oak Ridge National Laboratory. Oak Ridge, Tennessee. **TMI-2 Criticality Analysis: Analytical Models and Methods.** J.T. West; R.M. Westfall; G.E. Whitesides; and J.T. Thomas. Springfield, VA: NTIS, 1980. 7 p. American Nuclear Society International Conference. Washington, D.C., November 17, 1980. (CONF-801107-29; CONTRACT W-7405-ENG-26)
Calculations of core disruption are presented for TMI-2 reactor core as a result of the accident.
*Reactor core damage; Reactor core analysis

1419
Oak Ridge National Laboratory. Oak Ridge, Tennessee. **TMI-2 Criticality Analysis: Parametric Studies and Overall Results.** R.M. Westfall; J.T. West; G.E. Whitesides; and J.T. Thomas. Springfield, VA: NTIS, 1980. 7 p. American Nuclear Society International Conference. Washington, D.C., November 17, 1980. (CONF-801107-32; CONTRACT W-7405-ENG-26)
Studies on core damage probability were conducted with the thought that the knowledge gained could be applied to cleanup operations for TMI-2.
*Criticality; Cleanup; Reactor core damage

1420
Oak Ridge National Laboratory. Oak Ridge, Tennessee. Chemical Technology Division. **Emergency Actions Concerned with Effluent Control at TMI.** R.E. Brooksbank; D.O. Campbell; J.W. Snyder; and O.O. Yarbro. Springfield, VA: NTIS, 1980. 33 p. American Nuclear Society Session on the Public Safety Aspects of TMI. American Nuclear Society Annual Meeting. Las Vegas, Nevada, June 8-19, 1980. (CONF-800607-60; CONTRACT W-7405-ENG-26)
Observations on a TMI site transformed from an electrical power production facility to a radiochemical processing plant are made.
*Cleanup; Waste handling; Ion exchange

1421
Office of Naval Research. London, England. **Three Mile Island and Other Emergency Concerns.** J.R. Neighbors. Springfield, VA: NTIS, June 30, 1980 3 p. Included in *European Scientific Notes* 34(6):272-4, June 30, 1980. (PB81-139503)
Three Mile Island and other energy topics are discussed.
*Reactor accidents

1422
Sandia National Laboratories. Albuquerque, New Mexico. **Analysis of the TMI-2 Dome Radiation Monitor.** M.B. Murphy; G.M. Mueller; and W.C. Jernigan. Springfield, VA: NTIS, August 1985. 173 p. (18 refs.) illus. (GEND INF-063; CONTRACT AC07-761D01570)
Questions have arisen regarding the accuracy of in-containment radiation readings of the dome radiation monitor, HP-R-214, during the March 28, 1979 loss of coolant accident at Three Mile Island. This report discusses the accuracy of readings,

examines the monitor itself, and estimates radiation environment inside the containment during the accident.

*Radiation monitors; Instrument function; Reactor containment building; Radiation analysis

1423
Science Applications, Inc. Monroeville, Pennsylvania. **Review of Literature on the TMI Accident and Correlation to the LWR Safety Technology Program.** W.J. Miller. Springfield, VA: NTIS, May 1980. 220 p. (SAI-124-79-PGH; ALO-85; CONTRACT AC04-76DP00789)

This report is a collection of available literature concerning the TMI-2 accident and its correlation to the LWR Safety Technology Program.

*Bibliographies

1424
Science Applications, Inc. Rockville, Maryland. **Measurements of Exp 129 I and Radioactive Particulate Concentrations in the TMI-2 Containment Atmosphere During and After Venting.** J.E. Cline; P.A. Roy; J.N. Hollcroft; J. Houbaugh, Jr.; and T.L. McVey. Springfield, VA: NTIS, April 1981. 25 p. (GEND-009; CONTRACT AC07-761D01570)

Measurements of the radioactivity in the TMI-2 containment building before and after the June 28 to July 11, 1980, venting of Iodine 129 and Krypton 85 are reported. Krypton 85 was reduced significantly while the iodine returned to its earlier concentration in about 15 days.

*Reactor containment building; Venting; Iodine radioisotopes; Krypton; Radiation monitoring

1425
UKAEA. Atomic Energy Technical Unit. Harwell, England. **Three Three Mile Island Incident.** L.M. Davies. Springfield, VA: NTIS, 1979. 8 p. (INIS-mf-5449)

The TMI accident, sequence and core damage, is presented along with the engineering design of the TMI-2 pressurized water reactor.

*Reactor core damage; Accident sequence; Nuclear reactor design; Pressurized water reactors

1426
Union Carbide Corporation. Computer Sciences Division and Nuclear Division. Oak Ridge, Tennessee. **Criticality Analysis of Disrupted Core Models of Three Mile Island Unit 2.** R.M. Westfall; J.T. West; G.E. Whitesides; and J.T. Thomas. Springfield, VA: NTIS, December 1979. 111 p. (ORNL/CSD/TM-106; CONTRACT W-7405-ENG-26)

Three core model damage assessment reports were examined for the President's Commission Report. Data are presented on these including criticality analyses.

*Reactor core damage; Criticality; Reactor core analysis

1427
U.S. Ad Hoc Interagency Population Dose Assessment Group. **Population Dose and Health Impact of the Accident at the Three Mile Island Nuclear Station: (A Preliminary Assessment for the Period March 28, 1979, through April 7, 1979).** Prepared by an Interagency Team from the Nuclear Regulatory Commission, Environmental Protection Agency, and the Department of Health, Education and Welfare. Springfield, VA: NTIS, May 10, 1979. 99 p.

This interagency report was conducted to assess the health impact of the collective radiation dose received by the two million residents within the fifty-mile radius of Three Mile Island as a result of the March 28 accident. The period up to, and including, April 7, 1979 was examined. The conclusion, based on information gathered from thermoluminescent dosimeters, is that the health impact will be minimal.

*Radiation dosage; Population exposure; Health effects

1428
U.S. Ad Hoc Interagency Population Dose Assessment Group. **Summary and Discussion of Findings from Population Dose and Health Impact of the Accident at the Three Mile Island Nuclear Station: (A Preliminary Estimate for the Period March 28, 1979, through April 7, 1979).** Ad Hoc Interagency Population Dose Assessment Group. Bureau of Radiological Health. Springfield, VA: NTIS, 1979. 7 p.

This report contains the "preface" and "summary and discussion of findings" sections of the full report.

*Radiation dosage; Population exposure; Radioactive releases; Health effects

1429
U.S. Department of Health and Human Services. Bureau of Radiologic Health. **Use of Photographic Film to Estimate Exposure Near the Three Mile Island Nuclear Power Station.** R.E. Shuping. Springfield, VA: NTIS, February 1981. 26 p. (PB81-183402; FDA/BRH-81/51; DHHS/PUB/FDA-81-8142)

The use of photographic film to determine amount of radiation release from TMI-2 is reported.

*Film, photographic; Radiation monitoring; Radioactive releases

1430
U.S. Environmental Protection Agency. **Three Mile Island Nuclear Reactor Accident of March 1979. Environmental Radiation Data Update. A Report to the President's Commission on the Accident at Three Mile Island.** E.W. Bretthauer; R.F. Grossman; D.J. Thome; and A.E. Smith. Springfield, VA: NTIS, December 21, 1979. 328 page table. (EPA 600 4-79)

A table.

*Radiation monitoring; Environmental pollutants

1431
U.S. Environmental Protection Agency. **Three Mile Three Mile Island Nuclear Reactor Accident of March 1979. Environmental Radiation Data Update. 3 volumes. 1, 2, 3.** B.A. Hilton and R.F. Grossman. Springfield, VA: NTIS, March 1981. (EPA 600/4-81-014A; EPA 600/4-81-014B; EPA 600/4-81-014C; v.1 PB81-191165; v.2 PB81-191173; v.3 PB81-191181)

The original report contains a list of environmental radiation monitoring data collected in the TMI vicinity following the March 28, 1979 accident. The material was collected by the EPA, NRC, DOE, HHS, Commonwealth of Pennsylvania and Bethlehem Steel Corporation. This final update contains additional data not previously reported.

*Radiation monitoring; Environmental pollutants

1432
U.S. General Accounting Office. **Impact of Federal R & D Funding on Three Mile Island Cleanup Costs.** Washington, DC: GPO, January 15, 1982. 23 p. (EMD-82-28)

The various research and development programs of the Department of Energy are reviewed. The thrust of operations is two fold, one of data acquisition and one of research and development aimed at reactor evaluation and waste immobilization. Appendices list actual and projected/proposed expenditures.

*Research and development; Financing; Costs; Reactor core analysis; Radioactive waste; Waste handling

1433
U.S. General Accounting Office. **Three Mile Island: The Financial Fallout. Report by the Comptroller General of the United States.** Washington, DC: GPO, July 7, 1980. 85 p. 3 appendices. (EMD-80-89; B-199244) - preliminary pages

The GAO reviewed the financial situation of GPU with regard to loss of generating power of the TMI units, the cleanup costs and the necessity of buying power from the grid in order to meet customer needs. GAO concludes that the accident has created a severe financial situation with regard to: (1) borrowing, (2) cleanup costs, and (3) cash flow. It is also noted that federal regulatory agencies have "done little to expedite the system's recovery from the accident."

*Financing; Cleanup; Costs

1434
U.S. General Accounting Office. Energy and Minerals Division. **Greater Commitment Needed to Solve Continuing Problems at Three Mile Island.** Springfield, VA: NTIS, August 26, 1981. 113 p. (EMD-81-106)

This report examines GPU's financial situation with regard to the accident and cleanup operations at TMI. The GAO recommends, on the basis of its study, that multi-year funding to the Department of Energy be provided for research and development programs at Three Mile Island.

*Cleanup; Financing; General Public Utilities; Research and development

1435
U.S. General Accounting Office. Energy and Minerals Division. **Three Mile Island: The Most Studied Nuclear Accident in History. Summary. Report to Congress.** Springfield, VA: NTIS, September 9, 1980. 86 p. (PB81-156077; EMD-80-109)

This report summarizes the TMI accident to date and the effect it has had on the nuclear industry regarding safety precautions.

*Reactor accidents; Safety

1436
U.S. General Accounting Office. Resources Community and Economic Development Division. **Nuclear Regulatory Commission Should Report on Progress in Implementing Lessons Learned from the Three Mile Island Accident. Report to Congress.** Springfield, VA: NTIS, July 19, 1985. 105 p. (GAO/RCED-85-72, B-213365; PB85-242212/GAR)

Progress on the May 1980 Three Mile Island Action Plan (to promote more efficient operation of nuclear power plants) is reported.

*Safety; Operators; Recommendations; Lessons learned

1437
Verein Deutscher Ingenieure. Duesseldorf, Germany F.R. Gesselschaft Energietechnik. **Harrisburg Accident. A Documentation of the VDI - Gesselschaft Energietechnik.** Springfield, VA: NTIS, 1979. 16 p. (INIS-mf-5444)

A description of the TMI-2 accident is presented based on the known available facts.

*Reactor accidents; Reactor accident analysis

# 17

## Pennsylvania State Documents

1438
Pennsylvania. Adjutant General's Office. Department of Military Affairs. **After Action Report; Three Mile Island Nuclear Incident. 28 March - 5 April, 1979.** Annville, PA: The Adjustant General, 1979. 15 p. 10 "annexes" or appendices.

This report reviews the events of the accident and response leading up to the activation of the Pennsylvania National Guard. The actions of the Guard throughout the critical period are noted. A four-page summary of problems encountered, lessons learned, as well as recommendations, are a valuable part of this report.

*Reactor accidents; Emergency preparedness; Pennsylvania National Guard; Recommendations

1439
Pennsylvania. Commission to Study and Evaluate the Consequences of the Incident at Three Mile Island. **Pennsylvania Commission to Study and Evaluate the Consequences of the Incident at Three Mile Island.** Harrisburg: The Commission, June 5, 1979.

The press release naming the members of the Governor's Commission, Subcommittee assignments, and Lt. Governor Scranton's charge to the Commission are given.

*Accident investigations

1440
Pennsylvania. Commission to Study and Evaluate the Consequences of the Incident at Three Mile Island. **Report of the Governor's Commission on Three Mile Island.** Presented to the Honorable Dick Thornburgh, Governor, Commonwealth of Pennsylvania. Harrisburg: The Commission, February 26, 1980. 211 p.

The Pennsylvania Governor's Commission was formed to assess the Commonwealth's performance during the TMI-2 emergency, to look at the accident's consequences, and to determine what the state government could do to alleviate the impact of the accident on its citizens. Accordingly, this report addresses the environmental, health, economic, and legal issues, emergency preparedness, and recovery responses of the state and federal agencies. Recommendations in six broad areas are considered and a short bibliography completes the report.

*Accident investigations; Reactor accidents; Health effects; Environmental effects; Economic effects; Legal aspects; Emergency preparedness; Cleanup; Recommendations; Bibliographies

1441
Pennsylvania. Department of Health. Division of Epidemiological Research. **Cancer Mortality and Morbidity (Incidence) Around TMI: Includes News Release, Summary and Main Report.** G.K. Tokuhata and E. Digon. Harrisburg: Pennsylvania Department of Health, September 1985. 70 p. tables.

This report is the result of the Pennsylvania Department of Health examination of cancer incidence and deaths in the communities (within twenty miles) around Three Mile Island. Major findings of the report include the following: (1) there is no evidence that cancer mortality increased because of TMI; (2) there were no increases in cancer by site -- including leukemia; (3) there was not an increased incidence of newly diagnosed cancers; and (4) the Mother-Child registry reports no significant increase in cancer at this time.

*Health effects - cancer

1442
Pennsylvania. Department of Health. Division of Epidemiological Research. **Fetal and Infant Mortality and Congenital Hypothyroidism Around TMI**. G.K. Tokuhata and E. Digon. Harrisburg: Pennsylvania Department of Health, 1981. 6 p. tables. Paper presented at the International Symposium on Health Impacts of Different Sources of Energy. Nashville, Tennessee, June 22-26, 1981.

As a part of the Pennsylvania Department of Health studies on the health impact of the Three Mile Island accident, pregnancy outcome and infant health were determined for the area encompassed by a ten-mile radius of the damaged reactor. The years 1977, 1978, and 1979 are reviewed, and data analyzed by calendar quarters.

*Health effects; Pregnancy outcome; Infant mortality; Hypothyroidism

1443
Pennsylvania. Department of Health. Division of Epidemiological Research. **Health Related Economic Costs of the Three Mile Island Accident**. T.W. Hu and K.S. Slaysman. University Park, Pennsylvania State University. Harrisburg: Pennsylvania Department of Health, July 1981. 86 p. (9 refs.) (CONTRACT 641996)

Examination of the short term health-related economic costs of the TMI accident for the period immediately following is detailed. Decisions to evacuate or stay, cost of evacuation, and cost of medical services and supplies as well as alcohol and cigarettes are considered. It is estimated that for the fifteen-mile radius during the two-week period following the accident, economic cost ranged between $9 and $15 million.

*Economic effects; Health care; Local residents; Evacuation; Stress; Accident investigations

1444
Pennsylvania. Department of Health. Division of Epidemiological Research. **Health Studies in the Three Mile Island Area**. G.K. Tokuhata. Harriburg: Pennsylvania Department of Health, 1981. 11 p.

Having identified the population to be studied in the aftermath of the Three Mile Island incident, this report describes the various areas to be addressed by the Pennsylvania Department of Health in further researches. These include pregnancy outcome, psychological stress, congenital/neonatal hypothyroidism, radiation dose assessment, chromosome studies, mental health, infant mortality, population mobility, health economics, cardiac mortality and family unity studies.

*Health effects; Pregnancy outcome; Health effects - psychological; Stress; Hypothyroidism; Population mobility; Radiation dosage; Economic effects

1445
Pennsylvania. Department of Health. Division of Epidemiological Research. **Pregnancy Outcome Around Three Mile Island**. G.K. Tokuhata. Harrisburg: Pennsylvania Department of Health, 1981. 7 p. (16 refs.) Paper presented at a conference, "Linking Public Health Social Worker and Public Social Services for Comprehensive Care for Mothers and Children." University of Pittsburgh, Pittsburgh, Pennsylvania, March 29-April 2, 1981.

An examination of available data regarding the possible effects of the TMI accident on pregnancy outcome is reported. An overview of the possible effects radiation may have on the fetus, the importance of stress on pregnancy and other factors which may influence pregnancy are taken into account.

*Pregnancy outcome; Health effects; Pregnancy outcome; Reactor accidents; Stress

1446
Pennsylvania. Department of Health. Division of Epidemiological Research. **Three Mile Island Health Effects Research Program**. G.K. Tokuhata. Harrisburg: Pennsylvania Department of Health, 1980. 3 p.

An outline and brief description of the various studies that are a part of the "Health effects research program" is presented. These include: (1) TMI census, (2) pregnancy outcome, (3) congenital/neonatal hypothyroidism, (4) health behavioral (stress), (5) health economics, (6) radiation dose assessment, (7) cytogenetic (chromosome), (8) cardiac mortality, (9) family breakage (divorce), (10) mental health, and (11) long-term disease surveillance, especially cancer mortality.

*Health effects; Pregnancy outcome; Hypothyroidism; Stress; Economic effects; Radiation dosage; Health effects - cancer; Research and development

1447
Pennsylvania. Departmment of Health. Division of Epidemiological Research. Department of Epidemiology and Disease Prevention. **Impact of TMI Nuclear Accident upon Pregnancy Outcome, Congenital Hypothyroidism and Infant Mortality**. G.K. Tokuhata. Harrisburg: Pennsylvania Department of Health, 1981. 14 p. bibliography.

A study of pregnancy outcome for the twenty-mile radius around Three Mile Island for the period March 28, 1979 to March 27, 1980, (about 4000 deliveries), incidence of congenital hypothyroidism among infants, and infant mortality for a ten-mile radius were reviewed. There does not appear to be any additional adverse physical health effects as a result of the accident at Three Mile Island.

*Health effects; Pregnancy outcome; Hypothyroidism; Infant mortality

1448
Pennsylvania. Department of Health. Division of Epidemiology. **Crisis Evacuation During the Three Mile Island Nuclear Accident: The TMI Population Registry**. M.K. Goldhaber and J.E. Lehman. Harrisburg: Pennsylvania Department of Health, November 1982. updated May 1983. 42 p. (refs.)

A study of the population within five miles of the Three Mile Island power plant revealed that about 64% evacuated at some time during the crisis period following the accident. A number of interesting demographic variables are presented. The single most significant factor in evacuation appeared to be families with preschool children.

*Evacuation; Preschool children

1449
Pennsylvania. Department of Health. Division of Epidemiology Research. **Three Mile Island Mother-Child Registry**. J. Bratz. Harrisburg: Pennsylvania Department of Health, 1983. (also appears in *Pennsylvania Department of Health Newsletter: Health Reporter* 4(12) 1 and 4, December 1983)

A description of the methodology and data collection aspects of the TMI mother-child registry as it grew out of the TMI pregnancy outcome study is reported.

*Health effects; Pregnancy outcome; Population registry; Mothers; Children

1450
Pennsylvania. Department of Health. Division of Epidemiology Research. **The Three Mile Island Population Registry. Report One: A General Description**. M.K. Goldhaber. Harrisburg: Pennsylvania Department of Health, October 1981. 38 p. tables.

The Pennsylvania Department of Health survey of the population living within five miles of Three Mile Island was initiated soon after the accident in order to establish a baseline study group for future determinations of the effect of the accident. This report provides a description of the population and outlines plans for subsequent studies.

*Population registry; Research and development

1451
Pennsylvania. Department of Health. TMI Advisory Panel on Health Research Studies. **Mobility of the Population Within 5 Miles of Three Mile Island During the Period from August 1979 through July 1980**. M.K. Goldhaber; P.S. Houts; and R. DiSabella. Harrisburg: Pennsylvania Department of Health, September 1981. 50 p.

A study to determine whether local residents moved away from the TMI area (five-mile radius) after the accident was undertaken. Results indicate that the rate at which people moved remained the same pre- and post-accident and attitudes of "movers" and "non-movers" were virtually the same.

*Population mobility; Local residents

1452
Pennsylvania Department of Health. TMI Advisory Panel on Health Research Studies. **Health Related Behavioral Impact of the Three Mile Island Nuclear Incident. A Report Submitted to the TMI Advisory Panel on Health Research Studies. Part I.** P.S. Houts; R.W. Miller; G.K. Tokuhata; and K.S. Ham. Hershey: Pennsylvania State University College of Medicine, April 8, 1980. 5 chapters. variously numbered.

This report examines the psychological effect that the March 28, 1979, TMI incident had on the population surrounding the nuclear facility. "The material included here is part one of the total report, and will focus on: (1) characteristics of the population residing in the area of TMI, (2) description of the evacuation decisions and behavior during the crisis period, and (3) the extent, severity, and duration of distress experienced by persons living in the vicinity of TMI."

*Health effects - psychological; Evacuation; Stress; Population mobility

1453
Pennsylvania. Department of Health. TMI Advisory Panel on Health Research Studies. **Health Related Behavioral Impact of the Three Mile Island Nuclear Incident. A Report Submitted to the TMI Advisory Panel on Health Related Studies. Part II.** P.S. Houts; R.W. Miller; G.K. Tokuhata; and K.M. Ham. Harrisburg: Pennsylvania Department of Health. November 21, 1980. 2 chapters. variously numbered.

This second part of the study attempts to describe: "(1) characteristics of persons who were distressed during the initial crisis in March and April, 1979, and (2) characteristics of persons who increased or decreased distress over the nine month period from April 1979 to January 1980." Also contained in this part are summaries of information which might "have relevance for public policies," e.g. evacuation planning, social support, etc.

*Health effects - psychological; Stress

1454
Pennsylvania. Department of Health. TMI Advisory Panel on Health Research Studies. **Health Related Behavioral Impact of the Three Mile Island Nuclear Incident. A Report Submitted to the TMI Advisory Panel on Health Research Studies. Part III**. P.S. Houts; R.M. DiSabella; and M.K. Goldhaber. Harrisburg: Pennsylvania Department of Health, May 12, 1981. 27 p.

This paper presents the results of a study to determine whether distress levels had changed in comparison to early reports. Attitude questions, about how the TMI situation was being handled and what residents' thoughts were regarding krypton venting are reported.

*Health effects - psychological; Stress; Krypton; Venting

1455
Pennsylvania. Governor Thornburgh. **Governor Dick Thornburgh's Proposal to Finance the Cleanup of Three Mile Island**. Harrisburg: Government Office, July 1981. 20 p.

A proposal for financing the cleanup of TMI which divides and shares the responsibility among the federal, state, local governments and the utility is presented.

Also included in the proposal is the creation of a non-government agency, "National Energy Research Institute," to serve as a financial and technical support vehicle for the cleanup and for future energy concerns.

*Cleanup; Financing; Thornburgh, Richard

1456
Pennsylvania. Governor's Commission on Three Mile Island. Legal Subcommittee. **Legal Aspects of the Three Mile Island Accident: Report on the Legal Subcommittee to the Governor's Commission on Three Mile Island**. Harrisburg: The Commission, December 1979. 71 p.

This committee's work and recommendations were included in the Report of the Governor's Commission, but this separate report was produced for those persons who are specifically interested in the legal aspects of the nuclear issue. "Legal aspects" include proceedings in court and before regulatory bodies for both private citizens and public bodies. The Subcommittee's recommendations complete the report.

*Legal aspects; Accident investigations

1457
Pennsylvania. Governor's Office of Policy and Planning. **The Socio-Economic Impacts of the Three Mile Island Accident. Final Report**. Harrisburg: The Office, 1980. 229 p. Prepared in co-operation with the Department of Agriculture, Department of Community Affairs, Department of Commerce, Department of Insurance, Department of Labor and Industry and the Department of Revenue. (PB81-164196; EDA-81-028; Grant EDA-01-6-01984-40)

Under the direction of the Pennsylvania Governor's Office, the effect of the TMI accident on the economics of manufacturing in the twenty-mile radius surrounding the plant was assessed. The study covers the twelve months following the accident.

*Socioeconomic effects

1458
Pennsylvania. Governor's Office of Policy and Planning. **Three Mile Island Socio-Economic Impact Study**. Prepared by the Governor's Office of Policy and Planning in cooperation with the Departments of Agriculture, Community Affairs, Commerce, Insurance, Labor and Industry, and Revenue. Harrisburg: The Office, December 14, 1979. 105 p. map.

The study reviewed the short- and potential long-term economic impacts that the accident at TMI had on Pennsylvania. Short-term impact was related to interruption of routine business primarily because of evacuation. The potential longer impact centers around increased electricity costs, insurance and legal claims related to the accident.

*Socioeconomic effects

# 18

## Federal and State Hearings

1459
Pennsylvania. General Assembly. House of Representatives. Select Committee - Three Mile Island. **Three Mile Island Report: Hearings Concerning Three Mile Island.** James L. Wright, Jr., Chairman. Harrisburg: The Select Committee, 1980. 96 p.
*Hearings

1460
United States. Congress. House. Committee on Armed Services. Military Installations and Facilities Subcommittee. **Civil Defense and the Three Mile Island Nuclear Accident: Report of the Military Installations and Facilities Subcommittee.** 96th Congress, 1st session, December 18, 1979. Washington, DC: GPO, 1979. 11 p. Committee Print No. 13.
*Hearings

1461
United States. Congress. House. Committee on Armed Services. Military Installations and Facilities Subcommittee. **Civil Defense Aspects of the Three Mile Island Nuclear Accident: Hearings Before the Military Installations and Facilities Subcommittee.** 96th Congress, 1st session, May 16 - June 14, 1979. Washington, DC: GPO, 1979. 243 p. bibliography. Serial No. 96-11.
*Hearings

1462
United States. Congress. House. Committee on Energy and Commerce. Subcommittee on Energy Conservation and Power. **Financial Fallout from Three Mile Island: Hearings Before the Subcommittee on Energy Conservation and Power.** 97th Congress, 1st session on H.R. 2512, May 1 and September 22, 1981. Washington, DC: GPO, 1982. 389 p. illus. bibliography. Serial No. 97-163.
*Hearings

1463
United States. Congress. House. Committee on Energy and Commerce. Subcommittee on Oversight and Investigations. **Cleanup Efforts at Three Mile Island: Hearing Before the Subcommittee on Oversight and Investigations.** 97th Congress, 1st session, March 30, 1981. Washington, DC: GPO, 1982. 72 p. Serial No. 97-108.
*Hearings

1464
United States. Congress. House. Committee on Government Operations. Environment, Energy, and Natural Resources Subcommittee. **Nuclear Safety -- Three Years After Three Mile Island: Joint Hearing Before Certain Subcommittees of the Committee on Government Operations and Interior and Insular Affairs.** 97th

Congress, 2d session, March 12, 1982. Washington, DC: GPO, 1982. 58 p.
*Hearings

1465
United States. Congress. House. Committee on Interior and Insular Affairs. **Reporting of Information Concerning the Accident at Three Mile Island: A Report.** Prepared by the majority staff of the Committee on Interior and Insular Affairs. 97th Congress, 1st session, March 1981. Washington, DC: GPO, 1981. 150 p. illus.
*Hearings

1466
United States. Congress. House. Committee on Interior and Insular Affairs. Subcommittee on Energy and the Environment. **Accident at the Three Mile Island Nuclear Power Plant: Oversight Hearings Before a Task Force of the Subcommittee on Energy and the Environment.** 96th Congress, 1st session, May 9-15, 1979. Washington, DC: GPO, 1979 - . 11 volumes. Serial No. 96-8.
*Hearings

1467
United States. Congress. House. Committee on Interior and Insular Affairs. Subcommittee on Energy and the Environment. **Current Status of the Three Mile Island Nuclear Generating Station, Units 1 and 2: Oversight Hearing Before the Subcommittee on Energy and the Environment.** 98th Congress, 1st session, April 26, 1983. Washington, DC: GPO, 1983. 211 p. illus. Serial No. 98-10.
*Hearings

1468
United States. Congress. House. Committee on Interior and Insular Affairs. Subcommittee on Energy and the Environment. **Financial Implications of the Accident at Three Mile Island: Oversight Hearings Before the Subcommittee on Energy and the Environment.** 97th Congress, 1st session, May 4 and 5, 1981. Washington, DC: GPO, 1981. 917 p. illus. bibliography. Serial No. 97-15.
*Hearings

1469
United States. Congress. House. Committee on Interior and Insular Affairs. Subcommittee on Energy and the Environment. **Financing the Cleanup of Three Mile Island Unit 2 Nuclear Power Plant: Oversight Hearings Before the Subcommittee on Energy and the Environment.** 97th Congress, 2d session, April 23 and 27, 1982. Washington, DC: GPO, 1983. 603 p. illus. bibliography. Serial No. 97-36.
*Hearings

1470
United States. Congress. House. Committee on Interior and Insular Affairs. Subcommittee on Energy and the Environment. **Three Mile Island Cleanup and Rehabilitation: Oversight Hearing Before the Subcommittee on Energy and the Environment.** 96th Congress, 2d session, May 22, 1980. Washington, DC: GPO, 1980. 236 p. illus. bibliography. Serial No. 96-38.
*Hearings

1471
United States. Congress. House. Committee on Interstate and Foreign Commerce. Subcommittee on Energy and Power. **NRC's Response to the Report of the President's Commission on Three Mile Island: Hearing Before the Subcommittee on Energy and Power.** 96th Congress, 1st session, November 5, 1979. Washington, DC: GPO, 1980. 41 p. Serial No. 96-104.
*Hearings

1472
United States. Congress. House. Committee on Science and Technology. Subcommittee on Energy Research and Production. **Kemeny Commission Findings: Oversight Hearing Before the Subcommittee on Energy Research and Production.** 96th Congress, 1st session, November 14, 1979. Washington, DC: GPO, 1980. 129 p. bibliography. Serial No. 96-74.
*Hearings

1473
United States. Congress. House. Committee on Science and Technology. Subcommittee on Energy Research and Production. **Lessons Learned in Utility Management and the Status of the R and D Program Following the Accident at Three Mile Island: Hearing Before the Subcommittee on Energy Research and Production.** 98th Congress, 2d session, May 22, 1984. Washington, DC: GPO, 1985. 512 p. bibliography. No. 125
*Hearings

1474
United States. Congress. House. Committee on Science and Technology. Subcommittee on Energy Research and Production. **Nuclear Power Plant Safety After Three Mile Island: Report Prepared by the Subcommittee on Energy Research and Production.** Prepared by Gordon Chipman with assistance of Robert Civiak. 96th Congress, 2d session, March 1980. Washington, DC: GPO, 1980. 74 p. bibliography. Serial No. 96-JJ.
*Hearings

1475
United States. Congress. House. Committee on Science and Technology. Subcommittee on Energy Research and Production. **Plans for Improved Safety of Nuclear Power Plants Following the Three Mile Island Accident: Hearing Before the Subcommittee on Energy Research and Production.** 96th Congress, 1st session, September 19, 1979. Washington, DC: GPO, 1979. 255 p. Serial No. 96-81.
*Hearings

1476
United States. Congress. House. Committee on Science and Technology. Subcommittee on Natural Resources and the Environment. **Three Mile Island Nuclear Plant Accident: Hearing Before the Subcommittee on Natural Resources and the Environment.** 96th Congress, 1st session, June 2, 1979. Washington, DC: GPO, 1979. 251 p. illus. bibliography. Serial No. 96-54.
*Hearings

1477
United States. Congress. Senate. Committee on Energy and Natural Resources. **Financing the Cleanup of the Three Mile Island Nuclear Power Plant: Joint Hearing Before the Committee on Energy and Natural Resources, and the Subcommittee on Nuclear Regulation of the Committee on Environment and Public Works.** 97th Congress, 1st session, October 20, 1981. Washington, DC: GPO, 1982. 432 p. illus. bibliography. Energy Committee No. 97-52. Environment Committee No. 97-H33.
*Hearings

1478
United States. Congress. Senate. Committee on Environment and Public Works. **Nuclear Property Insurance Act of 1981: Hearing Before the Committee on Environment and Public Works.** 97th Congress, 2d session on S 1606, May 20, 1982. Washington, DC: GPO, 1982. 365 p. illus. bibliography. Serial No. 97-H52.
*Hearings

1479
United States. Congress. Senate. Committee on Environment and Public Works. **Supplemental Expenditures by the Committee on Environment and Public Works.** Report to accompany S. Res. 171. Washington, DC: GPO, 1979. 6 p. Serial No. 96-194.
*Hearings

1480
United States. Congress. Senate. Committee on Environment and Public Works. Subcommittee on Nuclear Regulation. **Nuclear Accident and Recovery at Three Mile Island: A Report.** Prepared by the Subcommittee on Nuclear Regulation. 96th Congress, 2d session, June 1980. Washington, DC: GPO, 1980. 423 p. illus. bibliography. Serial No. 96-14.
*Hearings

1481
United States. Congress. Senate. Committee on Environment and Public Works. Subcommittee on Nuclear Regulation. **Report of the President's Commission on the Three Mile Island Accident: Joint Hearing Before the Subcommittee on Nuclear Regulation, United States Senate, and the Subcommittee on Energy and the Environment of the Committee on Interior and Insular Affairs, House of Representatives.** 96th Congress, 1st session, October 31, 1979. Washington, DC: GPO, 1980. 154 p. Serial No. 96-H34.
*Hearings

1482
United States. Congress. Senate. Committee on the Environment and Public Works. Subcommittee on Nuclear Regulation. **Staff Studies, Nuclear Accident and Recovery at Three Mile Island.** Subcommittee on Nuclear Regulation. 96th Congress, 2d session, July 1980. Washington, DC: GPO, 1980. 564 p. illus. bibliography. Serial No. 96-14A.
*Hearings

1483
United States. Congress. Senate. Committee on the Environment and Public Works. Subcommittee on Nuclear Regulation. **Three Mile Island Nuclear Power Plant Accident: Hearings Before the Subcommittee on Nuclear Regulation.** 96th Congress, 1st session, April 10, 23, 30, 1979; October 2 and 3, 1979; November 8 and 9, 1979. Washington, DC: GPO, 1979 - . 3 parts, illus. bibliography. Serial No. 96-H12.
*Hearings

1484
United States. Congress. Senate. Committee on Governmental Affairs. Subcommittee on Energy, Nuclear Proliferation, and Federal Services. **Impact Abroad of the Accident at the Three Mile Island Nuclear Power Plant: March - September 1979.** Prepared for the Subcommittee on Energy, Nuclear Proliferation, and Federal Services by the Congressional Research Service, Library of Congress. Edited by Warren H. Donnelly. 96th Congress, 2d session, May 1980. Washington, DC: GPO, 1980. CRS-81 p. bibliography. Committee Print.
*Hearings

1485
United States. Congress. Senate. Committee on Labor and Human Resources. Subcommittee on Health and Scientific Research. **Three Mile Island Nuclear Accident, 1979: Hearing Before the Subcommittee on Health and Scientific Research.** 96th Congress, 1st session, April 4, 1979. Washington, DC: GPO, 1979. 140 p. bibliography.
*Hearings

# 19

## Popular Literature

1486
Alpern, David M.; with Lord, Mary, and Hager, Mary. "Nuclear Fallout." **Newsweek** 94(November 12,1979):38-39.
*President's Commission; Recommendations

1487
Alpern, David M.; Hewitt, Bill; and Lord, Mary. "Three Mile Island's Legacy." **Newsweek** 95(April 7, 1980):33-34.
*Protest; TMI - anniversary (1); Lessons learned

1488
"Although the Full Consequences." **Environment** 21(May 1979):21.
*Nuclear industry

1489
Arlen, Michael J. "The Air: The Opening of the Baseball Season." **New Yorker** 55(April 16, 1979):153-154.
*Media; Press coverage

1490
Asher, Linda. "Behavioral Fallout-Watching After Three Mile Island." **Psychology Today** 13(June 1979):31.
*Health effects - psychological; Research and development

1491
"Atomic Mishap -- Setback for Nuclear Power." **U.S. News & World Report** 86(April 9, 1979):23.
*Overview; Nuclear industry

1492
"Back From the Brink; Atomic-Age Pioneers Start Their Lives Again." **Time** 113(April 16, 1979):24-26.
*Local residents; Nuclear workers; Cooldown; Nuclear Regulatory Commission

1493
Bagne, Paul. "After the China Syndrome." **Omni** 6(September 1984):56-58, 60, 62, 116-118.
*Cleanup; China Syndrome

1494
Balk, A. "The Editor's Corner." **Atlas World Press Review** 26(May 1979):4.
*International reaction

1495
Beck, Melinda; with Hager, Mary. "$380 Million and Counting." **Newsweek** 102(October 10, 1983):30, 37.
*Cleanup; Costs

1496
Begley, Sharon L.; with Lord, Mary. "Three Mile Island: A Hot Attraction." **Newsweek** 94(October 1, 1979):12, 14.
*Tourism

1497
Behr, Peter. "Three Mile Island: What Happened." **Environment** 21(May 1979):2-4.
*Accident sequence; Accident - causes; Human error; Nuclear reactor design

1498
Bertin, Leonard. "Harrisburg." **Ascent** 1(Summer 1979):18-26.
*Overview; Canada

1499
Booker, Simeon. "Black Pa. Mayor's Key Role Overlooked in Nuclear Crisis." **Jet** 56(April 19, 1979):6-7.
*Reid, Robert; Middletown, PA

1500
Boslough, John. "3 Mile Island Casts a Pall on the Future." **U.S. News & World Report** 88(March 31, 1980):62.
*Nuclear industry; Public opinion; Politics; TMI - anniversary (1)

1501
Bower, Bruce. "Chronic Stress Still Lingers Near TMI." **Science News** 127(April 6, 1985):214.
*Health effects - psychological; Stress, chronic; TMI - anniversary (5)

1502
Brand, Stewart. "The Real News from Three Mile Island." **CoEvolution Quarterly** 22(Summer 1979):2-11.
*Accident sequence; Schweickart, Russell

1503
Brightsen, R.A. "The Way to Save Nuclear Power." **Fortune** (September 10, 1979):126-128, 130, 132.
*Nuclear power; Energy policy

1504
Brown, Paul B. "Brother Can You Spare $1 Billion?" **Forbes** 127(May 25, 1981):154.
*General Public Utilities; Cleanup; Financing

1505
Brown, Paul B. "Contrarians, Beware." **Forbes** 128(October 26, 1981):171, 174.
*Stock market; General Public Utilities; Financing

1506
"Capital Fallout; Kudos for the Kemeny Report." **Time** 114(November 12, 1979):34.
*President's Commission; Recommendations

1507
"Carter Acts on TMI: A Flood of Changes." **Science News** 116(December 15, 1979):405-406.
*President's Commission; Recommendations; Carter administration

1508
Cary, Bruce. "`People Played Like There Was no Tomorrow.'" **Progressive** 44(June 1980):18-19.
*Local residents

1509
Chaze, William L. "Curious Fallout at Three Mile Island." **U.S. News & World Report** 87(September 24, 1979):55.
*Tourism

1510
Church, George J. "Looking Anew at the Nuclear Future." **Time** 113(April 16, 1979):32-33.
*Nuclear industry

1511
Church, George J.; reported by Lee, Gary, and Stoler, Peter. "Radiation Sickness." **Time** 118(October 26, 1981):18-20.
*Nuclear industry; Cleanup

1512
Churcher, Sharon. "Documents Suggest that Utility Knew About Three Mile Island Danger." **New York** 17(November 12, 1984):15.
*General Public Utilities; Public relations; Reactor core

1513
Clark, Tim. "`All the Things We Do He Does Better.'" **Yankee** 44(March 1980):66-71, 120-136.
*Kemeny, John G.

1514
"Cleaning Up." **Financial World** 149(September 1, 1980):46-47.
*Cleanup; Decontamination; Chem-Nuclear Systems; Water contamination, radioactive

1515
Copulos, Milton R. "Is This What They Want? I. What Did Happen? What Have We Learned?" **National Review** 31(May 11, 1979):608-610.
*Accident sequence; Lessons learned

1516
Cory, Christopher T. "The Weekend Nuclear Exodus." **Psychology Today** 13(January 1980):26-27.
*Evacuation; Health effects - psychological; Telephone surveys

1517
"Court Stalls TMI-1 Restart." **Science News** 121(January 23, 1982):55.
*Restart; Court cases; Health effects - psychological

1518
Cousins, Norman. "An Exchange with Edward Teller." **Saturday Review** 7(March 1, 1980):10.
*Health effects - cancer; Price-Anderson Act; Teller, Edward

1519
Cousins, Norman. "Q & A That Raises More Q's Than A's." **Saturday Review** 6(September 29, 1979):10.
*Price-Anderson Act; Health effects - cancer; Nuclear industry; Teller, Edward

1520
Critchley, Jay M. "A Public Proposal: Three Mile Island Historic Nuclear Park and Planned Community." **Whole Earth Review** no. 8(Fall 1985):l.
*TMI Historical Nuclear Park; Satire

1521
"The Critical Problems of Nuclear Power." **Business Week** no. 2581(April 16, 1979):27-28.
*Nuclear industry

1522
Cunningham, Ann Marie. "Is There a Seismograph for Stress?" **Psychology Today** 16(October 1982):46-50, 52.
*Health effects - psychological; Stress

1523
"Decontaminating Three Mile Island." **Physics Today** 33(February 1980):90.
*Decontamination; Cleanup

1524
DeMott, John S.; reported by Redman, Christopher, and Van Voorst, Bruce. "An Industry Still in Disarray." **Time** 121(April 11, 1983):72-74.
*Nuclear industry; TMI - anniversary (4)

1525
Diamond, Edwin, and Passman, Leigh. "Three Mile Island: How Clear was TV's Picture?" **TV Guide** 27(August 4, 1979):4-5, 8, ll-l2.
*Network news; News coverage; Television

1526
Diamond, Stuart. "Three Mile Island: A Multibillion Dollar Three-Ring Circus." **Mechanix Illustrated** 78(August 1982):64-66, 92-93.
*Cleanup; Financing

1527
"Dim Bailout Hopes for GPU." **Business Week** no. 2680(March 23, 1981):43.
*General Public Utilities; Financing; Insurance; Cleanup; Costs

1528
Dworetzky, Tom. "Taking the Cover off a Dangerous Pressure Cooker." **Discover** 5(October 1984):29-32.
*Cleanup

1529
Edwards, Sandra. "TMI's Worst Casualty: Stress." **Science Digest** 87(May 1980):74-76.
*Health effects - psychological; Stress

1530
Ehrlich, Anne, and Ehrlich, Paul. "The Lessons of Three Mile Island." **Mother Earth News** 62(March/April 1980):148-149.
*Lessons learned

1531
Eppinger, Josh. "Nuclear Energy on the Brink." **Science Digest** 93(February 1985):56, 58-59, 80, 82.
*Nuclear industry

1532
"Fallout from an Accident." **Fortune** 100(December 17, 1979):15.
*General Public Utilities; Financing; Kuhns, William G.; Pennsylvania Public Utilities Commission

1533
"False Start at TMI." **Science News** 128(September 7, 1985):150.
*Restart; Court cases

1534
Faltermayer, Edmund. "Nuclear Power after Three Mile Island." **Fortune** 99(May 7, 1979):114-118, 120, 122.
*Nuclear industry; Lessons learned

1535
"The Financial Fallout at GPU." **Business Week** no. 2596(July 30, 1979):64-65.
*General Public Utilities; Financing

1536
"Findings at Three Mile Island." **Newsweek** 104(November 19, 1984):62.
*Reactor core; Iodine radioisotopes

1537
Fonda, Jane, and Hayden, Tom. "Mobilizing Against the Nukes." **New York** 12(April 16, 1979):43.
*Activism

1538
Ford, Daniel. "A Reporter at Large: Three Mile Island I. - Class Nine Accident." **New Yorker** 57(April 6, 1981:49-120.
*Overview

1539
Ford, Daniel. "A Reporter at Large: Three Mile Island. II. - The Paper Trail." **New Yorker** 57(April 13, 1981):46-109.
*Nuclear Regulatory Commission; Regulation

1540
Fox, Catherine. "A Brown-out for Nuclear Energy." **Maclean's** 92(November 5, 1979):36.
*President's Commission; Recommendations

1541
Fox, Catherine, and Lowther, William. "Harrisburg Syndrome." **Macleans** 92(April 9, 1979):32-34.
*Overview

1542
Friedman, Sharon M. "Before and After TMI." **Environment** 26(December 1984):4-5,34.
*Press coverage; Newspapers

1543
"Further Fallout and Further Lessons." **Time** 113(May 14, 1979):59.
*Preliminary investigations; Lessons learned

1544
Gaffron, Margie. "`A Fear We are Just Learning to Live With.'" **Progressive** 44(June

1980):24.
*Local residents

1545
Gendlin, Frances, and Gancher, David. "Three Mile Island: Life Imitating Art." **Sierra** 64(May/June 1979):71.
*Politics; Public relations; Media

1546
Gibbons, Don. "TMI Venting Controversy." **SciQuest** 53(July/August 1980):21.
*Venting; Krypton; Health effects - cancer

1547
"A Global Shock Wave; Samplings from a Worldwide Reaction." **Atlas World Press Review** 26(June 1979):39-41.
*International reaction; Media

1548
Gold, Michael. "Inside Three Mile Island." **Science 82** 3(October 1982):16.
*Cleanup; Reactor core analysis; Videocamera use

1549
Goldman, Peter; with Agrest, Susan; Borger, Gloria; Lord, Mary; and Martin, David C. "In the Shadow of the Towers." **Newsweek** 93(April 9, 1979):29.
*Local residents; Evacuation

1550
Gray, Mike. "What Really Happened at Three Mile Island?" **Rolling Stone** (May 17, 1979):47-51.
*Accident sequence

1551
Greenberg, Joel. "Human Error: The Stakes are Raised." **Science News** 117(February 23, 1980):122-123, 125.
*Human error; Operators; Training, occupational; Safety; Nuclear reactor design

1552
Guild, Gerald A. "The Fallout from Three Mile Island." **Financial World** 148(May 1, 1979):51.
*Stocks and bonds; Nuclear industry; Financing; Virginia Electric

1553
Gwynne, Peter. "The Nuclear Power Dilemma -- Which Horn Do We Choose?" **Science Digest** 86(August 1979):6-10.
*Nuclear industry

1554
"Happy Birthday, TMI." **National Review** 32(April 18, 1980):454.
*International reaction; Nuclear industry; TMI - anniversary (1)

1555
Harding, Jim. "Three Mile Island; Three Years After." **Sierra** 67(March/April 1982):20-3.
*Cleanup; Nuclear industry; TMI - anniversary (3)

1556
"Harrisburg." **National Review** 31(April 27, 1979):522-23.
*Journalism; China Syndrome

1557
"The Heat Builds Up Again Over Three Mile Island." **Business Week** no. 2823(January 9, 1984):95-96.
*Restart; General Public Utilities; Management; Financing

1558
Herbert, W. "TMI: Uncertainty is Causing Chronic Stress." **Science News** 121(May 8, 1982):308.
*Health effects - psychological; Stress, chronic

1559
Hertsgaard, Mark. "There's Life After T.M.I." **Nation** 230(March 8, 1980):269-271.
*Nuclear industry

1560
Holland, Jessica. "Fear and Trembling: Some Myths Exploded." **New York** 12(July 2, 1979):60.
*Nuclear workers; Employment

1561
"The Hot River Valley Syndrome." **Nation** 228(April 14, 1979):385.
*Nuclear industry; Risk; Government

1562
"`Hot' Water Cleanup Begins at TMI." **Science News** 120(October 17, 1981):247.
*Cleanup; Decontamination; Cleanup; Costs

1563
"How it Works." **Time** 113(April 9, 1979):10-11.
*Nuclear reactor design

1564
"Human Three Mile Island Survey." **Science News** 115(April 28, 1979):281.
*Radiation dosage; Health effects

1565
"In the Wake of Three Mile Island." **Science News** 115(May 5, 1979):292.
*Cooldown; Babcock & Wilcox; Accident investigations

1566
"In the World of Nuclear Power: Crisis." **Life** 2(May 1979):22-30.
*Overview; Nuclear industry

1567
Jacobs, M.E. "Action on Kemeny Recommendations." **Physics Today** 33(February 1980):90-92.
*President's Commission; Recommendations

1568
Jacobs, Michael E. "Study Blames `People-Related Problems' for TMI Accident." **Physics Today** 33(February 1980):89-90.
*President's Commission; Human error; Operators

1569
Jaffe, Susan. "Labor Divided on Three Mile Island." **Rolling Stone** (April 30, 1981):8-9.
*TMI - anniversary (2); Labor; Protest

1570
Jones, Peter M. "Incident at Three Mile Island." **Senior Scholastic** 112(September 6, 1979):6-9.
*Overview; Juvenile literature

1571
Keerdoja, Eileen; LaBrecque, Ron; and Namuth, Tessa. "Three Mile Island is Still a Problem." **Newsweek** 99(March 1, 1982):8.
*Cleanup; Restart

1572
Keerdoja, Eileen; Monroe, Sylvester; and Lord, Mary. "The Lessons Learned at Three Mile Island." **Newsweek** 101(June 27, 1983):13.
*Lessons learned; Safety; Denton, Harold

1573
Keisling, Bill, and Perrone, Ed. "Corporate Meltdown; The Lessons of Three Mile Island." **Progressive** 43(June 1979):12-17.
*Accident sequence; Lessons learned

1574
Kelly, Orr. "Is U.S. Prepared for a New Three Mile Island?" **U.S. News & World Report** 96(March 19, 1984):55, 58.
*Simulation; Safety; Evacuation; Emergency planning

1575
Kemeny, John G. "Political Fallout." **Society** 18(July/August 1981):5-9.
*President's Commission; Politics

1576
Ketchum, R.M. "Letter from the Country." **Blair & Ketchum's Country Journal** 6(July 1979):18-19.
*Local residents; Amish

1577
Kinsley, Michael. "Mental Cases." **Harper's** 265(July 1982):8-9.
*Stress; People Against Nuclear Energy v. U.S. Nuclear Regulatory Commission; Court cases

1578
Krass, Allan S. "The Killer Beyond Your Lightswitch." **Mechanix Illustrated** 76(May 1980):35-37, 122-124, 126.
*Evacuation; Emergency planning; Nuclear industry

1579
Langdon, Dolly; Neuhaus, Cable; and Walter, Greg. "Three Mile Island Revisited: Amid a Fallout of Fear and Distrust, the Crisis Goes on." **People** 13(March 24, 1980):43-44, 47-48.
*TMI - anniversary (1); Local residents; Denton, Harold

1580
Latham, Aaron. "Hollywood vs. Harrisburg." **Esquire** 91(May 22, 1979):77-82, 84, 86.
*China Syndrome

1581
"Lessons from Atomic Scare." **U.S. News & World Report** 86(April 16, 1979):21-23.
*Overview; Preliminary investigations

1582
"Lessons Learned in a Year." **Time** 115(March 24, 1980):62.
*Lessons learned; Safety; TMI - anniversary (1)

1583
Levi, B.G. "What Went Wrong with the Three Mile Island Reactor?" **Physics Today** 32(June 1979):77-78.
*Accident sequence

1584
Lewis, Carolyn D. "Taking Nuclear Risks." **Newsweek** 94(November 12, 1979):32-33.
*President's Commission; Risk

1585
"Loans for Firms Near Three Mile Island." **Nation's Business** 67(July 1979):17.
*Local business; Economic effects

1586
Luta, Lynne. "`We Packed a Lifetime of Hopes.'" **Progressive** 44(June 1980):22.
*Local residents; Evacuation

1587
McCracken, Samuel. "The Harrisburg Syndrome." **Commentary** 67(June 1979):27-39.
*Nuclear industry; China Syndrome

1588
McCracken, Samuel. "Two Years Later, Most Systems Go!" **National Review** 33(March 20, 1981):278, 280-281.
*Nuclear industry; TMI - anniversary (2)

1589
McGowan, Alan. "The Editorial Page." **Environment** 21(April 1979):cover.
*Public opinion

1590
McGowan, Alan. "The Editorial Page." **Environment** 21(May 1979):cover.
*Public opinion; Credibility

1591
McKenna, David L. "Science Says `Excuse Me' After the Radioactive Burp." **Christianity Today** 23(September 21, 1979):13.
*Science; Philosophy; Theology

1592
MacPherson, Myra. "Something Happened at the Plant!" **McCall's** 107(November 1979):132-133, 166-171.
*President's Commission; Trunk, Ann; Local residents; Middletown, PA

1593
McQuaig, Linda. "The Legacy of Three Mile Island." **Maclean's** 97(March 12, 1984):10, 12.
*TMI - anniversary (5); Cleanup; Restart; General Public Utilities; Management

1594
Mallowe, Mike. "Fantasy Island." **Philadelphia Magazine** 75(March 1984):86-89, 109, 112-127.
*Public relations; General Public Utilities

1595
Marrett, Cora Bagley. "Accident Analysis." **Society** 18(July/August 1981):66-72.
*President's Commission; Public policy; Organizational behavior

1596
Marshall, Eliot. "Matching Wits with a Nuclear Reactor." **Smithsonian** 11(June 1980):107-108, 110-111.
*Computer simulation

1597
Mathews, Tom; with Agrest, Susan; Borger, Gloria; Lord, Mary; Marbach, William D.; and Cook, William J. "Nuclear Accident." **Newsweek** 93(April 9, 1979):24-28, 30, 33.
*Overview; Accident sequence

1598
Mathews, Tom; with Cook, William C.; LaBrecque, Ron; Manning, Richard; Willenson, Kim; Lord, Mary; and Martin, David C. "Cooling off the Crisis." **Newsweek** 93(April 16, 1979):28, 33, 35, 38.
*Cooldown; Preliminary investigations

1599
Mayer, Allan J.; with Lord, Mary. "The Kemeny Blast." **Newsweek** 94(November 5, 1979):54.
*President's Commission; Recommendations

1600
"Memories of a Near Meltdown." **Time** 123(February 13, 1984):41.
*Local residents; Cleanup; General Public Utilities; Management

1601
Mielke, William C. "Facing Nuclear Terror in Harrisburg." **Christian Century** 96(April 18, 1979):427-428.
*Health effects - psychological; Theology

1602
Miller, Ellen Motoviloff. "I Fled Three Mile Island." **Senior Scholastic** 112(September 6, 1979):10-11.
*Local residents; Evacuation; Juvenile literature

1603
Mills, Susan. "Voices from Three Mile Island." **Progressive** 44(June 1980):16-24.
*Local residents

1604
Mitchell, Robert Cameron. "Nuclear Power and Public Opinion; Before and After Three Mile Island." **Current** 223(June 1980):34-39.
*Nuclear power; Public opinion

1605
"Money Trouble Lingers at Three Mile Island." **Business Week** no. 2592(July 2, 1979):22-23.
*General Public Utilities; Financing

1606
"More Evidence on Three Mile Island." **Consumer's Research Magazine** 63(July 1980):15-16.
*Accident investigations; Recommendations

1607
"More Harrisburg Studies." **Science News** 115(June 9, 1979):375.
*Health effects

1608
"More TMI Questions." **Science News** 119(February 14, 1981):102.
*Metropolitan Edison; Communications; Accident investigations

1609
Morganthau, Tom; with Cook, William J.; Lord, Mary; and Malamud, Phyllis. "Inside the NRC." **Newsweek** 93(April 16, 1979):35.
*Nuclear Regulatory Commission

1610
Morganthau, Tom; with Lord, Mary. "Trying to Clean Up Three Mile Island." **Newsweek** 95(March 24, 1980):39.
*Cleanup

1611
Morland, Howard. "The Meltdown that Didn't Happen; Optimism and Inertia at Three Mile Island." **Harpers** 259(October 1979):16-18, 22-23.
*Accident sequence; Radiation, low level

1612
Morrissey, Robert B. "Three Mile Island: Sifting Out the Facts." **America** 141(September 8, 1979):91-93.
*Accident sequence; Radiation terminology

1613
Nelkin, Dorothy, and Pollak, Michael. "Nuclear Protest and National Policy." **Society** 18(July/August 1981):34-38.
*Public policy; Protest; International reaction; Energy policy

1614
Nelson, Lin. "The Last Mile." **Progressive** 48(June 1984):16.
*Health effects; Restart

1615
Nevans, Ronald. "Has Public Opinion Killed the Nuclear Industry?" **Financial World** 148(October 1, 1979):28-31.
*Investments; Financing; Public opinion; Risk; Nuclear industry

1616
Norton, Boyd "Supercritical: A Nuclear Excursion." **Audubon** 82(May 1980):80-105.
*Accident sequence; TMI - anniversary (1)

1617
"Now Comes the Fallout; The White House and Congress Reapprise Nuclear Safety." **Time** 113(April 16, 1979):22-24.
*Safety; Congress; Carter administration; Public policy; Politics

1618
"Now for Operation Teakettle; A Long, Arduous Cooldown at Three Mile Island." **Time** 113(April 30, 1979):60-61.
*Cooldown

1619
"Nuclear Accident Underscores Need for County Preparedness." **American City &**

**County** 94(September 1979):16.
*Disaster planning; Evacuation

1620
"Nuclear Energy: How Safe Is It?" **Consumers' Research Magazine** 63(May 1980):9-14.
*President's Commission; Recommendations; Nuclear industry; Safety

1621
"A Nuclear Nightmare; Confusion and Fear Spew from a Damaged Reactor in Pennsylvania." **Time** 113(April 9, 1979):8-12, 15-16, 19.
*Overview; Accident sequence

1622
"Nuclear Power Anyone?" **National Review** 32(October 17, 1980):1242-1243.
*Nuclear industry; President's Commission

1623
"The Nuclear Think Tank at Three Mile Island." **Business Week** no. 2582(April 23, 1979):35.
*Nuclear industry; Engineers; Volunteers; General Public Utilities

1624
Olson, McKinley. "The Unacceptable Risk." **Nation** 228(April 14, 1979):387-388.
*Risk; Nuclear power

1625
Olson, McKinley C. "After T.M.I. - A Meltdown of Trust." **Nation** 230(April 19, 1980):449, 466-468.
*Public opinion; TMI - anniversary (1)

1626
Osborne, John. "Something Out There." **New Republic** 180(April 28, 1979):6, 8, 10.
*Communications

1627
O'Toole, Lawrence. "Unforgettable Anniversary." **Maclean's** 94(April 6, 1981):32-33.
*TMI - anniversary (2); Protest; Cleanup

1628
Pauly, David, with Abraham, Pamela Lynn. "Fallout for B&W." **Newsweek** 93(May 7, 1979):82.
*Babcock & Wilcox

1629
Perrow, Charles. "Normal Accident at Three Mile Island." **Society** 18(July/August 1981):17-26.
*Accident prevention; Risk; Nuclear reactor design; Human error; Human factors

1630
Peterson, I. "A Fresh Start at Three Mile Island?" **Science News** 127(June 8, 1985):359.
*Restart; Court rulings; General Public Utilities; Management; Activism

1631
Peterson, I. "Getting to the Core of the TMI Accident." **Science News** 124(September 17, 1983):183.
*Cleanup; Reactor core analysis

1632
Peterson, I. "Getting to the Core of Three Mile Island." **Science News** 121(April 17, 1982):263.
*Cleanup; Reactor core analysis; Research and development

1633
Peterson, Russ W. "Russ Peterson and the Nukes." **Audubon** 82(May 1980):5-6.
*President's Commission; Nuclear power

1634
Petzal, David E. "In Mi Opinion." **Mechanix Illustrated** 77(March 1981):8.
*Cleanup; Financing

1635
"Poll Finds Residents Against TMI Restart." **Industry Week** 213(May 3, 1982):34, 39.
*Restart; Referendum; Local residents

1636
Pollard, Robert. "At N.R.C. It's Safety Last." **Nation** 236(May 7, 1983):569-572.
*Nuclear Regulatory Commission; Lessons learned; Safety; Licensing; Politics; TMI - anniversary (4)

1637
Posner, Michael. "Mopping Up at Three Mile Island." **Maclean's** 95(March 29, 1982):22.
*Cleanup; Decontamination; Restart; TMI - anniversary (3)

1638
"Post-TMI Miscarriages May Show Stress." **Science News** 124(August 6, 1983):92.
*Health effects; Abortion, spontaneous; Pregnancy outcome; Stress

1639
Purcell, Arthur H. "Three Mile Island's Three Fateful Dates ... And the Fate of Nuclear Energy." **Science Digest** 87(April 1980):44-48.
*Overview; TMI - anniversary (1); Nuclear industry

1640
"Questioning All." **Time** 113(June 11, 1979):58.
*Health effects

1641
Quickel, Stephen W. "Nuclear Postscript." **Financial World** 149(February 1, 1980):4.
*Kemeny, John G.; Regulation

1642
Raloff, J. "`Ill Health' Still Plagues Nuclear Industry." **Science News** 121(June 5, 1982):374.
*Health effects - psychological; Court rulings; Nuclear industry

1643
Raloff, J. "TV at TMI: Hard-Core Rubble. **Science News** 122(July 31, 1982):68.
*Cleanup; Reactor core; Videocamera use

1644
Randall, Nan. "What if...?" **New York** 12(April 16, 1979):cover, 39-41.
*Evacuation; Emergency planning

1645
"A Ray of Sunshine for Nuclear Power." **U.S. News & World Report** 99(October 14, 1985):16.
*Restart; Nuclear industry; Licensing

1646
"Reagan Appeals TMI Stress Ruling." **Science News** 122(September 11, 1982):168.
*Health effects - psychological; Reagan administration

1647
Rein, Richard K. "The Man in the Hot Seat at Three Mile Island was a Cool Troubleshoter Named Denton." **People** 11(April 23, 1979):34-35.
*Denton, Harold

1648
Rice, Berkeley. "Unemployment, Not Radiation, Scares Three Mile Islanders." **Psychology Today** 13(July 1979):25.
*Unemployment; Nuclear workers; Stress

1649
Richardson, Jack. "A Disastrous Calm." **New York** 12(April 16, 1979):40, 42-43.
*Local residents; Journalism

1650
"Rivers and Reactors." **SciQuest** 55(January 1982):4-5.
*Environmental effects; Susquehanna River; Peach Bottom plant

1651
Rosen, Ira. "Carter Panel Puts Heat on Nukes." **Rolling Stone** (October 18, 1979):17.
*President's Commission; Carter administration

1652
Rosen, Ira D. "Countdown to Meltdown: Grace Under Pressure in Harrisburg." **Nation** 228(April 21, 1979):428-430.
*Local residents; Media; Press coverage

1653
Rost, Peter. "Three Mile Island Trip Produced Hitherto Concealed Facts." **Geographical Magazine** 52(September 1980):847.
*Satire; Energy policy

1654
Sandler, Ross. "Mixed Rulings on Nuclear Power." **Environment** 25(July/August 1983):2-3.
*Restart; Court rulings

1655
"Scathing Look at Nuclear Safety." **Time** 114(November 5, 1979):38, 43.
*President's Commission; Recommendations

1656
Schardt, Arlie; with Howard, Lucy, and Foote, Donna. "Covering Three Mile Island." **Newsweek** 93(April 16, 1979):93-94.
*Media; Press coverage

1657
Scherer, Ron. "Life Inside a `Hot' Reactor." **U.S. News & World Report** 97(November 12, 1984):82.
*Cleanup; Decontamination; Nuclear workers

1658
Schotsch, Linda. "Living with a Nuclear Neighbor; the Beckers Cope with Three Mile Island." **Farm Journal** 103:L-1-L-2.
*Local residents; Agriculture

1659
"A Scrap over Nuclear Safety." **Newsweek** 94(September 3, 1979):32.
*President's Commission; Nuclear Regulatory Commission; Licensing; Safety

1660
Senders, John W. "The Ballad of Three Mile Island." **Psychology Today** 13(April 1980):56.
*Human error; Poetry

1661
Sharpe, Tom H. "Coming Up for Air." **Progressive** 45(September 1981):21.
*Radiation dosage; Iodine radioisotopes; Local residents

1662
Sheils, Merrill; with Cook, William J.; Lord, Mary; LaBrecque, Ron; Reese, Michael; and Henkoff, Ronald. "Atom Plant Safety -- The Big Question." **Newsweek** 93(April 9, 1979):33-34, 36, 39.
*Safety

1663
"Signs of Distress Near TMI." **Science News** 119(March 21, 1981):185.
*Health effects - psychological; Stress, chronic; Telephone surveys

1664
Simmons, Henry T. "TMI Plus Three Years." **SciQuest** 55(February 1982):15-21.
*TMI - anniversary (3); Safety; Reagan administration; Politics; Public policy; Nuclear industry

1665
Starr, Roger. "The Three Mile Shadow." **Commentary** 70(October 1980):48-55.
*Nuclear industry; Public policy; Accident - causes

1666
"Steam Tubes Further Stay TMI-1 Restart." **Science News** 121(April 24, 1982):283.
*Restart, Steam tubes

1667
Sternglass, Ernest. "The Whole Truth About Three Mile Island; NRC Ignored the Full Hazard." **Progressive** 43(June 1979):18.
*Radiation dosage; Population exposure

1668
Sternglass, Ernest J. "The Invisible Death (Part I): The First Casualty at T.M.I." **Nation** 232(February 28, 1981):225, 240-244.
*Radiation dosage; Health effects; Infant mortality

1669
Sternglass, Ernest J. "The Invisible Death (Part II): The Lethal Path of T.M.I. Fallout." **Nation** 232(March 7, 1981):267-269, 272-273.
*Health effects; Infant mortality; Radiation, low level; Radiation effects

1670
Sternglass, Ernest J. "T.M.I. Update." **Nation** 232(April 25, 1981):485-486.
*Health effects; Infant mortality

1671
Stoler, Peter. "Legacy of Three Mile Island." **Time** 115(March 24, 1980):58-59, 61.
*TMI - anniversary (1); Lessons learned; Middletown, PA

1672
Stoler, Peter; reported by Branegan, Joy; and Nash, J. Madeleine. "Pulling the Nuclear Plug." **Time** 123(February 13, 1984):34-39, 41-42.
*Nuclear industry

1673
Stranahan, Susan Q. "Three Mile Island; It's Worse Than You Think!" **Science Digest** 93(June 1985):54-57, 82-83, 87.
*Reactor core analysis; Cleanup

1674
"The Talk of the Town: Notes and Comments." **New Yorker** 55(April 16, 1979):27-28.
*General

1675
"The Talk of the Town: Notes and Comments." **New Yorker** 55(April 23, 1979):35.
*Public opinion; Nuclear industry

1676
"Talk of the Town: Notes and Comments." **New Yorker** 55(June 25, 1979):21.
*General

1677
"Talk of the Town: Notes and Comments." **New Yorker** 55(November 12, 1979):44-46.
*President's Commission

1678
"The Three Mile Accident." **Science News** 115(April 7, 1979):227-228.
*Accident sequence; Hydrogen bubble; Cleanup

1679
"Three Mile Island." **Progressive** 43(May 1979):4.
*Activism; Nuclear power

1680
"Three Mile Island: A Costly Cleanup." **Newsweek** 98(September 21, 1981):90.
*Cleanup; Costs; Financing

1681
"Three Mile Island: Fallout of Fear." **Time** 121(April 11, 1983):74.
*Local residents; Health effects

1682
"Three Mile Island: Rogovin Blasts NRC." **Science News** 117(February 2, 1980):68.
*Rogovin report; Recommendations; Safety

1683
"Three Mile Island: Tolerable Limits." **America** 141(November 17, 1979):293.
*President's Commission; Recommendations; Nuclear power; Risk

1684
"Three Mile Island: The Verdicts Begin." **Science News** 116(November 3, 1979):309-310.
*Nuclear Regulatory Commission; Metropolitan Edison; Management; President's Commission; Recommendations

1685
"Three Mile Island Dose Estimates." **Science News** 115(June 2, 1979):360.
*Radiation dosage

1686
"Three Mile Island Fuels Church Debate." **Christianity Today** 23(May 4, 1979):48.
*Theology

1687
"Three Mile Island in Albany." **SciQuest** 53(March 1980):4.
*Radiation dosage; Albany, NY

1688
"Three Mile Island One Year Later." **Science News** 117(April 5, 1980):216.
*General Public Utilities v. Babcock & Wilcox; TMI - anniversary (1)

1689
"Three Mile Island Verdict: Human Error is to Blame." **Time** 114(August 13, 1979):50.
*Accident investigations; Human error; Accident - causes

1690
"Three Mile Island's Lingering Ills." **Business Week** no. 2608(October 22, 1979):75.
*Pennsylvania Public Utilities Commission; Restart decision; Nuclear Regulatory Commission; General Public Utilities; Financing; Regulation

1691
"Three Mile Island's Missing Watchdogs." **Business Week** no.2612(November 19, 1979):44-45.
*Pennsylvania Public Utilities Commission; Restart decision; General Public Utilities; Nuclear Regulatory Commission; Regulation

1692
"Three Mile Pile." **Forbes** 124(September 17, 1979):218, 220.
*Cleanup; Financing

1693
Tischler, Nancy M. "Three Mile Island." **Christianity Today** 23(May 4, 1979):15.
*Theology

1694
"TMI - One Year Later." **Progressive** 44(April 1980):7-8.
*TMI - anniversary (1)

1695
"TMI + 2." **National Review** 33(April 3, 1981):336-337.
*TMI - anniversary (2)

1696
"TMI Fallout: Demoralization." **Science News** 116(November 10, 1979):326.
*Health effects - psychological; President's Commission; Mental health

1697
"TMI Stress: The Long Goodbye." **Science News** 117(April 12, 1980):230.
*Health effects - psychological; Stress; TMI - anniversary (1)

1698
"TMI Tests Trashed." **Science News** 128(September 21, 1985):186.
*Safety tests; Nuclear workers

1699
"TMI-1 Restart Halted." **Science News** 127(June 15, 1985):375.
*Restart; Court rulings

1700
"TMI-1 Restart Underway." **Science News** 128(October 12, 1985):229.
*Restart; Court cases

1701
Topolski, E. Anthony. "`We Can't Sell our House.'" **Progressive** 44(June 1980):23.
*Local residents; Real estate

1702
Trachtman, Paul. "Phenomena, Comment and Notes." **Smithsonian** 10(July 1979):14, 16.
*Nuclear Regulatory Commission; Politics

1703
Tucker, Carll. "Accident-Prone." **Saturday Review** 6(May 26, 1979):76.
*Energy consumption; Literature

1704
Tucker, William. "Is This What They Want? II. Nightmare or Bad Dream?" **National Review** 31(May 11, 1979):610-611.
*Public opinion; Nuclear industry

1705
"TV Images Suggest a TMI Core Melt." **Science News** 127(March 9, 1985):148.
*Reactor core; Fuel melting; Videocamera use

1706
Underwood, Nora. "Recalling a Nightmare." **Macleans** 99(May 12, 1986):31.
*Local residents

1707
Van Boskirk, Andrew S. "`We Went to a Motel and Waited...'" **Progressive** 44(June 1980):17.
*Local residents; Evacuation

1708
Vargo, Mike. "Innocence Lost." **Pennsylvania Illustrated** 3(August 1979):22-33, 47-48, 62-64.
*Local residents; Reid, Robert; Garver, John; Middletown, PA

1709
Vollers, Maryanne, and Keisling, Bill. "Dress Rehearsal for the Apocalypse." **Rolling Stone** (May 17, 1979):51-52.
*Local residents

1710
[Waggoner, D.] "Life in the Shadow of Nuclear Disaster: Fear, Anger, Courage and Eroding Faith." **People** 11(April 16, 1979):36-41.
*Local residents

1711
Walinsky, Adam. "Nuclear (T)error." **Newsweek** 93(May 7, 1979):26-27.
*Accident prevention; Human error

1712
Walker, Dale Elizabeth. "The Media and T.M.I." **Nation** 232(April 25, 1981):482.
*News coverage; Television; Health effects; Infant mortality

1713
Walter, Greg. "Two Gung-Ho Nuclear Engineers Brave the Unknown to Clean Up Three Mile Island." **People** 14(April 11, 1980):30-31.
*Cleanup; Nuclear workers

1714
Wasserman, Harvey. "What is Happening to the Animals?" **Progressive** 44(June 1980):25-28.
*Environmental effects; Ecology

1715
"Watching the Watchdogs." **Time** 113(April 16, 1979):31.
*Nuclear Regulatory Commission

1716
"We All Live in Harrisburg." **Atlas World Press Review** 26(June 1979):37-39.
*International reaction; Europe; Politics; Economic effects

1717
"We're Fighting for Our Lives." **Time** 115(March 24, 1980):61-62.
*Nuclear industry; TMI - anniversary (1)

1718
Weaver, Erma. "`Love May Yet Be Stronger Than Death.'" **Progressive** 44(June 1980):20-21.
*Local residents

1719
"What Next for Nuclear Power." **U.S. News & World Report** 87(November 12, 1979):33-35.
*President's Commission; Recommendations

1720
"Will Kemeny Report Sway Other Probes?" **Industry Week** 203(November 12, 1979):128-129.
*President's Commission

1721
Williams, Dennis A.; with Kasindorf, Martin; Lubenow, Gerald C.; and LaBrecque, Ron. "Beyond `The China Syndrome'." **Newsweek** 93(April 16, 1979):31.
*China Syndrome

1722
Williams, Dennis A.; with Lord, Mary, and LaBrecque, Ron. "The Nuclear Tapes." **Newsweek** 93(April 23, 1979):30, 35.
*Nuclear Regulatory Commission; Accident investigations

1723
"Windscale, Three Mile Island -- and Chernobyl." **New Statesman** 111(May 2, 1986):3-4.
*Nuclear industry

1724
Woerthwein, Wicky. "`Three Mile Island Happened to Us.'" **Redbook Magazine** 154(April 1980):46-47, 49, 146.
*Local residents; Evacuation

# 20

# Proceedings

1725
Akers, D.W. "TMI-2 Core Debris Chemistry and Radionuclide Behavior." **Abstracts of Papers of the American Chemical Society** 189(April 1985):82.

1726
Akers, D.W., and Hobbins, R.R. "Elemental and Radionuclide Content of TMI-2 Core Debris Grab Samples." **Transactions of the American Nuclear Society**, 50(November 1985):222-223.

1727
Alcorn, F.M., and Gudorf, M.R. "Criticality Safety Considerations for TMI-2." **Transactions of the American Nuclear Society** 35(November 1980):270-271.

1728
Alvares, N.J. "Fire Damage to Polymeric Materials in the Three Mile Island Unit 2 Reactor Building." **Abstracts of Papers of the American Chemical Society** 189(April 1985):60.

1729
Alvares, N.J.; Beason, D.G.; Eidam, G.R.; and Reeder, D.L. "Investigation of Hydrogen Burn Damage in the TMI-2 Reactor Building." **Transactions of the American Nuclear Society** 43(1982):11-12.

1730
Augustine, F., and Fricke, V. "Scattered Dose Rates During the TMI-2 Plenum Lift." **Transactions of the American Nuclear Society** 50(November 1985):481-482.

1731
Augustine, F., and Worku, G. "Contamination in the TMI-2 Plenum." **Transactions of the American Nuclear Society** 50(November 1985):461-462.

1732
Austin, W.E.; Baston, V.F.; Hofstetter, K.J.; and Owen, D.E. "Experimental Examinations of TMI-2 Debris Samples Indicate Absence of Pyrophoric Characteristics." **Transactions of the American Nuclear Society** 46(1984):482-483.

1733
Auxier, John A. "The Accident at Three Mile Island." **Health Physics** 39(December 1980):1057.

1734
Ayers, A.L., and Lilburn, B.J. "Preparations to Receive and Store the TMI-2 Core Debris." **Transactions of the American Nuclear Society** 50(November 1985):110.

1735
Babel, P.J.; Brosey, B.H.; and Distenfeld, C.H. "Reactor Fuel Detection and Distribution in the TMI-2 Nuclear Auxiliary Systems." **Health Physics** 50(S1, 1986):S2.

1736
Bacher, P.; Gery, A.; and Capel, R. "How Electricite de France has Addressed Post-TMI Problems." **Transactions of the American Nuclear Society** 35(November 1980):412-413.

1737
Bahn, A.; Tokuhata, G.; King, M.; and Stein, G. "Registration and Long-Term Health Surveillance of the Three Mile Island Population." **American Journal of Epidemiology** 112(September 1980):442.

1738
Baratta, A.J.; Jester, W.A.; Daugherty, N.M.; Kelchner, A.; and Reilly, M.A. "Nuclear Radiation Concepts and TMI-2 Purging -- A Pre-Purge Seminar for TMI Area Journalists." **Transactions of the American Nuclear Society** 38(June 1981):76-77.

1739
Baratta, A.J.; Jester, W.A.; Gricar, B.G.; Reilly, M.A.; and Eidam, G. "A Citizens' Radiation Monitoring Program for the TMI Area." **Transactions of the American Nuclear Society** 35(November 1980):40-42.

1740
Barkhurst, R.P. "Effects of 3 Mile Island on Nuclear Plant Staffing." **Transactions of the American Nuclear Society** 34(June 1980):686.

1741
Barrett, L.H. "TMI Short-Term Lessons Learned Radiation Protection Recommendations -- NRC Perspective." **Transactions of the American Nuclear Society** 35(November 1980):452.

1742
Baston, V.F., and Hofstetter, K.J. "Adherent Activity on TMI-2 Internal Surfaces." **Abstracts of Papers of the American Chemical Society** 189(April 1985):81.

1743
Baston, V.F., and Hofstetter, K.J. "Long Term Appearance Rate of Radionuclides in TMI-2 Coolant." **Transactions of the American Nuclear Society** 47(1984):111-113.

1744
Baston, V.F.; Hofstetter, K.J.; Bain, G.M.; and Elrick, R.M. "A Comparison of TMI-2 and Laboratory Results for Cesium Activity Retained on Reactor Material Surfaces." **Transactions of the American Nuclear Society** 50(November 1985):223-225.

1745
Battist, Lewis. "Environmental Measurement Requirements Resulting from the TMI-2 Accident." **IEEE Transactions on Nuclear Science** NS-28(February 1981):231-233.

1746
Bedell, Douglas. "Releases of On-Site Information -- Public Relations Associated with TMI Decontamination." **Transactions of the American Nuclear Society** 38(June 1981):624-625.

1747
Behling, U.H. "Health Effects of the TMI-2 Accident." **Health Physics** 50(S1, 1986):S55.

1748
Belda, Ing W. "Analysis of the Three Mile Island Accident." **Health Physics** 44(February 1983):187-188.

1749
Beller, L.S., and Martin, M.R. "Ultrasonic Mapping of the Postaccident TMI-2 Core Configuration." **Transactions of the American Nuclear Society** 46(1984):482.

1750
Bengel, P.R. "TMI-2 Head Removal." **Transactions of the American Nuclear Society** 47(1984):106-107.

1751
Bergeron, P.A.; Chapman, J.R.; Turnage, J.C.; Husain, A.; and Szymczak, W.J. "Post-TMI Analysis at Yankee Atomic Electric Company." **Transactions of the American Nuclear Society** 34(June 1980):866-867.

1752
Bixby, W.W.; Reilly, J.K.; Grant, P.J.; Quinn, G.J.; and McIntosh, T.W. "Transuranic Waste Management Program at the TMI-2." **Transactions of the American Nuclear Society** 46(1984):110-111.

1753
Bland, W.M. "TMI Rare Event -- Lessons to be Learned." **Proceedings of the Annual Reliability and Maintainability Symposium** 1982(1982):17-20.

1754
Bond, G.R.; Devine, J.C.; Kulynych, G.E.; and Rider, R.L. "Plans for the Examination of the TMI-2 Core." **Transactions of the American Nuclear Society** 35(November 1980):200.

1755
Bond, W.D.; King, L.J.; Knauer, J.B.; Hofstetter, K.J.; and Thompson, J.D. "Cleanup of the TMI-2 Demineralizer Resins." **Abstracts of Papers of the American Chemical Society** 189(April 1985):102.

1756
Bradbury, P. "TMI-2 Defueling System Design." **Transactions of the American Nuclear Society** 50(November 1985):616-618.

1757
Brasher, J.W. "Radiological Protection Program at Three Mile Island, Unit 2." **Health Physics** 43(July 1982):95.

1758
Breen, R.J. "Defense-in-Depth Approach to Safety in Light of TMI- 2." **Transactions of the American Nuclear Society** 34(June 1980):593-594.

1759
Bretthauer, E.W.; Grossman, F.; Efurd, W.; Douglas, G.; Bunce, L.; and Bills, M. "The Environmental Protection Agency's Radiation Monitoring and Surveillance Activities During the Purging of Three Mile Island Unit 2." **Transactions of the American Nuclear Society** 35(November 1980):59-60.

1760
Broughton, J.M.; Tolman, E.L.; Allison, C.M.; Behling, S.R.; Hagrman, D.L.; and Taylor, D.J.N. "TMI-2 -- Core Condition and Postulated Accident Scenario." **Transactions of the American Nuclear Society** 50(November 1985):210-211.

1761
Brownson, D.A.; Baratta, A.J.; Robinson, G.E.; Kim, D.; and Huegel, D.S. "Pressurizer Size and Surge Line Impedance -- Effects on the TMI Accident." **Transactions of the American Nuclear Society** 44(1983):345-347.

1762
Bryan, G.H., and Siemens, D.H. "Development and Demonstration of a Process for Vitrification of TMI Zeolite." **Transactions of the American Nuclear Society** 39(1981):174-176.

1763
Bryan, G.H.; Siemens, D.H.; and Knox, C.A. "Vitrification of High Activity TMI Zeolite." **Transactions of the American Nuclear Society** 45(1983):130-131.

1764
Buhl, A.R., and Robinson, J.C. "Reactor Diagnostics Passed the TMI Test." **Transactions of the American Nuclear Society** 34(June 1980):795-796.

1765
Buhl, A.R.; Robinson, J.C.; Oakes, L.C.; and Cain, D.G. "Core Conditions Derived from Instrumentation Responses During the Three Mile Island Accident." **Transactions of the American Nuclear Society** 34(June 1980):557-558.

1766
Burchill, W.E. "The Effects on Operation of Combustion Engineering Reactors Caused by Three Mile Island." **Transactions of the American Nuclear Society** 44(S1, 1983):83-84.

1767
Burman, D.L.; Hochreiter, L.E.; Jacobs, S.E.; Paddleford, D.F.; Shell, S.L.; and Yeh, H.C. "Degraded Core Cooling Calculations for TMI-2." **Transactions of the American Nuclear Society** 34(June 1980):869-870.

1768
Burton, H.M., and Bixby, W.W. "Technology Transfer at Three Mile Island Unit." **Transactions of the American Nuclear Society** 42(1982):24-25.

1769
Cain, D.G. "Function and Role of TMI-2 Instrumentation in Core Damage Assessment." **Transactions of the American Nuclear Society** 34(June 1980):793-795.

1770
Campbell, D.O.; Brooksbank, R.E.; and King, L.J. "Off-Gas Handling at TMI -- Escape of Radioactivity and Implications for the Future." **Transactions of the American Nuclear Society** 38(June 1981):499-500.

1771
Cannon, N.S.; Helbert, H.J.; Trijillo, R.; Hecker, L.A.; and Cannon, C.P. "Effect of the Three Mile Island LOCA on Cables and Connector Components." **Transactions of the American Nuclear Society** 45(1983):239-240.

1772
Card, C.J.; Divine, J.R.; Munson, L.F.; and Naughton, M.D. "Chemical Decontamination Methods Applicable to the TMI-2 RCS." **Transactions of the American Nuclear Society** 43(1982):20.

1773
Carlson, E.R.; Broughton, J.M.; and Eidam, G.R. "TMI-2 Core Examinations -- Techniques and Results." **Journal of Metals** 37(11, 1985):21.

1774
Carlson, E.R., and Cook, B.A. "Characterization of TMI-2 Core Debris Materials." **Transactions of the American Nuclear Society** 50(November 1985):212-214.

1775
Catlow, F. "Impact of the Accident at Three Mile Island on Plant Control and Instrumentation Philosophy." **Transactions of the South African Institute of Electrical Engineers** 74(August 1983):210-220.

1776
Childress, P.C., and McGuinn, E.J. "Defueling Canister Designs for the TMI-2 Core Removal Program." **Transactions of the American Nuclear Society** 47(1984):108-110.

1777
Childs, R.L.; Selby, D.L.; Robinson, J.C.; Canada, R.G.; and Monday, R.L. "Calculation of the TMI-2 Source Range Monitor Reading for Several Core Water Levels." **Transactions of the American Nuclear Society** 34(June 1980):558-559.

1778
Chu, T.L. "Probabilistic Analysis of TMI-Type Accidents Including Human Actions." **Transactions of the American Nuclear Society** 45(1983):377-378.

1779
Chung, H.M. "Zircaloy Reaction in Hydrogen Steam Mixtures -- Implications for TMI-2 Core Damage." **Transactions of the American Nuclear Society** 35(November 1980):198-199.

1780
Cline, J.E.; Thomas, C.D., Jr.; Cameron, D.S.; Barefoot, E.D.; Daniel, J.A.; Schlomer, E.A.; Cooper, J.R.; and Tarpinian, J.E. "Measurements of Surface and Volume Contamination on Components and Floors of Three Mile Island-2 Reactor Containment Building Using a Shielded and Collimated Germanium Detector." **Health Physics** 43(July 1982):97-98.

1781
Cole, R.K.; Baskes, M.I.; and Rumble, E.T. "Thermal-Hydraulic Analysis of the 1st 16 Hours of the TMI-2 Accident." **Transactions of the American Nuclear Society** 34(June 1980):562-563.

1782
Colletti, J.P.; Easly, J.A.; and Luoma, J.D. "TMI-2 Post-Accident Criticality Analyses." **Transactions of the American Nuclear Society** 35(November 1980):271-272.

1783
Collins, E.D.; Campbell, D.O.; King, L.J.; Knauer, J.B.; and Wallace, R.M. "Development of the Flowsheet Used for Decontaminating High Activity Level Water at TMI-2." **Abstracts of Papers of the American Chemical Society** 189(1985):99.

1784
Collins, Harold E. "Emergency Planning and Preparedness Before and After Three Mile Island." **IEEE Transactions on Nuclear Science** NS-27(February 1980):919-923.

1785
Connell, C.W., and Brazill, G.J. "Changes to the B-and-W 205 Fuel Assembly NSSS Since TMI-2." **Transactions of the American Nuclear Society** 39(1981):820-821.

1786
Cook, B.A.; Stoyack, J.E.; and Owen, D.E. "Metallurgical Analysis of the TMI-2 Core Damage." **Journal of Metals** 37(11, 1985):21.

1787
Corcoran, W.R.; Burchill, W.E.; and Pfeifer, J.W. "CE Response to TMI Regulatory Requirements." **Transactions of the American Nuclear Society** 34(June 1980):596-597.

1788
Cox, T.E. "Fission Product Transfer in the TMI-2 Purification System." **Transactions of the American Nuclear Society** 43(1982):14-15.

1789
Cronenberg, A.W.; Behling, S.R.; Domenico, W.F.; and Broughton, J. "Analysis of Damage Potential to the TMI-2 Lower Head by Core Debris." **Transactions of the American Nuclear Society** 50(November 1985):317-319.

1790
Cummings, Garth E. "Operator/Instrumentation Interactions During the Three Mile Island Incident." **IEEE Transactions on Nuclear Science** NS-27(February 1980):931-934.

1791
Cutter, S. et al. "Special Session on Three Mile Island: Risk Evaluation and Evacuation Responses." **Proceedings, Middle States Division, Association of American Geographers** 13(1979):80-88.

1792
Dam, A.S. "Reactor Plant Pressure and Temperature Control During TMI-2 Initial Recovery." **Transactions of the American Nuclear Society** 34(June 1980):574-576.

1793
Daniel, J.A., and Cunnane, J.C. "Characterization of Fission Product Deposition in the TMI-2 Reactor Coolant and Auxiliary Systems." **Transactions of the American Nuclear Society** 43(1982):14.

1794
Daniel, J.A., and Schlomer, E.A. "Identification of Fuel Debris in TMI-2 Systems by Gamma-Ray Spectroscopy." **Transactions of the American Nuclear Society** 44(1983):240-242.

1795
Daniel, J.A., and Schlomer, E.A. "Post-Accident Sampling and Measurement Experience at TMI-2." **Transactions of the American Nuclear Society** 45(1983):260-261.

1796
Daniels, R.S. "Three Mile Island Assessments." **AJR; American Journal of Roentgenology** 141(November 1983):1090.

1797
Daniels, R.S.; McIntosh, T.W.; and Reilly, J.K. "Overview and Update on Waste Management Issues at TMI-2." **Transactions of the American Nuclear Society** 43(1982):141-142.

1798
Davidson, L.M., and Baum, A. "Physiological Symptoms of Post-Traumatic Stress at Three Mile Island." **Psychosomatic Medicine** 47(January/February 1985):80-81.

1799
Dempsey, R.H. "The Effects of the TMI Accident -- Reassessment of Simulator Training Needs." **Transactions of the American Nuclear Society** 34(June 1980):714-715.

1800
Denton, Harold R. "NRC Action Plans." **Transactions of the American Nuclear Society** 34(June 1980):591.

1801
Devine, J.C. "Three Mile Island-2 Progress Report." **Transactions of the American Nuclear Society** 45(1983):679-680.

1802
Devine, J.C., and Fillnow, R.H. "Three Mile Island Unit-2 -- Status of the Recovery." **Transactions of the American Nuclear Society** 44(1983):505.

1803
Devine, J.C., and Fillnow, R.H. "TMI-2 Recovery Strategy and Cleanup Progress." **Journal of Metals** 37(11, 1985):20.

1804
Diamond, S. "Nuclear Information Since Three Mile Island -- A Little Effort, Not Much Success." **Transactions of the American Nuclear Society** 38(June 1981):494-495.

1805
Dickerson, Marvin H., and Gudiksen, Paul H. "Atmospheric Release Advisory Capability (ARAC) Response to the Three Mile Island Accident." **IEEE Transactions on Nuclear Science** NS-27(February 1980):927-930.

1806
Distenfeld, C. "Instrument Performance and Measurements at TMI: Projected Needs for Other Nuclear Accidents." **Health Physics** 39(December 1980):1037.

1807
Distenfeld, C.; Bokharee, S.; Brosey, B.; and Greenborg, J. "A Search for Fuel in the TMI-2 Primary System External to the Reactor Vessel." **Transactions of the American Nuclear Society** 49(June 1985):252-254.

1808
Distenfeld, C.; Brosey, B.; Greenborg, J.; and Babel, P. "Exposure Environment in the TMI-2 Reactor Building." **Health Physics** 49(July 1985):126.

1809
Distenfeld, C., and Klemish, J. "A Low Cost Radioiodine Air Sampling System for Nuclear Incidents: Experience at Three Mile Island." **IEEE Transactions on Nuclear Science** NS-27(February 1980):675-681.

1810
Distenfeld, C.H., and Bokahree, S.A. "A Method of Determining Fuel Deposition in the TMI-2 Pressurizer." **Health Physics** 50(S1, 1986):S2-S3.

1811
Dornsife, W.P. "The 3-Mile Island Accident, What Really Caused the Crisis." **Transactions of the American Nuclear Society** 34(June 1980):79.

1812
Doyle, T.E., and Alvarez, J.L. "Analysis of TMI-2 Samples Using the Molecular Optical Laser Examiner." **Transactions of the American Nuclear Society** 43(1982):16-17.

1813
Dubiel, R.W. "Health Physics Problems at TMI." **Transactions of the American Nuclear Society** 34(June 1980):637-638.

1814
Eidam, G.R. "TMI-2 Accident -- Visual Inspection." **Abstracts of Papers of the American Chemical Society** 189(April 1985):61.

1815
Eidam, G.R. "TMI-2 Reactor Characterization Program." **Transactions of the American Nuclear Society** 47(1984):305-306.

1816
Eidam, G.R., and Devine, J.C. "TMI-2 Current Status and Plans." **Abstracts of Papers of the American Chemical Society** 189(April 1985):103.

1817
Eidam, G.R., and Leigh, D.W. "Dose Reduction and Contamination Control in the TMI-2 Reactor Building." **Transactions of the American Nuclear Society** 43(1982):18-19.

1818
Ferguson, R.L. "A Reliable Plant is a Safe Plant. Lessons Learned from TMI." **Proceedings of the 1980 Annual Reliability and Maintainability Symposium** 1980(1980):76-78.

1819
Fillnow, R.H., and Devine, J.C. "TMI-2 Defueling -- Techniques and Progress." **Journal of Metals** 37(11, 1985):40.

1820
Fleming, R.; Baum, A.; and Gatchel, R.J. "Psychophysiological Concomitants of Chronic Stress at Three Mile Island." **Psychophysiology** 19(May 1982):316.

1821
Franke, Bernd. "What Release and Doses Can Go Undetected by Environmental Radiation Monitoring Systems at Three Mile Island?" **Health Phsyics** 50(S1, 1986):S90.

1822
Franz, W.A.; Rider, R.L.; Austin, W.A.; and Cole, N. "TMI-2 Quick Look Examination." **Transactions of the American Nuclear Society** 43(1982):4-5.

1823
Fricke, V.R., and Peterson, H.K. "Modeling the TMI-2 Reactor Radiation Source Before Removing the Head." **Transactions of the American Nuclear Society** 45(1983):618-619.

1824
Gardner, H.R., and Allen, R.P. "Decontamination of TMI-2 Leadscrew Sections with Pressurized Water." **Transactions of the American Nuclear Society** 46(1984):719.

1825
Gardner, H.R.; Allen, R.P.; and Polentz, L.M. "Development of Nonchemical Decontamination Techniques for Use at TMI-2." **Transactions of the American Nuclear Society** 46(1984):716-717.

1826
Gardner, H.R.; Allen, R.P.; Polentz, L.M.; Skiens, W.E.; Wolf, G.A.; and Anderson, L.R. "Evaluation of Non-Chemical Decontamination Techniques for TMI-2 RCS Application." **Transactions of the American Nuclear Society** 43(1982):20-21.

1827
Giefer, D.L. "Remote Equipment Development at TMI-2." **Transactions of the American Nuclear Society** 50(November 1985):620-621.

1828
Gold, R.; Roberts, J.H.; Ruddy, F.H.; Preston, C.C.; McElroy, W.N.; Rao, S.V.; Greenborg, J.; and Fricke, V.R. "Neutron Dosimetry with Solid State Track Recorders in the Three Mile Island Unit-2 Reactor Cavity." **Nuclear Tracks and Radiation Measurements** 10(3, 1985):447-460.

1829
Gold, R.; Roberts, J.H.; Ruddy, F.H.; Preston, C.C.; McElroy, W.N.; Rao, S.V.; Greenborg, J.; and Fricke, V.R. "Results and Interpretation of SSTR Neutron Dosimetry in the Three Mile Island Unit-2 Reactor Cavity." **Transactions of the American Nuclear Society** 47(1984):114-115.

1830
Goldman, M.I.; Davis, R.J.; Strahl, J.F.; Tonkay, D.W.; and Shawn, L.W. "Development of Preliminary Radionuclide Mass Balance for the TMI-2 Accident." **Transactions of the American Nuclear Society** 44(1983):211-212.

1831
Gonzalez, L.C., and Gagliardo, R.S. "Electrical Modifications to Three Mile Island - Unit 2 During the Initial Stages of the Accident." **IEEE Transactions on Power Apparatus and Systems** PAS-99(November/December 1980):2327-2333.

1832
Greenborg, J., and Babel, P.J. "Location and Quantification of Fuel Lines in the TMI-2 Coolant System." **Transactions of the American Nuclear Society** 43(1982):15-16.

1833
Greenwood, R.H. "TMI-2 Reactor Building Sump Water Level Measurement and Removal." **Transactions of the American Nuclear Society** 43(1982):145-146.

1834
Gur, D.; Rao, G.R.; and Gibbs, E. "Population Dose During the TMI Incident in Perspective." **Health Physics** 39(December 1980):1039.

1835
Gur, D.; Rao, G.; Good, W.; Miller, D.; and Hollis, R. "Radiation Dose Assignment to Individuals Residing Near TMI." **Health Physics** 43(July 1982):138-139.

1836
Halbig, J.K.; Bosler, G.E.; Phillips, J.R.; and Rinard, P.M. "A System for Measuring TRU Materials in a Three Mile Island Demineralizer." **Transactions of the American Nuclear Society** 46(1984):314.

1837
Harms, A.A., and Murray, A.M. "A Topological Characterization of Public Response to Three Mile Island." **Transactions of the American Nuclear Society** 39(1981):725-726.

1838
Harrison, J.F., Jr.; Trikouros, N.G.; and Irani, A.A. "RETRAN Simulation of the Three Mile Island Unit 1 Turbine Trip Test." **Transactions of the American Nuclear Society** 32(1979):455-458

1839
Hayner, G.O., and Caldwell, C.S. "Microanalysis of a Core Debris Sample Removed from TMI-2." **Transactions of the American Nuclear Society** 50(November 1985):214-217.

1840
Henderson, Oran K. "Commonwealth of Pennsylvania Emergency Preparedness and Response -- The Three Mile Island Incident." **IEEE Transactions on Nuclear Science** NS-27(February 1980):924-926.

1841
Henninger, R.J., and DeMuth, N.S. "TRAC-PF1 Feed-and-Bleed Calculations for TMI-1." **Transactions of the American Nuclear Society** 45(1983):469-471.

1842
Henrie, J.O., and Postma, A.K. "Analysis of the TMI-2 Hydrogen Burn." **Transactions of the American Nuclear Society** 43(1982):12- 13.

1843
Henrie, J.O.; Quinn, G.J.; and Greenborg, J. "Special Handling of Three Mile Island Submerged Demineralizer System Vessels." **Transactions of the American Nuclear Society** 45(1983):49-50.

1844
Hildebrand, J.E. "TMI-2 Cleanup Update." **Health Physics** 49(July 1985):173.

1845
Hildebrand, James E., and Schmidt, James W. "TMI Experience with Beta Dosimetry Problems." **Health Physics** 43((July 1982):120.

1846
Hitz, C.G., and Deltete, C.P. "TMI-2 Decontamination Technology Development." **Journal of Metals** 37(11, 1985):40.

1847
Hlubek, W. "3 Mile Island -- Impacts on the Operation of German Nuclear Power Plants." **Transactions of the American Nuclear Society** 34(June 1980):688.

1848
Hofstetter, K.J. "TMI-2 Water Chemistry." **Abstracts of Papers of the American Chemical Society** 189(April 1985):80.

1849
Hofstetter, K.J.; Baston, V.F.; Bain, G.M.; and Hayner, G.O. "Decontamination Barrier on TMI-2 Leadscrew." **Transactions of the American Nuclear Society** 46(1984):384-385.

1850
Hofstetter, K.J., and Hitz, C.G. "Zeolite Processing of the TMI-2 Reactor Building Sump and the Reactor Coolant System." **Transactions of the American Nuclear Society** 43(1982):146-147.

1851
Hofstetter, K.J.; Loewenschuss, H.; and Baston, V.F. "Adherent Activity on TMI-2 Leadscrew." **Transactions of the American Nuclear Society** 45(1983):278-279.

1852
Houts, Peter S.; Miller, Robert W.; Ham, Kum Shik; and Tokuhata, George K. "Extent and Duration of Psychological Distress of Persons in the Vicinity of Three Mile Island." **Proceedings of the Pennsylvania Academy of Science** 54(1980):22-28.

1853
Hsu, C.J.; Shier, W.G.; and Levine, M.M. "OTSG Modeling for the Analysis of the TMI Incident." **Transactions of the American Nuclear Society** 34(June 1980):573-574.

1854
Hull, A.P. "A Consideration of the Need for Evacuation to Protect Public Safety at Three Mile Island." **Transactions of the American Nuclear Society** 34(June 1980):91-92.

1855
Hull, Andrew P. "A Critical Evaluation of Source and Environmental Radiation Measurements at Three Mile Island." **Health Physics** 39(December 1980):1037.

1856
Hull, Andrew P. "A Critique of Source Term and Environmental Measurement at Three Mile Island." **IEEE Transactions on Nuclear Science** NS-27(February 1980):664-668

1857
Ireland, J.R. "TRAC Analysis of Three Mile Island Accident." **Transactions of the American Nuclear Society** 34(June 1980):501-503.

1858
Irving, Bruce. "Three Mile Island Hands-On Decontamination Auxiliary Building." **Transactions of the American Nuclear Society** 38(June 1981):625.

1859
Jester, W.A.; Baratta, A.J.; Rudy, K.E.; Ford, B.C.; Bonner, J.J.; and Okyere, E.W. "Monitoring Kr-85 During TMI-2 Purging Using the Penn State Noble Gas Monitor." **Transactions of the American Nuclear Society** 35(November 1980):61-62.

1860
Karrasch, B.A., and Smotrel, J.R. "A Summary of Natural Circulation Alternatives for Long Term Core Cooling at TMI-2." **Transactions of the American Nuclear Society** 34(June 1980):565-566.

1861
Keaten, R.W., and Walsh, P.S. "The Role of Instrumentation in the TMI-2 Accident." **Transactions of the American Nuclear Society** 34(June 1980):789.

1862
Kettler, D.J. "Economics of Future Generation -- Nuclear Versus Coal Post-TMI." **Transactions of the American Nuclear Society** 38(June 1981):40-41.

1863
King, Bruce L. "Industry Response to Three Mile Island." **Proceedings of the Pennsylvania Academy of Science** 54(1980):29-30.

1864
Knief, R.A. "Nuclear Criticality Safety Training for TMI-2 Operators." **Transactions of the American Nuclear Society** 44(1983):306-307.

1865
Knief, R.A.; Boltz, D.J.; and Irizarr, C.A. "Behavioral Learning Objectives for Plant Simulation at TMI." **Transactions of the American Nuclear Society** 39(1981):281-282.

1866
Knief, R.A.; Long, R.L.; and Newton, S.L. "Training Requirements at TMI -- Harbinger for the Nuclear Industry." **Transactions of the American Nuclear Society** 45(1983):195-196.

1867
Knox, Joseph B.; Dickerson, Marvin H.; Greenly, George D.; Gudiksen, Paul H.; and Sullivan, Thomas J. "Utilization of the Atmospheric Release Advisory Capability (ARAC) During and After the Three Mile Island Accident." **Health Physics** 41(December 1981):850.

1868
Kripps, L.J.; Center, B.M.; and Long, R.L. "TMI Radioactive Material Release Pathways." **Transactions of the American Nuclear Society** 34(June 1980):633.

1869
Lapides, M.E. "Evaluation of Radiographic Techniques for Assessing TMI-2 Fuel Debris." **Transactions of the American Nuclear Society** 43(1982):9.

1870
Larsen, Diane K.; Graham, J. A. Bill; and West, Charles P. "The Impact of Three Mile Island on Radiation Emergency Planning in Tennessee." **Health Physics** 39(December 1980):1014-1015.

1871
Larson, J.E. "The Uniqueness of TMI-2 Technical Specifications and their Potential Applicability to Other Plants." **Transactions of the American Nuclear Society** 49(June 1985):302-303.

1872
Lee, J.Y. "TMI-2 Containment Building Accident Water Sampling and Reactor Primary Coolant Accident Source Term." **Transactions of the American Nuclear Society** 39(1981):837-838.

1873
Lee, N.; Meyer, P.E.; Schwarz, W.R.; and Tauche, W.D. "Development of Notrump to Meet Post-TMI Needs." **Transactions of the American Nuclear Society** 46(1984):801-802.

1874
Lefkowitz, S.; Sliter, G.; and Winkleblack, R.K. "Overview of Progress and Plans for EPRIs TMI-2 Mechanical Component Survivability Program." **Transactions of the American Nuclear Society** 43(1982):9-10.

1875
Legendre, Allen L., Jr. "The Financial Impact of the TMI Action Plan on Florida Power Corporation." **Transactions of the American Nuclear Society** 38(June 1981):539-540.

1876
Leung, J.C.M. "Assessment of Hydrogen Release Into the Containment During the TMI-2 Accident." **Transactions of the American Nuclear Society** 38(June 1981):435-436.

1877
Lodde, G.M.; Peterson, H.K.; Bondick, J.M.; and Surgeoner, D.M. "Health Physics Aspects of the Reactor Building Airborne Concentrations and Effluents for TMI Unit-2 from 3/28/79 to 5/31/84." **Health Physics** 47(July 1984):181.

1878
Logan, J.A.; Hultman, C.W.; and Lewis, T.J. "TMI-2 Reactor Vessel Head Removal and Damaged Core Removal Planning." **Transactions of the American Nuclear Society** 43(1982):3.

1879
Long, R.L. "Utility-University Cooperation in the Educational Response to TMI-2." **Transactions of the American Nuclear Society** 35(November 1980):34.

1880
Long, R.L.; Jones, J.D.; and Giacobbe, F.S. "Failure Analysis of the TMI-1 Once Through Steam Generator Tubes." **Transactions of the American Nuclear Society** 43(1982):327-328.

1881
McConnell, J.W., and Sanders, R.D. "Degradation of Resins in Epicor-II Prefilters from Three Mile Island." **Transactions of the American Nuclear Society** 50(November 1985):109.

1882
McIntosh, T.W.; Reilly, J.K.; and Quinn, G.J. "Update on High Activity Waste Disposition from TMI-2." **Transactions of the American Nuclear Society** 47(1984):110-111.

1883
McIsaac, C.V. "TMI-2 Reactor Building Source Term Measurements." **Transactions of the American Nuclear Society** 50(November 1985):225-226.

1884
McIsaac, C.V., and Hetzer, D.C. "Surface Deposition Measurements of the TMI-2 Gross Decontamination Experiment." **Transactions of the American Nuclear Society** 43(1982):19.

1885
Marotta, C.R. "Re-Criticality Potential of TMI-2 Core." **Transactions of the American Nuclear Society** 35(November 1980):272-274.

1886
Miller, A.D. "Radiation Source Terms and Shielding at TMI-2." **Transactions of the American Nuclear Society** 34(June 1980):633-635.

1887
Miller, C.W.; Cotter, S.J.; Little, C.A.; and Moore, R.E. "Comparison of Observed and Predicted Doses from the TMI Incident." **Transactions of the American Nuclear Society** 34(June 1980):90-91.

1888
Mills, L.M. "Impact of TMI -- A Change in Attitude." **Transactions of the American Nuclear Society** 34(June 1980):591-592.

1889
Mochizuki, K. "Effects of TMI on Reactor Licensing in Japan." **Transactions of the American Nuclear Society** 35(November 1980):414.

1890
Morrell, M.P., and Bohn, L. "Remote Camera Radiation Monitoring Experiments in TMI-2 Containment Building." **Transactions of the American Nuclear Society** 34(June 1980):635-636.

1891
Muller, H. Arnold; Trautlein, J.J.; Donovan, J. Ward; and Miller, Kenneth. "Things That Go Bump in the Night -- Rumination on a Medical Center's Response to Nuclear Catastrophe." **Clinical Research** 28(April 1980:297A.

1892
Muller, W. "Radiation Protection Aspects of the Three Mile Island Nuclear Power Plant Accident." **Health Physics** 40(February 1981):276.

1893
Munson, L.F.; Hartz, R.; Lo, R.; Travers, W.D. "Possible Options for Reducing Occupational Dose From the TMI-2 Basement." **Health Physics** 50(S1, 1986):S1-S2.

1894
Munson, L.F.; Lo, R.; Oliu, W.; and Hoenes, G.R. "Informing the Public About the Occupational Dose for Cleanup Activities at TMI-2." **Health Physics** 47(July 1984):144.

1895
Murphy, M.B. "In-Containment Radiation Monitoring at TMI-2." **Transactions of the American Nuclear Society** 43(1982):8-9.

1896
Murphy, M.B., and Thome, F.V. "TMI-2 Instrument Analysis Results." **Transactions of the American Nuclear Society** 39(1981):803-804.

1897
Nitti, D.A.; Hicks, J.H.; and Harner, K.L. "Fission Product Release and Transport Following the TMI-2 Accident." **Abstracts of Papers of the American Chemical Society** 181(March 1981):5.

1898
Osgood, N.L.; Hopkins, W.C.; and Peterson, H.K. "Analysis of Radiation Measurements Taken Inside the TMI-2 Reactor Vessel." **Health Physics** 47(July 1984):205.

1899
Osgood, N.L.; Williams, D.S.; Cooper, W.J. "Radiation Protection Considerations in TMI-2 Defueling Systems Design." **Health Physics** 49(July 1985):126-127.

1900
Osterhoudt, T.R.; Quinn, G.J.; and Reilly, J.K. "Remote Decontamination of the TMI-2 Makeup and Purification Demineralizers." **Transactions of the American Nuclear Society** 47(1984):467-468.

1901
Owen, D.E.; Bower, J.M.; and Fee, D.S. "The Acquisition of TMI-2 Core Debris Samples." **Transactions of the American Nuclear Society** 45(1983):798-799.

1902
Owen, D.E., and Martin, M.R. "TMI-2 Core Examination -- 1st Results." **Transactions of the American Nuclear Society** 43(1982):5.

1903
Owen, D.E., and Thomas, G.R. "Plenum Assembly Damage at TMI-2." **Transactions of the American Nuclear Society** 50(November 1985):219-220.

1904
Parklyn, J.D. "Emergency Planning in the Post-TMI Era -- A Change of Direction at the La Crosse Boiling Water-Reactor." **Transactions of the American Nuclear Society** 45(1983):546.

1905
Pasedag, W.F., and Postma, A.K. "The Impact of TMI on Future Licensing." **Abstracts of Papers of the American Chemical Society** 189(April 1985):104.

1906
Pasquenza, J.P. "Plant Performance in the Aftermath of TMI." **Transactions of the American Nuclear Society** 34(June 1980):599.

1907
Patti, F.J., and Dam, A.S. "Radiation Protection Considerations in TMI-2 Recovery Systems Design." **Transactions of the American Nuclear Society** 34(June 1980):638-639.

1908
Pavelek, M.D. "TMI-2 Postaccident Dose Reduction Through Coating Removal." **Transactions of the American Nuclear Society** 49(S2, 1985):32-33.

1909
Pelletier, C.A.; Cox, T.E.; Reeder, D.L.; Voilleque, P.G.; and Thomas, C.D. "Preliminary Results of the TMI-2 Radioactive Iodine Mass Balance Study." **Transactions of the American Nuclear Society** 43(1982):17-18.

1910
Perrone, P.E., and Lexa, A.F. "The Reactimeterc-TMI-2 Success and Guide to Future Nuclear Data Systems." **Transactions of the American Nuclear Society** 34(June 1980):790-791.

1911
Peterson, Henry K., and Frey, Steven R. "TMI-II Reactor Building Radiation Source Term Identification." **Health Physics** 45(July 1983):211.

1912
Peterson, Russell W. "Three Mile Island -- Lessons for America." **IEEE Transactions on Aerospace and Electronic Systems** AES-17(March 1981):229-233.

1913
Pettus, W.G., and Hassler, L.A. "Criticality Calculations for TMI-2 Defueling Canisters." **Transactions of the American Nuclear Society** 47(1984):108.

1914
Pickard, P.S.; Philbin, J.S.; Nelson, W.E.; and Straker, E. "Neutronic Sensitivity Analysis of the SRM Response for the TMI-2 Accident." **Transactions of the American Nuclear Society** 34(June 1980):559-560.

1915
Picklesimer, M.L. "An Estimate of the Damage to the TMI-2 Core." **Transactions of the American Nuclear Society** 35(November 1980):196-197.

1916
Porter, S.W., Jr., and Sherbini, S.S. "Beta Dosimetry -- Lessons Learned from TMI." **Health Physics** 39(December 1980):1017-1018.

1917
Powers, D.A. "Plausible Conditions of the TMI-2 Core." **Transactions of the American Nuclear Society** 34(June 1980):563-565.

1918
Quinn, G.J. "Offsite Shipment and Disposal of TMI-2 Core and Wastes." **Journal of Metals** 37(11, 1985):40.

1919
Quinn, G.J.; Henrie, J.O.; and Greenborg, J. "Problems and Solutions Associated with Combustible Gas Generation in TMI-2 Waste." **Transactions of the American Nuclear Society** 50(November 1985):142.

1920
Rainisch, R.; Fricke, V.R.; Warren, H.D.; and Caldwell, C.S. "Gamma Scanning of the TMI-2 Lower Reactor Vessel Head via in-Core Detectors." **Transactions of the American Nuclear Society** 50(November 1985):460-461.

1921
Rao, A.S.; Sawyer, C.D.; and Paradiso, F.M. "A Post-TMI Review of Small Loss of Coolant Accidents (LOCAs) for a BWR." **Transactions of the American Nuclear Society** 35(November 1980):338-340.

1922
Rao, G.; Good, W.; Herron, J.; and Gur, D. "The TMI Population: A Closer Look." **Health Physics** 41(December 1981):850.

1923
Rao, G.R. et al. "The TMI Population: A Closer Look." **Proceedings of the Pennsylvania Academy of Science** 56(1982)44.

1924
Reed, Cordell. "Impact of Three Mile Island on Regulatory Requirements." **Transactions of the American Nuclear Society** 34(June 1980):592-593.

1925
Reilly, Margaret A. "The Measurement of Krypton-85 by a Community Monitoring Program." **Transactions of the American Nuclear Society** 35(November 1980):60-61.

1926
Rider, R.L., and Kelly, D.F. "Preparing TMI-2 for Core Removal from the Reactor Vessel." **Transactions of the American Nuclear Society** 49(June 1985):343-346.

1927
Riethle, William E. "Metropolitan Edison Company's Radiation Environmental Monitoring During Venting of TMI-Unit 2 Containment." **Transactions of the American Nuclear Society** 35(November 1980):58-59.

1928
Ritchie, J.C., and Carter, M.W. "A Description of the System and Process Designed to Treat High-Activity Primary Coolant Water at Three Mile Island, Unit 2." **Transactions of the American Nuclear Society** 35(November 1980):506-507.

1929
Ritchie, John C., and Carter, Mel. "A Description of the System and Process Designed to Treat High Activity Primary Coolant Water at Three Mile Island, Unit II." **Health Physics** 39(December 1980):1033.

1930
Roberts, J.T.A., and Schwartz, F.L. "TMI-2 Technology Transfer Program for United States Utilities." **Journal of Metals** 37(11, 1985):21.

1931
Rogan, R.E. "Site Emergency Preparedness Programs and In-Plant Emergency Response -- A New Dimension in the Aftermath of TMI-2." **Transactions of the American Nuclear Society** 41(1982):536-537.

1932
Ruddy, F.H.; Roberts, J.H.; Gold, R.; and Preston, C.C. "Applications of Solid State Track Recorder Neutron Dosimetry for Fuel Debris Location in the Three Mile Island Unit 2 Makeup and Purification Demineralizers." **Nuclear Tracks and Radiation Measurements** 8(1-4, 1984):473-482.

1933
Sadauskas, G.J., and Carscadden, J.R. "3 Mile Island Recovery Radiation Monitoring Systems." **Transactions of the American Nuclear Society** 34(June 1980):796.

1934
Sanders, M., and Maslo, R.M. "A Proposed Engineering Enhancement of Power Plant Training Simulators in Light of Experiences Observed at TMI-2." **Transactions of the American Nuclear Society** 34(June 1980):717.

1935
Schaeffer, M.A.; Baum, A.; and Gatchel, R.J. "Adrenal Gland Secretion as a Function of Stress at Three Mile Island." **Psychophysiology** 19(September 1982):584.

1936
Schaeffer, M.A.; McKinnon, W.; Baum, A.; Reynolds, C.P.; Rikli, P.; Davidson, L.M.; and Fleming, I. "Immune Status as a Function of Chronic Stress at Three Mile Island." **Psychosomatic Medicine** 47(January/February 1985):85.

1937
Schmitt, R.C., and Quinn, G.J. "Preparations to Ship the TMI-2 Damaged Reactor Core." **Transactions of the American Nuclear Society** 50(November 1985):132-133.

1938
Schultz, M.A. "High Neutron Detector Readings at TMI After the Accident." **Transactions of the American Nuclear Society** 34(June 1980):82-83.

1939
Schwarz, W.F. "The Instrumentation and Electrical Program at TMI-2." **Transactions of the American Nuclear Society** 43(1982):7.

1940
Sehgal, B.R.; Duffey, R.B.; and Loewenstein, W.B. "Post-TMI Thermal Hydraulic Aspects of LWR Safety Analysis." **Transactions of the American Nuclear Society** 34(June 1980):865-866.

1941
Shanstrom, R.T. "Analysis of SPND Response During the TMI-2 Accident." **Transactions of the American Nuclear Society** 34(June 1980):561-562.

1942
Shaw, H.E.; Sandberg, R.O.; and Hovey, G.K. "TMI-2 -- Status of Recovery and Future Plans." **Transactions of the American Nuclear Society** 35(November 1980):412.

1943
Sheff, J.R.; Baston, V.F.; and Buchanan, D.R. "Particle Debris Source Term for the TMI-2 Primary System." **Transactions of the American Nuclear Society** 49(June 1985):251-252.

1944
Shepard, R.L, and Anderson, J.L. "PWR Instrumentation -- Recommendations for Improvement from TMI-2 Experience." **Transactions of the American Nuclear Society** 34(June 1980):792-793.

1945
Shults, W.D.; Carter, J.A.; Emery, J.F.; and Laing, W.R. "Analytical Chemistry Response and Three Mile Island." **Transactions of the American Nuclear Society** 34(June 1980):152.

1946
Shuping, R.E. "An Independent Estimation of Exposure Near the Three Mile Island Nuclear Power Station Using Photographic Film." **Health Physics** 39(December 1980):1037-1038.

1947
Simon, P.M. "Three Mile Island Unit-1 -- Control Room Human Factors Efforts." **Transactions of the American Nuclear Society** 44(1983):171-172.

1948
Skillman, G.R., and Elam, B.D., Jr. "TMI-2 Cooldown and Cleanup." **Proceedings of the American Power Conference** 42(1980):771-775.

1949
Snyder, B.J. "Regulatory Perspective on the Cleanup of TMI-2." **Transactions of the American Nuclear Society** 43(1982):2.

1950
Standerfer, F.R. "TMI-2: How the Cleanup is Coming." **Proceedings of the American Power Conference** (1985):738-742.

1951
Stephens, M. "Press Coverage and Public Opinion of Nuclear Safety Since Three Mile Island." **Transactions of the American Nuclear Society** 38(June 1981):493.

1952
Stone, G.F.; Maxwell, R.B.; and Hufham, J.W. "Radiological Emergency Planning in TVA -- Effects from 3 Mile Island." **Transactions of the American Nuclear Society** 34(June 1980):688.

1953
Storton, J.M., and Kramer, J.F. "TMI-2 Defueling Filter Test." **Abstracts of Papers of the American Chemical Society** 189(April 1985):101.

1954
Stratton, William R. "Fission Product Chemistry and the Accident at Three Mile Island." **Health Physics** 43(July 1982):171.

1955
Subbaratnam, T. "Environmental Impact of the Three Mile Island Nuclear Accident." **Health Physics** 40(April 1981):574.

1956
Sumpter, K.C.; Trickeit, K.A.; Feinauer, E.; Owen, D.E.; and Martin, M.R. "An Overview of the TMI-2 Core Examination Plan." **Transactions of the American Nuclear Society** 43(1982):5-6.

1957
Tagart, S.W., and Moore, K.E. "Thermal Analysis of the TMI-2 Primary Pressure Boundary and Potential Impact on Requalification Plans." **Transactions of the American Nuclear Society** 43(1982):10- 11.

1958
Tarpinian, J.E. "Dose Reduction Program for the TMI-2 Reactor Building." **Health Physics** 47(July 1984):163.

1959
Tarpinian, J.E. "General Area Dose Rate Reduction of the TMI-2 Reactor Building." **Transactions of the American Nuclear Society** 47(1984):376-377.

1960
Tarpinian, James E., and Hollenbeck, Peter. "A Summary of Radiological and Physical Conditions of the TMI-II Reactor Containment and Their Impact on Recovery." **Health Physics** 41(December 1981):879.

1961
Taylor, J.H. "The Impact of TMI-2 on Regulatory Requirements -- An Industry Perspective." **Transactions of the American Nuclear Society** 34(June 1980):595-596.

1962
Theising, J.W. "TMI-2 Containment Decontamination." **Transactions of the American Nuclear Society** 38(June 1981):625-626.

1963
Thomas, G.R. "Description of the TMI-2 Accident." **Abstracts of Papers of the American Chemical Society** 189(April 1985):57.

1964
Thomas, G.R. "Overview -- Significant Aspects of TMI-2 Core Material Performance." **Transactions of the American Nuclear Society** 35(November 1980):194-196.

1965
Thompson, J.D.; Knauer, J.B.; Bond, W.D.; and King, L.J. "TMI-2 Plant Demineralizer Sample Analysis." **Transactions of the American Nuclear Society** 47(1984):113-114.

1966
Thorpe, J.R. "Impacts of TMI on Regulatory Requirements -- The GPU Position." **Transactions of the American Nuclear Society** 34(June 1980):593.

1967
Thorpe, J.R. "TMI-1 Preparation for Restart." **Transactions of the American Nuclear Society** 44(1983):424-425.

1968
Tokuhata, G.K. "Health Studies in the Three Mile Island Area." **Transactions of the American Nuclear Society** 38(June 1981):82-83.

1969
Tolman, B.; Allison, O.; Davis, C.; and Polkinghorn, S. "Thermal Hydraulic Aspects of the TMI Accident." **Abstracts of Papers of the American Chemical Society** 189(April 1985):58.

1970
Trautlein, J.J.; Kales, J.D.; Martin, E.D.; and Bixler, E.O. "Three Mile Island -- Medical Issues." **Clinical Research** 28(April 1980):301A.

1971
Trikouros, N.G.; Broughton, T.G.; and Harrison, J.F. "Application of RETRAN for Three Mile Island Analysis." **Transactions of the American Nuclear Society** 34(June 1980):79-82.

1972
Vigil, M.G. "High Integrity Container Design Requirements for TMI Low Level Waste." **Transactions of the American Nuclear Society** 39(1981):211-212.

1973
Vigue, R.P. "The Cost Impact of Three Mile Island on Nuclear Projects." **AACE Transactions** (1982):J.3.1-J.3.6.

1974
Vinjamuri, K., and Hobbins, R.R. "Response of 17-4 PH Stainless Steel from TMI-2 Leadscrews." **Transactions of the American Nuclear Society** 50(November 1985):220-221.

1975
Voglewede, J.C. "Lessons Learned from TMI In-Reactor Instrumentation -- An NRC Viewpoint." **Transactions of the American Nuclear Society** 34(June 1980):791-792.

1976
Voillegue, P.G.; Cox, T.E.; and Pelletier, C.A. "Fission Product Behavior at TMI-2." **Abstracts of Papers of the American Chemical Society** 189(April 1985):59.

1977
Wald, N. "Radiation Health Aspects of the Three Mile Island Power Plant Accident." **International Journal of Nuclear Medicine and Biology** 9(2,1982):155.

1978
Walker, E. "Evaluation of Portable Survey Instruments for TMI-2 Containment Entry." **Health Physics** 41(December 1981):865.

1979
Walker, E., and Morrell, M.P. "TMI-2 Containment Assessment Program -- Radiological Measurements of the Environment Above the Operating Floor Using Spare Penetration." **Transactions of the American Nuclear Society** 38(June 1981):587-588.

1980
West, J.T.; Westfall, R.M.; Whitesides, G.E.; and Thomas, J.T. "TMI-2 Criticality Analysis Analytical Models and Methods." **Transactions of the American Nuclear Society** 35(November 1980):274-276.

1981
Westfall, R.M.; West, J.T.; Whitesides, G.E.; and Thomas, J.T. "TMI-2 Criticality Analysis Parametric Studies and Overall Results." **Transactions of the American Nuclear Society** 35(November 1980):276-277.

1982
Wethy, C.M. "The Effects of Three Mile Island on Nuclear Plant Operations." **Transactions of the American Nuclear Society** 44(S1, 1983):82.

1983
Whittaker, R. "Remote Operating Vehicle for TMI-2 Basement Recovery." **Transactions of the American Nuclear Society** 50(November 1985):621-622.

1984
Williams, Edward F., Jr. "Comparison of TMI-2 Offsite Emergency Monitoring Experiences with Present Federal Guidance." **IEEE Transactions on Nuclear Science** NS-27(February 1980):669-674.

1985
Wilson, W.B., and England, T.R. "TMI-2 Radionuclide Inventory and Decay Properties." **Transactions of the American Nuclear Society** 34(June 1980):787-788.

1986
Yancey, M.E. "TMI-2 SPND and Thermocouple Data Analysis." **Transactions of the American Nuclear Society** 43(1982):7-8.

1987
Yochheim, E.D.; Hicks, J.H.; and Frederick, K.F. "Radiochemistry Experience at TMI-2." **Transactions of the American Nuclear Society** 34(June 1980):636-637.

1988
Zebroski, E.L. "Implementing the Five Main Lessons of Three Mile Island." **IEEE Transactions on Nuclear Science** NS-29(February 1982):27-34.

# Author Index

# Subject Index

# Government Document Number Index

## About the Compilers

M. SANDRA WOOD is Head Reference Librarian, the George T. Harrell Library, The M.S. Hershey Medical Center of Pennsylvania State University. She is the editor of *Cost Analysis, Cost Recovery, Marketing, and Free-Based Services: A Guide for the Health Sciences Librarian* and the coeditor of *End User Searching in the Health Sciences.* She has contributed articles to *Reference Librarian* and *Bulletin of the Medical Library Association.*

SUZANNE M. SHULTZ is Director of the Library at the Polyclinic Medical Center of Harrisburg, Pennsylvania. She served as Book Review Editor of *Medical Reference Services Quarterly* and has contributed articles to *Transactions and Studies of the College of Physicians of Philadelphia, Pennsylvania Medicine,* and *Medical Reference Services Quarterly.*